Television
and
Video Engineering

Television
and
Video Engineering

A. Veera Lakshmi
(M.B.A.)

Senior Lecturer in Adhiparasakthi Engineering College, Melmaruvathur

and

R. Srivel

Faculty Member in the Electronics and Communication Engineering Department in P.B. College of Engineering, Chennai

Ane Books Pvt. Ltd.

New Delhi ♦ Chennai

Television and Video Engineering

A. Veera Lakshmi and R. Srivel

Published by

Ane Books Pvt. Ltd.

4821, Parwana Bhawan, 1st Floor, 24 Ansari Road, Darya Ganj,
New Delhi - 110 002, Tel.: +91(011) 23276843-44, Fax: +91(011) 23276863
e-mail: kapoor@anebooks.com, Website: www.anebooks.com

Branch Office:

- Avantika Niwas, 1st Floor, 19 Doraiswamy Road, T. Nagar,
  Chennai - 600 017, Tel.: +91(044) 28141554, 28141209
  e-mail: anebookschennai@gmail.com, rathinam@anebooks.com

ISBN : 978-93-8015-669-9

Printed at R.K. Offset Process, New Delhi

PREFACE

The importance of television is well known in the various fields of communication. Therefore it is necessary for an Electronic Engineer to know the Fundamentals of Television Circuits. From this point of view, it gives us a great pleasure to release this book which covers the entire syllabus of the subject "***Television and Video Engineering***" all the concepts of black and white and colour television are described for the easy understanding of the subject.

The view in this text book is therefore, more on Television Circuit Functions and block diagrams. Numerous new features that are now available have made present day TV receiver more improved than the older form. Topics like Satellite communication, Cable TV, 3D-TV, EDTV, VCR-recordings, Tele-text displays, Flat panel displays and recent developed equipments have also been introduced in this book.

This book uses simple and everyday language to explain the subject. Detailed explanation of each topic makes it easy for the students to understand the complicated circuits and blocks.

ACKNOWLEDGEMENTS

We wish to express our profound thanks to our Managing Director Mr. G.B. Senthil Kumar and Principal, Dr. S. Jayashri, Adhiparasakthi Engineering College for their kind support and encouragement.

We like to express our profound thanks to Chairman, Mr. G. Venkata Raman and Director, Dr. G. Subba Lakshmi, P.B. College of Engineering for their sustaining encouragement.

We wish to express thanks to Dr. G.K. Dwarakanath, Principal, P.B college of Engineering for their kind support and constant encouragement.

Last but by no means least, as we always, wish to thank Professor S. Srinivasan HOD/ECE, P.B.C.E and Professor M. Malathi, HOD/ECE, A.P.E.C and all our colleague staff members, friends R. Soundher and D. Balaji, for helping us on numerous occasions.

Finally we wish to thank entire team of ANE publications who have taken immense pain to get this book in time with quality printing.

Any suggestions for the improvement of this book will be acknowledged and well appreciated.

SYLLABUS

Chapter 1: FUNDAMENTALS OF TELEVISION

Geometry form and Aspect Ratio - Image Continuity - Number of scanning lines - Interlaced scanning - Picture resolution - Camera tubes - Image orthicon - vidicon-plumbicon-silicon diode array vidicon-solid state image scanners - monochrome picture tubes - composite video signal-video signal dimension - horizontal sync. Composition - vertical sync. Details - functions of vertical pulse train - scanning sequence details. Picture signal transmission - positive and negative modulation - VSB transmission sound signal transmission - standard channel bandwidth.

Chapter 2: MONOCHROME TELEVISION TRANSMITTER AND RECEIVER

TV transmitter - TV signal propagation - Interference - TV transmission Antennas - Monochrome TV receiver - RF tuner - UHF, VHF tuner - Digital tuning techniques - AFT-IF subsystems - AGC - Noise cancellation - Video and sound inter carrier detection - vision IF subsystem - video amplifiers requirements and configurations - DC re-insertion - Video amplifier circuits - Sync separation - typical sync processing circuits - Deflection current waveform - Deflection Oscillators - Frame deflection circuits - requirements - Line Deflection circuits - EHT generation - Receiver Antennas.

Chapter 3: ESSENTIALS OF COLOUR TELEVISION

Compatibility - colour perception - Three colour theory - luminance, hue and saturation-colour television cameras - values of luminance and colour difference signals - colour television display tubes - delta - gun-precision - in-line and Trinitron colour picture tubes - purity and convergence - purity and static and dynamic convergence adjustments - pincushion correction techniques - automatic degaussing circuit - grey scale tracking - colour signal transmission - bandwidth - modulation of colour difference signals - weighting factors - Formation of chrominance signal.

Chapter 4: COLOUR TELEVISION SYSTEMS

NTSC colour TV system - NTSC colour receiver - limitations of NTSC system - PAL colour TV system - cancellation of phase errors - PAL-D colour system - PAL coder - PAL-D colour receiver - chromo signal amplifier - separation of U and V signals - colour burst separation - Burst phase Discriminator - ACC amplifier - Reference Oscillator - Ident and colour killer circuits - U and V demodulators - Colour signal Matrixing - merits and demerits of the PAL system - SECAM system - merits and demerits of SECAM system.

Chapter 5: ADVANCED TELEVISION SYSTEMS

Satellite TV technology - Cable TV - VCR - Video Disc recording and playback - Tele Text broadcast receiver - digital television - Transmission and reception - projection television - Flat panel display TV receiver - Stereo sound in TV - 3D TV - EDTV - Digital equipments for TV studios.

CONTENTS

CHAPTER 1

TELEVISION

KEY TOPICS

- ✓ Image Continuity
- ✓ Geometry form and Aspect ratio
- ✓ Scanning and its types
- ✓ Interlaced scanning
- ✓ Picture resolution
- ✓ Camera tubes and its types
- ✓ Monochrome picture tube
- ✓ Composite video signal and its functions
- ✓ Picture signal transmission
- ✓ Positive modulation & negative modulation
- ✓ VSB transmission & reception
- ✓ Inter carrier sound modulation
- ✓ TV broadcast Channel allocations
- ✓ Standard channel bandwidth

LEARNING OBJECTIVES

After reading this chapter, the learner should be able to:

- ♦ The aim of a television system is to extend the sense of sight beyond its natural limit and to transmit sound associated with the scene.
- ♦ Picture transmission is optical character and its assemblage of a large number of bright and dark areas representing picture details.
- ♦ The aspect ratio of picture produced on the receiver screen and televised picture must be same if it is not, the picture will be too thin or too wide.
- ♦ The infinite number of pieces of information is needed to be picked up simultaneously for transmitting the picture details, but it is not practicable. Because it is not feasible to provide separate (signal path channel) for each picture signal, this problem can be solved by scanning.
- ♦ The major function of the camera tube is to convert an optical information of the scene into electrical form and it may be called as the eye of the TV system.
- ♦ Based on the target plate we can classify the following types of camera tubes.
    - Vidicon
    - Plumbicon
    - Saticon
    - New icon
    - Chalinicon
    - Silicon diode array
- ♦ A television transmitter transmit both audio and video signal using a channel bandwidth and two transmitters are employed one for picture

transmission which is amplitude modulated and uses a vestigial side band (VSB). Other is used for audio signal which is frequency modulated.

- The composite video signal waveform has two distinct polarity for black and white levels in picture. Hence two types of modulation is possible. They are positive modulation and negative modulation.
- The composite video signal is formed by the electrical signal corresponding to the picture information in the lines scanned in the TV camera pickup tube and sync details introduced in it.
- The monochrome picture tube is used to convert the video (electrical) signal's into optical signal.
- TV broadcast channel allocations and Standard channel bandwidth is discussed briefly.

1.1 TELEVISION BROADCASTING

Television means "tele + vision" *i.e.*, television is used to see the picture telecast from long distance and 'broadcasting' means to send in all direction.

In other words, television is used to extend the sense of sight beyond its natural limits and to transmit. Sound is associated with the scene. Figure 1.1 shows the block diagram of television system.

The television broadcasting equipment is usually quite sophisticated and expensive one. The service area for a commercial television broadcasting station is about 40 to 120 km in all direction from the transmitter. The transmitted wave contains two RF carriers, modulated by the desired information. Amplitude Modulation (AM) is used for the picture signal and Frequency Modulation (FM) is used for the sound signal.

1.1.1 Sound and Picture Transmission

The picture information is optical in nature. To transmit that signal into long distance we need to convert that optical signal into electrical signal. Here the camera tube are used to convert the optical natured picture information into a corresponding electrical signal, the amplitude of electrical signal varies in accordance with the variations of brightness. The electrical signal is amplified and then it is fed to the modulator for necessary modulation.

Here Amplitude Modulation (AM) is used to transmit the picture signal. Before modulation, the sync pulses and blanking pulses is added to keep camera and picture tube beams in synchronization. After proper amplification by the power amplifier it is fed to the combining network.

Similarly to transmit the audio signal FM modulation is used. After proper amplification it is fed to the combining network. This combining network is used to transmit both the picture and sound signal using a single

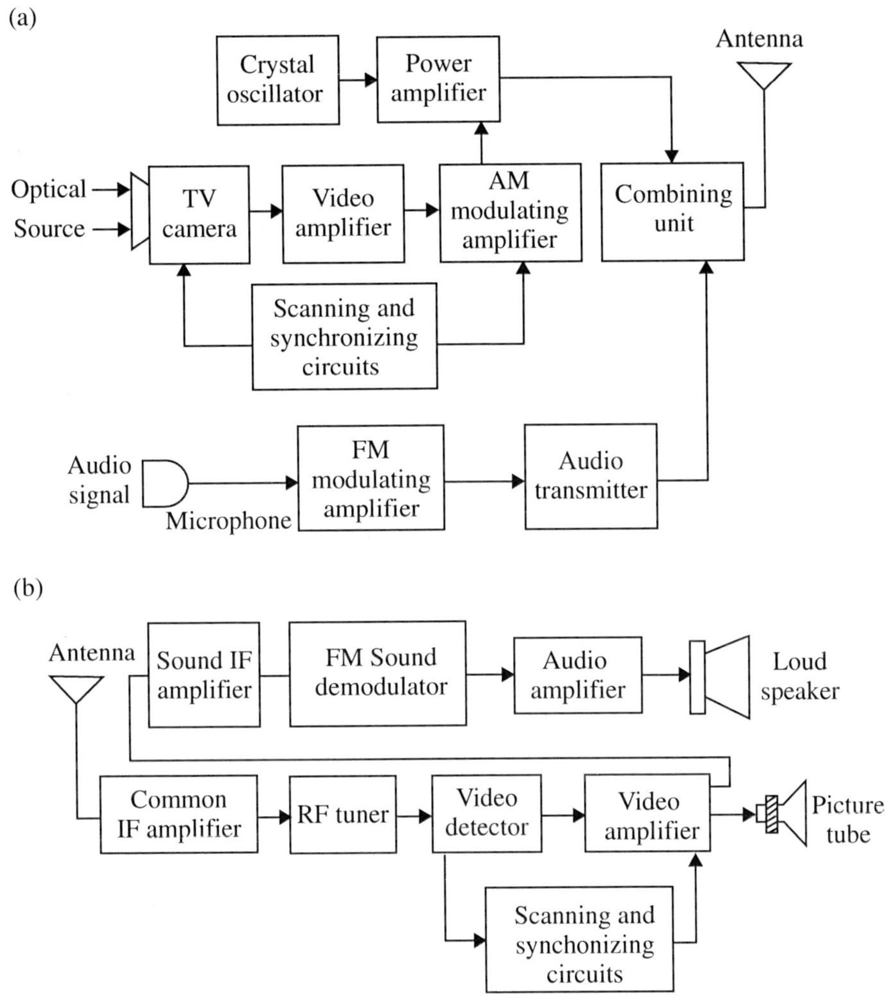

Figure 1.1 (a) Basic monochrome television transmitter. (b) Basic monochrome television receiver.

antenna without mixing each other. Diplexer is most commonly used for this purpose.

Separate carrier waves are used for the picture and sound signal, but they are transmitted by one transmitting antenna. The picture and sound signals are included in the broadcast channel for each station. For a commercial broadcasting station, the channel is made 7 MHz wide to include both the picture and sound.

1.1.2 Reception

The received RF signal is fed to the video detector. This video detector is used to separate the sound and video signal. After separation the video and sound signals are amplified and fed to the picture tube and loudspeaker. The picture tube which converts the electrical signal back into optical signal. The picture tube is also called as 'Kinescope'.

1.2 IMAGE CONTINUITY

While televising picture elements by the principle of scanning, it is necessary to present the picture to the eye in such a way that an illusion of continuity and any motion pictures appears on the screen as a smooth and continuous change.

- To achieve this it should follow the persistence of vision or storage characteristics of the human eye.
- The sensation produced when the nerves of the eye's retina and stimulated by an incident light does not cease immediately after the light is removed but persists for about $\frac{1}{16}^{\text{th}}$ of a second.
- If the scanning rate per second is made greater than the 16 or number of pictures shown per second is more then 16, the eye is able to integrate the change of brightness.
- Each picture or frame is projected individually as a still picture, but they are shown one after another in a rapid manner to produce the illusion of continuity in the motion pictures.
- In present day, 24 frames per one second is transmitted at a same rate in motion pictures in order to make an 'illusion of image continuity'.

1.2.1 Brightness Gradation

Signal transmission path should have proper transfer characteristics to preserve the same brightness, gradation as eye perceive while viewing the picture scene directly.

If non-linearities occurs necessary corrections have to be done both in the picture tube and camera tube. The picture tube emphasis white in reproducing the image. This correction is called 'gamma correction' which is used to obtain contrast ratio.

Factors that affect the total gradation of reproduced picture.

(a) Brightness

The overall or average intensity of illumination and it determines the background light in reproducing picture.

Brightness control is varied to obtain optimum average illumination of the scene.

(b) Contrast

Contrast it is the difference in light intensity between black and white parts of the picture below and above the average brightness level.

It is great enough to produce a strong picture with bright white and dark black for extreme intensity values.

Too much of contrast makes the picture hard, while too little contrast gives the impression of a washed out image.

(c) Viewing distance

The viewing distance from the screen of the TV receiver should not be so large because the eye cannot resolve the details of the picture.

The distance should not be so small that picture elements become separately visible. The above conditions are met when the vertical picture tube size subtends an angle at 15° at the eye approximately.

1.3 SYNCHRONIZATION AND SCANNING PROCESS

The camera tube converts the picture information into video signal by dissecting the image into a series of horizontal lines. Similarly the picture tube reassembles the image by line by line. These horizontal lines are produced by making the electron beam scan across the screen. There are 625 lines per picture frame. In addition to this, horizontal scanning vertical scanning is necessary to spread the lines from top to bottom of the screen. There are 25 complete picture frames per second. The scanning at the camera tube and picture tube must be synchronised with respect to the video signal to reassemble the picture information on the correct lines.

In India 625 line system is used for both monochrome and colour TV signal *i.e.*, each frame consists of 625 horizontal lines. Each frame is scanned at a rate of 25 times per second.

1.3.1 Need for Scanning

The whole picture is divided into a number of small elements called pixels. This infinite number of pieces of information are need to be picked up simultaneously for transmitting picture information but it is not practicable because, it is not feasible to provide separate signal (path channel) for each picture signal. This problem can be solved by a method known as "Scanning".

1.3.2 Scanning Process

The optical information is converted into electrical signal is carried out element by element, one at a time in a sequential order to cover the entire picture.

Scanning is the process which is done both in the camera tube (transmitter) and in the picture tube (receiver). During scanning electron beam is moved from left to right and bottom to top as shown in the Figure 1.2 the left to right movement of the electron beam is called 'Trace'. And from right to left movement of the electron beam at very faster rate is called 'Retrace'. In camera tube no video signal will be obtained during the retrace period. Similarly in the picture tube no image is traced during the retrace period because there is no video signal input to the picture tube cathode.

Thus, in the scanning process the electron beam is deflected from left to right and top to bottom to scan each and every picture element (pixel). Thus, during scanning in the camera tube each picture element (pixel) is converted into a video signal.

In picture tube during the scanning process the video signal reproduce each and every picture element on the picture tube screen. As the scanning process is done at much faster rate, we can see the full picture frame at a time.

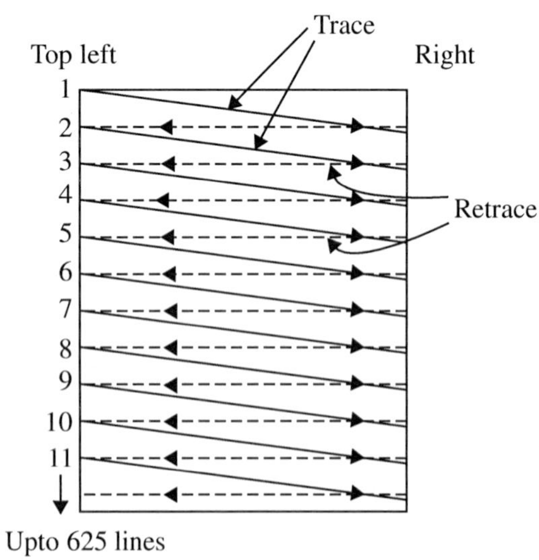

Figure 1.2 Scanning process.

1.3.3 Horizontal Scanning

Horizontal scanning is done by moving the electron beam from left to right and right to left as shown in the Figure 1.3. To produce the deflection, a saw tooth current is passed through the horizontal deflection coils in the deflection yoke.

In the Figure 1.3, an arrow on the lines indicates the path of the scanning beam. The solid lines shows the trace period of the scanning lines. The dotted line indicates the retrace of the beam.

Figure 1.3 shows the waveform of saw tooth current in horizontal deflection coils. As shown in the waveform, a linear rise of current in the deflection coil shift the electron beam from left to right of the screen this is called trace. As the maximum current is reached, the electron beam is at right side. The trace period is a 52μ seconds.

At the peak value, current reverses its direction and decreases rapidly to its initial value. This fast reversal moves an electron beam from right

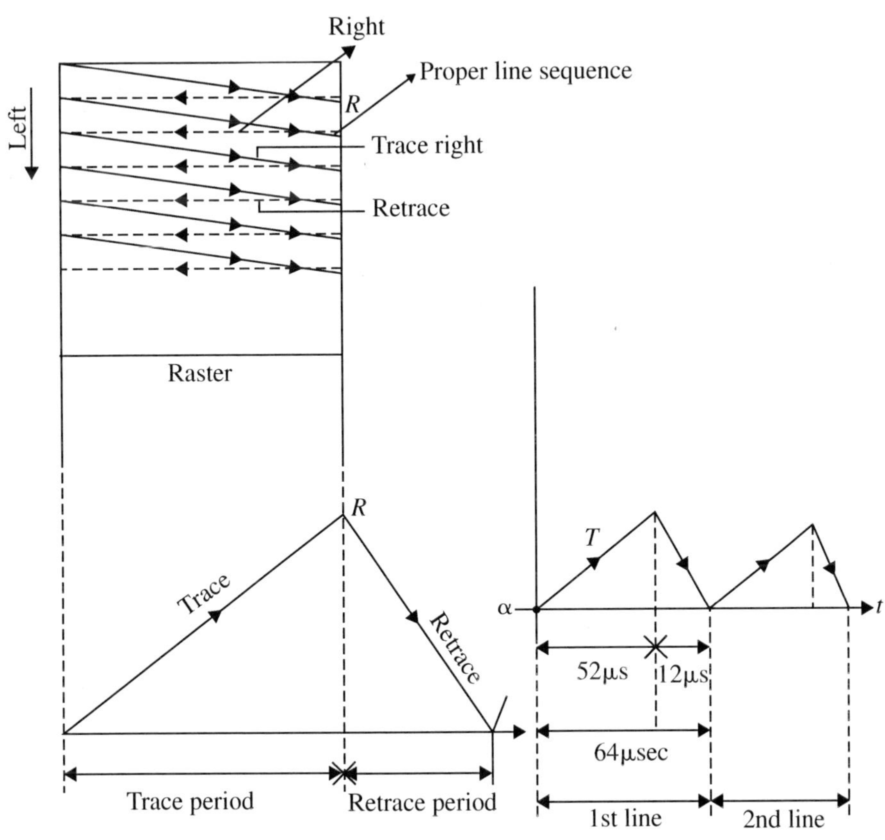

Figure 1.3 Horizontal scanning.

to left of the screen. This is called retrace or fly back period. The retrace period is 12μsec. The period one scanning line is equal to the period of saw tooth current waveform in horizontal deflection coils. Thus the period of one horizontal scanning line includes trace and retrace periods, hence, totally to scan a single horizontal line 64μsec time is needed.

$$\begin{aligned}\text{Horizontal line scanning frequency} &= \frac{1}{T} \\ &= \frac{1}{64\,\mu\text{sec}\ (52\,\mu\text{s} + 12\,\mu\text{s})} \\ &= 15{,}625\text{ Hz}\end{aligned}$$

So the horizontal line scanning frequency is equal to

$$\begin{aligned}\begin{pmatrix}\text{Number of lines}\\ \text{in a frame}\end{pmatrix} \times \begin{pmatrix}\text{Number of frames}\\ \text{transmitted per second}\end{pmatrix} &= 625 \times 25 \\ &= 15{,}625\text{ Hz.}\end{aligned}$$

1.3.4 Vertical Scanning

Vertical scanning is done by shifting the electron beam from top to bottom and again bottom to top as shown in the Figure 1.4 (a) and (b). To produce the deflection, a saw tooth current is passed through the vertical deflection coils in the deflection yoke.

In the Figure 1.4(a), arrows on the lines indicate the path of the scanning beam. The solid lines shows the trace period of the scanning lines. The dotted line indicates the retrace of the beam. Figure 1.4(c) shows the waveform of saw tooth current in vertical deflection coils.

As shown in the waveform, a linear rise of current in the deflection coil shift the electron beam from top to bottom of the screen. This is called vertical trace. As the maximum current is reached, the electron beam is at bottom of the screen. The trace period is a 18.722 ms.

At the peak value, current reverses its direction and decreases rapidly goes to minimum value. This fast reversal moves an electron beam from bottom to top of the screen. This is called vertical retrace or fly back period. The retrace period is 1.27 ms.

The vertical retrace period is very short during the retrace period the electron beam is at cut-off completely. Since horizontal deflection of the beam continues in vertical retrace period also, so many horizontal lines are scanned during the vertical retrace period. But these lines are not visible on the screen because the electron beam is at cut-off. Thus many horizontal lines are wasted during the vertical retrace period.

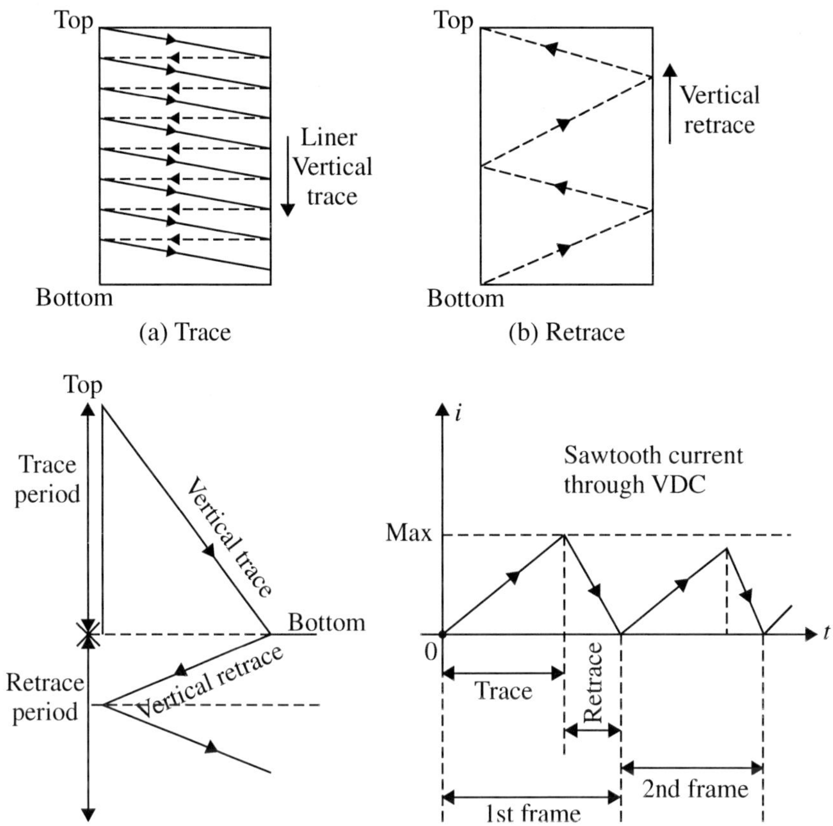

(c) Current wave forms in vertical deflection coils

Figure 1.4 Vertical scanning.

As shown in the Figure 1.4, one complete frame is scanned for one cycle of saw tooth current. Thus 25 frames are scanned in a second so the vertical scanning frequency is 50 Hz.

$$\begin{aligned}
\text{Vertical time} &= \frac{1}{t} = \frac{1}{50\,\text{Hz}} \\
&= 20\,\text{ms} \\
\text{Vertical trace time} &= 18.722\,\text{ms} \\
\text{Vertical retrace time} &= 1.27\,\text{ms} \\
\text{Number of horizontal lines lost} &= \frac{\text{vertical retrace time}}{\text{one horizontal time}} \\
\text{During the vertical retrace} &= \frac{1.28\,\text{ms}}{64\mu\,\text{sec}} = 20\ \text{lines.}
\end{aligned}$$

1.3.5 Sequential or Progressive Scanning

In sequential scanning is the process in which both horizontal and vertical directions are scanned simultaneously to provide complete pictures. Here Horizontal and vertical deflection current are given to the deflection coil simultaneously. So the electron beam is deflected left to right and top to bottom at the same time.

So, the complete scanning is done by line by line. After scanning the first line, second line is scanned. The third line is scanned and so on up to the last line.

1.3.5.1 Disadvantage of Sequential Scanning

The main drawback is that the time taken to complete the cycle is high. And also flicker will be produced, because of the low scanning rate.

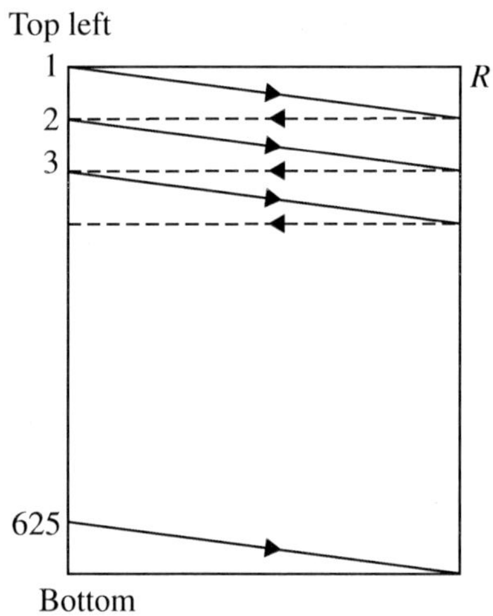

Figure 1.5 Progressive scanning.

1.3.6 Persistence of Vision

The image falling on the retina of the eye remains sensitive for $\frac{1}{16}^{\text{th}}$ of a second. This is called persistence of vision. The impression made by any light, seen by the eye persists (stored) for a small fraction of a second after the light source is removed. If a many views are presented to an eye during all the images at the same time. When the elements are scanned rapidly enough, they appear to eye as complete picture. This is used in cinema and television in obtaining the illusion of continuity by means of rapidly flashing picture frame.

1.3.7 Flicker

The rate of 24 picture per second in motion pictures and 25 frames per second in television picture, is enough to cause an illusion of continuity. So its is enough to scan 16 frames per second. But we are scanning 25 frames per second. This is not rapid enough to allow the brightness of 1st picture frame to blend smoothly into a next picture frame. This results in a definite flicker of light that is very annoying to the observer when the screen is made alternately bright and dark.

This problem is solved in motion picture by showing each picture twice, so that 48 views of the scene are shown per second although there are still the same 24 picture fames per second. In television, the picture frame scanning is done at the same rate as the frequency which is 50 Hz or 60 Hz. As a result of increased blanking rate, flicker is eliminated.

1.3.8 Interlaced Scanning

The flickering effect can be minimized by using interlaced scanning. From the point of view of flicker, it is observed that 50 picture frames per second is the minimum requirement in television scanning. For example, horizontal line scanning frequency of (15,625 + 15,625 = 31,625 Hz) is required for a 625 line system with a period of ($64\mu s/2 = 32\mu s$). This leads to a very high bandwidth of 10 MHz for a desired resolution of 546/2 alterations in the horizontal line if the scanning is the simple sequential way.

To reduce the bandwidth requirement with effective scan rate of 50 Hz 'Interlaced Scanning' is employed. Figure 1.6(a) illustrates the method of interlaced scanning. In this method, the picture is divided into two sets of field, each containing half of other fractional number of interlaced lines and the fields are scanned sequentially.

In Figure 1.6(b) interlaced, the 625 lines are divided into two sets of 312½ lines each. The first 'odd field' of 312½ odd number lines are first scanned sequentially. Half way through the 313th line,the electron spot is returned to the top of the screen (point C) and the remaining 312½ even number lines called the second field the 'even field' are then scanned in between the lines of the first set.

The scanning is done by operating the vertical field scans at 50 Hz so that the two successive interlaced scans each at 25 Hz rate make up the complete picture frame. This keeps the line scanning speed down, as only 312½ lines are scanned in 1/50 sec. The bandwidth requirement is reduced, since the 625 lines of the full picture are scanned 25 times per second, here the area of screen is covered in an interlaced fashion at twice the rate.

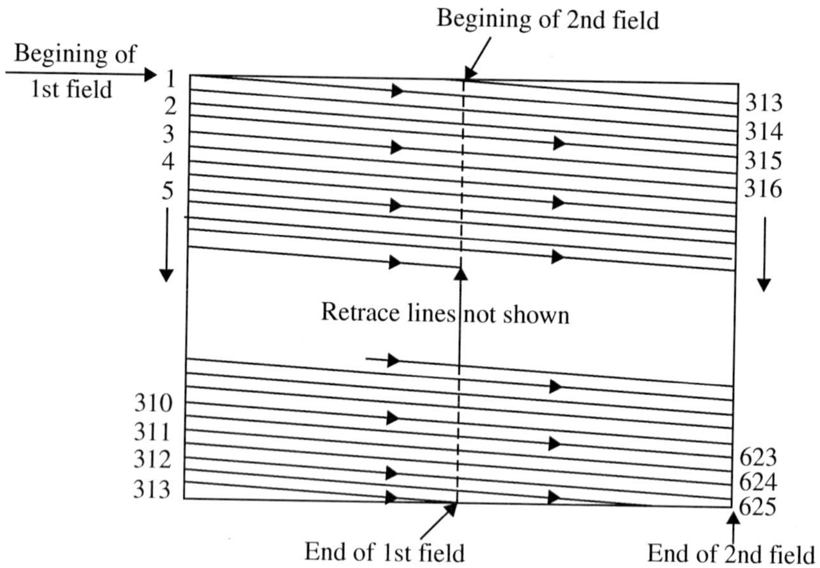

Figure 1.6 (a) Fields of interlaced scanning.

1.3.8.1 Scanning Sequence

During the vertical trace the electron beam starts at point A it is shifted from left to right by the saw tooth current in the deflection coil. The electron beam is deflected downward simultaneously. So at the end of 1st vertical trace the beam is at the point B. Thus during this vertical trace period 292.5 lines are scanned.

From the point B the beam is deflected upward to the point C during the first vertical retrace during this retrace period 20 horizontal lines are lost. Because during vertical retrace period horizontal (trace and retrace) scanning takes place continuously. Hence these 20 lines are in active. During the second vertical trace period the electron beam is at the point C at the centre of the top position.

Now the remaining even lines to be scanned. So, 20 lines are lost per field and in total 40 horizontal lines are lost during the scanning of each frame.

In practice, the fly back from the bottom to the top takes a finite time equal to several line periods. Usually up to 20 lines are allowed for vertical fly back after each of the two fields. This means that out of 625 lines only 585 lines actually bear picture information. These lines are called the 'active lines'. For interlaced scanning, the picture lines must be odd and it is then necessary to supply two regularly spaced synchronising pulses to the field time base for every picture frame.

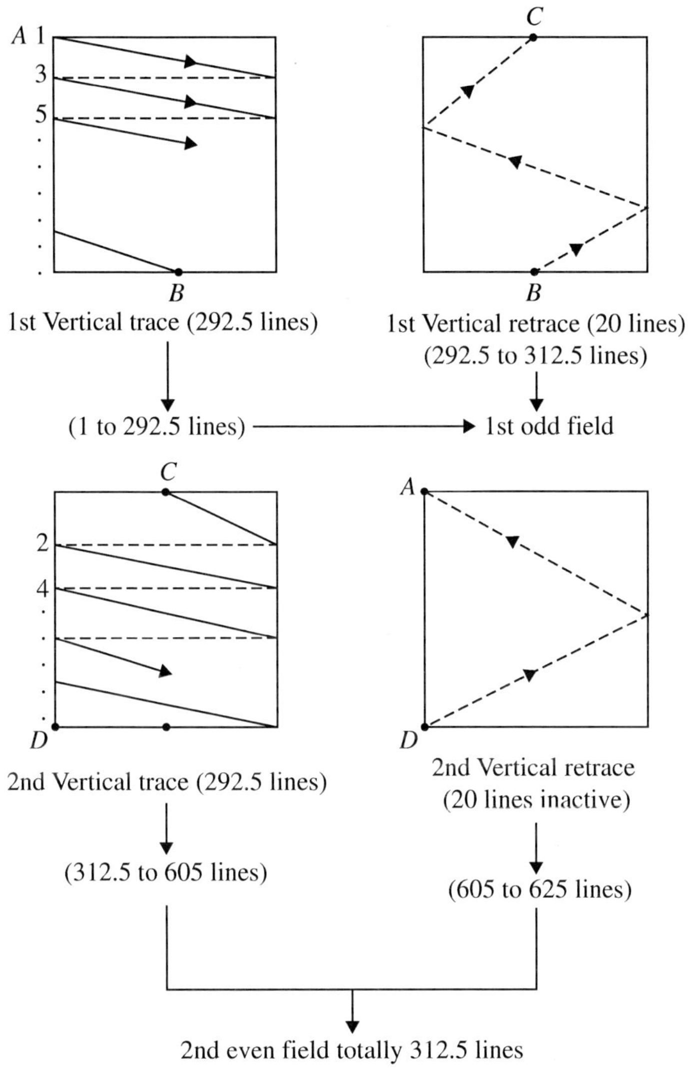

Figure 1.6 (b) Interlaced scanning process.

The odd-line field starts at the top left and ends at the bottom centre, while the even-line field begins at the top centre and ends at the bottom right. This interleaves the two fields exactly to give a complete scanning. The fly back times for both odd and even fields, must be exactly the same.

In the interlaced scanning,

$$\begin{pmatrix}\text{The horizontal}\\ \text{scanning frequency}\end{pmatrix} = \begin{pmatrix}\text{number of lines}\\ \text{in a field}\end{pmatrix} \times \begin{pmatrix}\text{number of frames}\\ \text{transmitted per second}\end{pmatrix}$$
$$= 312.5 \times 50 = 15{,}625\ \text{Hz}$$

$$\text{Vertical scanning frequency} = 50\,\text{Hz}$$

$$\text{Vertical time} = \frac{1}{t} = \frac{1}{50\,\text{Hz}} = 20\,\text{ms}$$

$$\text{Vertical trace time} = 18.722\,\text{ms}$$

$$\text{Vertical retrace time} = 1.27\,\text{ms}$$

$$\text{Number of horizontal lines lost} = \frac{\text{vertical retrace time}}{\text{one horizontal time}}$$

$$\text{During the vertical retrace} = \frac{1.28\,\text{ms}}{64\mu\,\text{sec}} = 20\text{ lines}$$

So, 20 lines are lost per field and in total 40 lines are lost during the scanning of each frame.

$$\begin{pmatrix}\text{Active number}\\ \text{of lines}\end{pmatrix} = \begin{pmatrix}\text{total number of}\\ \text{lines in a frame}\end{pmatrix} - \begin{pmatrix}\text{total number of horizontal lines}\\ \text{lost during even and odd fields}\end{pmatrix}$$

$$N_a = 312.5 + 312.5 - 20 + 20$$

$$N_a = 585\text{ lines}$$

In active number of lines = 40.

1.4 NEED FOR SYNCHRONISING PULSES

Synchronising pulses are transmitted during the blanking period along with the original video signal from the transmitter through the receiver. These pulses are used to keep the scanning sequence coincidence with each other in the camera and picture tube, these pulses make the horizontal and vertical scanning in synchronization both at camera tube and picture tube. Otherwise distorted picture will be seen on the screen. These pulses are sent during the blanking period (retrace period), so that sync pulse will not be visible to the eyes.

The sync pulses are given to the horizontal and vertical deflection oscillators in the receiver side. Identified, separated and used for triggering. So it can control exact timing coordination (starting and ending) of the scanning lines in the picture tube. Then only the image focused on the camera tube will be reproduced on the picture tube screen.

1.4.1 Horizontal and Vertical Synchronization

At the receiver the picture tube scanning beam must reassemble picture elements on each horizontal line with the same left-right position as the camera tube. Similarly, as the beam scans vertically the successive scanning lines on the picture tube screen must present the same picture elements as corresponding lines at the camera tube.

Hence for each horizontal line, a horizontal sync pulse is transmitted to keep horizontal synchronization. Similarly a vertical sync pulse is transmitted for each field to synchronize the vertical scanning motion. The horizontal sync pulse has frequency of 15,625 Hz and frequency of the vertical sync pulses is 50 Hz.

The synchronising pulses are transmitted as a part of the picture during the blanking period when no picture information is transmitted. The sync signals are combined with the picture signal in such a way that part of the modulated picture signal amplitude is used for the sync pulse and the reminder for the camera signal. It can be observed that all the pulses have the same amplitude but differ in pulse width. The sync pulses shown in Figure 1.7 include:

(i) the horizontal pulses from left to right.
(ii) a group of five equalising pulses.
(iii) a serrated vertical pulse.
(iv) five additional equalising pulses.

This is followed by three more horizontal pulses.

For every field must be one wide vertical pulses which is actually composed of five individual pulses separated from each other at the receiver. The five separations are inserted in the vertical pulse at half-line intervals. The equalization pulses are also spaced at half-line intervals. Thcsc half-line pulses are used for horizontal synchronization, while alternate pulses are used for even and odd fields.

The equalising pulses are used to provide identical wave shapes in the separated vertical sync signal for even and odd fields to obtain constant timing for good interlace, their repletion rate is twice 15,625 or 31,250 Hz. The sync enables the picture information reproduced on the raster to hold still in the correct position. If there is no vertical sync, the similarly if

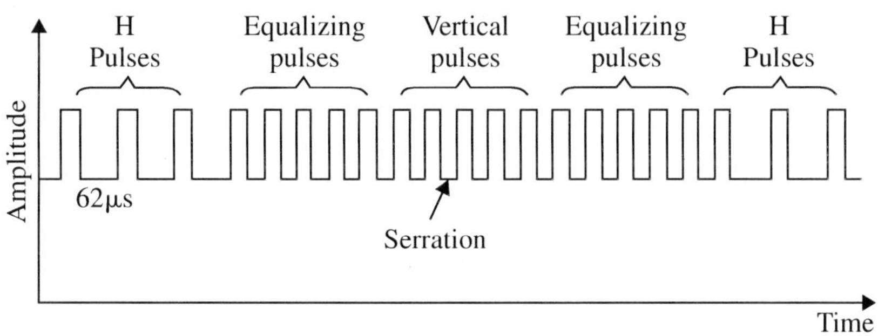

Figure 1.7 Synchronising pulses

there is no horizontal sync, the picture tilts to the left or right and the line structure breaks into diagonal segments.

1.5 KELL FACTOR

The maximum scanning lines should be around 860 for distinct resolution. This is true when the scanning beam passes over a single line. And its scans all the line from top to bottom.

Because of practical limitations the vertical resolution is reduced than the theoretical value. Around 70% of total lines get scanned separately and remaining 30% of lines get merged with other lines due to finite beam size and alignment errors thus the effective number of lines distinctly resolved are:

$$n_r = K \times n_v$$

where n_r is the number of distinctly resolved lines, n_v is the number of horizontal scanning lines and Kell factor (K) or resolution factor.

Thus the K gives the effective number of horizontal lines those can scanned because of practical limitations. The value of Kell factor lies between 0.65 to 0.75. Thus Kell factor multiples the theoretical maximum lines to give 'effective' scanning lines.

For $n_v = 585$, we get n_r as

$$n_r = 0.69 \times 585 = 400 \text{ lines.}$$

Deciding factors for scanning lines:

(i) The effective number of scanning lines can be resolved around 602 (*i.e.*, $n_r = K \times n_v = 0.69 \times 860 = 593.4$).

(ii) The value of Kell factor lies between 0.65 to 0.75 thus Kell factor multiples the theoretical maximum lines to give 'effective' scanning lines.

(iii) The channel bandwidth increases with increase in scanning lines. This reduces the number of channels those can be transmitted in a particular band.

(iv) The number of scanning lines increased means the vertical resolution is better but the maximum numbers of scanning lines depend upon the resolving capability of the human eye.

(v) The human eye can resolve maximum 860 scanning lines theoretically. But practically we observed that the improvement in sharpness detected by the eye is not match for scanning lines greater than 500.

1.6 NEED FOR BLANKING PULSE

During scanning process of every horizontal line the electron beam is deflected from right to left for retrace. During retrace period no video signal is applied to the cathode of the picture tube. But, the electron beam from the cathode will strike the picture tube screen to produce a bright white horizontal line. Similarly at the end of the every field there is a vertical retrace.

During this period also electron beam can strike the screen to produce bright vertical line. So, to avoid this retrace period. The retrace line should be made invisible. For this purpose we are sending blanking pulses along with the video signal, from the transmitter to the receiver. There are two types of blanking pulses they are:

(i) Horizontal blanking.
(ii) Vertical blanking.

1.6.1 Horizontal and Vertical Blanking Pulse

The term blanking means going to black. Blanking voltage is the part of the video signal voltage at the black level. This voltage cuts off the beam current in the picture tube to blank out light from the screen. The blanking pulses are utilised to make invisible the retraces required in scanning. The horizontal blanking pulses at 15,625 Hz black out the retrace from right to left for each line. Similarly vertical blanking pulses 50 Hz blank out the retrace from bottom to top for each field.

1.6.2 Horizontal Blanking

These pulses increase the signal amplitude to about 75% *i.e.*, above the black level. And also during this period the control grid of the picture tube is made more negative with respect to cathode. So the electron beam cannot strike the picture tube screen. Thus the horizontal retrace is made invisible.

Approximately 19% of each horizontal line is the period of time for horizontal blanking. The total time for H is 64 s including trace and retrace. Hence the blanking time for each line is $64 \times 0.19 = 12.16\mu s \approx (12\mu s)$ as shown in the Figure 1.8.

1.6.3 Vertical Blanking

These pulses increase the signal amplitude to about 75%, *i.e.*, above the black level. And also during this period the control grid of the picture tube

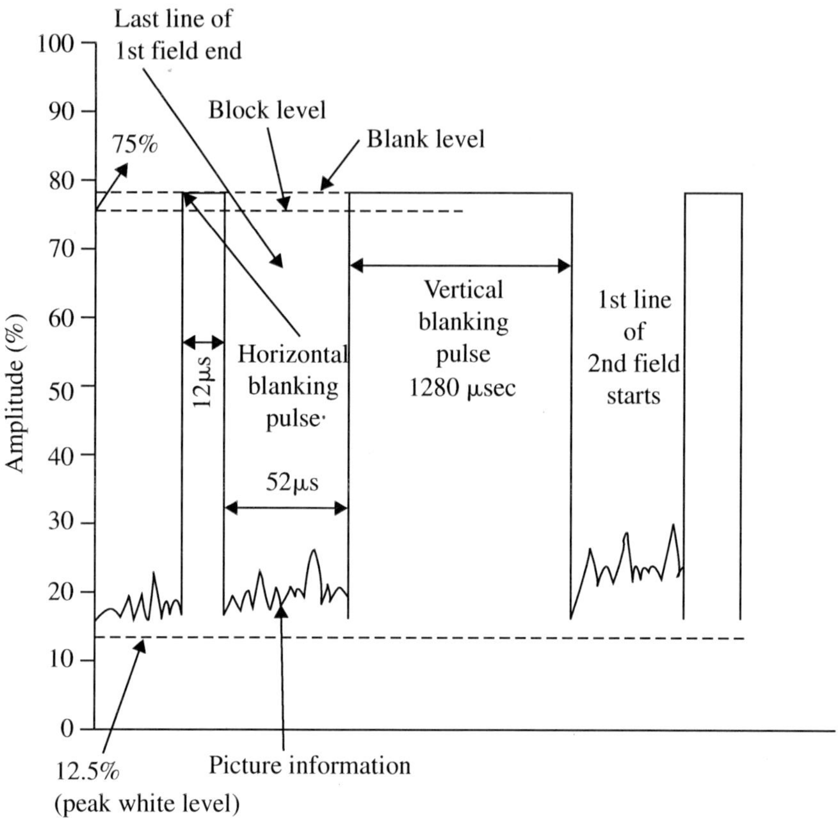

Figure 1.8 Blanking pulse.

is made more negative with respect to cathode. So the electron beam cannot strike the picture tube screen. Thus the vertical retrace is made invisible.

The time for vertical blanking approximately 8% of each vertical field. The total time for V is $1/50\mu$ sec including downward trace and retrace. Hence the blanking time for each field is given by $1/50 \times 0.08 = 0.0016$ sec, *i.e.*, 1600μ s. This means that within 1420μ s, the vertical retrace must be completed from bottom to top the picture.

During the blanking time, the retrace occurs because of synchronization of the scanning. The sync pulses coincide with the start of the retraces. Each horizontal sync pulse is inserted in the video signal within the time of the horizontal blanking pulse. This means that a blanking pulse comes first to put the video signal at blank level and then a sequence applies to blanking horizontal and vertical retraces. A typical blanking signal is illustrated in Figure 1.8.

1.7 ASPECT RATIO

Practical test have indicated that the best viewing comfort and artistic appreciation are obtained when the picture screen has a rectangular format with the aspect ratio *i.e.*, the width to height ratio of 4.3. This is represented in point wise:

(i) The movement of both the eyeballs are less restricted In the horizontal plane as compared to that in the vertical plane.
(ii) Also the region of the maximum resolution at the width than along the height.
(iii) The rectangular frame with aspect ratio of 4:3 enables direct television transmission of film programmes without wastage of any film area.
(iv) In motion picture format with an aspect ratio of 4:3 is commonly employed as it is most pleasing artistically and less fatiguing to the eye. It is given as:

$$\text{Aspect ratio} = \frac{\text{Width of the picture}}{\text{Height of the picture}}$$

1.8 RESOLUTION

The ability of the image reproducing system, to resolve the fine details of the picture, distinctly both in horizontal and vertical direction is called resolution of the system. By increasing the number of picture elements, the resolution is increased. Also the quality of reproduced picture elements is increased. So it is always good to have high resolution. Resolution is classified into two types. (i) vertical resolution and (ii) horizontal resolution.

1.8.1 Vertical Resolution

The ability of a scanning system to resolve vertical details in a scene depends upon the number of scanning lines used per frame. This is generally referred to as the scan ratio of a television system. The realistic limit to the number of lines is set by the resolving capability of the human eye, *viz*, about one minute of visual angle. For comfortable viewing, an angle of about 10 to 15 can be taken as the optimum visual angle. Hence the best viewing distance for watching television is about 10° as shown in Figure 1.9.

The maximum number of dark and white elements which can be resolved by the human eye in the vertical direction in a screen of height H is given by, the number of lines of vertical resolution, according to the

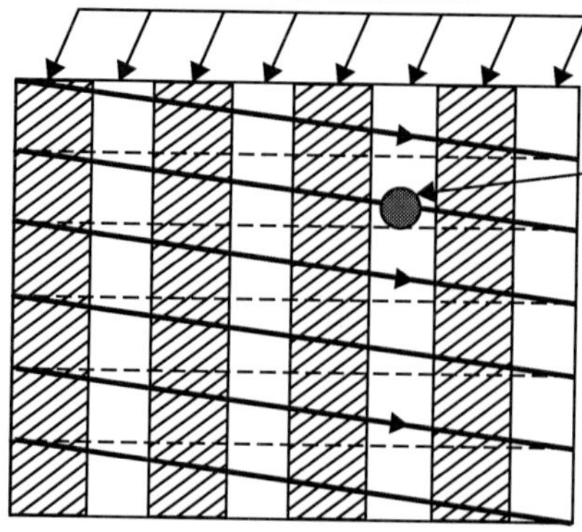

Figure 1.9 Vertical resolution.

relation:

$$\frac{H}{d} = n_v \times \alpha_0$$

where n_v is the number of black and white lines of resolution, α_0 the minimum angle of resolution in radians, and D the distance of the viewer. Thus, for $d/H = 6$, *i.e.*, visual angle about 10°

$$n_v = \frac{H}{d}\alpha_0 = \frac{1}{6} \times \frac{60}{1} \times \frac{180}{\pi} = \text{approximately 600 lines}$$

for $d/H = 4$, *i.e.*, visual angle about 15°

$$n_v = \frac{1}{4} \times \frac{60}{1} \times \frac{180}{\pi} = \text{approximately 900 lines.}$$

The maximum numbers of alternate dark and white elementary horizontal lines which can be resolved by the eye are thus 600 for a 10°.

1.8.1.1 Visual Angle or 900 for a 15° Visual Angle

In the 625 line system, the number of active lines left after deducing lines lost in vertical blanking are (625 – 40) = 585 lines. With a Kell factor of 0.7, the vertical resolution is $(0.69 \times 585) = 400$ lines. The horizontal resolution may not exceed this value, multiplied by the aspect ratio.

$$\text{Vertical resolution} = K \times n_a = 0.69 \times 585 = 400 \text{ lines}$$

1.8.2 Horizontal Resolution

The horizontal resolution of a television system is the ability of the scanning system to resolve the horizontal details. That is, changes in brightness levels of elements along a horizontal scanning line. The horizontal resolution can be expressed as a measure of the ability to reproduce horizontal information or lines of resolution as shown in Figure 1.10.

The horizontal resolution in a scanning system depends upon the rate at which the scanning spot is able to change brightness level as it passes through a horizontal line across the vertical lines of resolution shown in the Figure 1.10.

In a 625 line system, there are effectively about 400 lines of vertical resolution. The horizontal resolution should be the same order. Because of the aspect ratio of 4:3, the number of vertical lines for equivalent horizontal resolution will be $(400 \times 4/3) = 533/2 \approx 266$ cycles of black and white alterations of elementary areas. For the 625 line system the horizontal scan or line frequency (fh) is given by:

$$fh = \frac{\text{trace period of each horizontal line}}{\text{number of cycles}}$$

$$= \frac{52\mu s}{266}$$

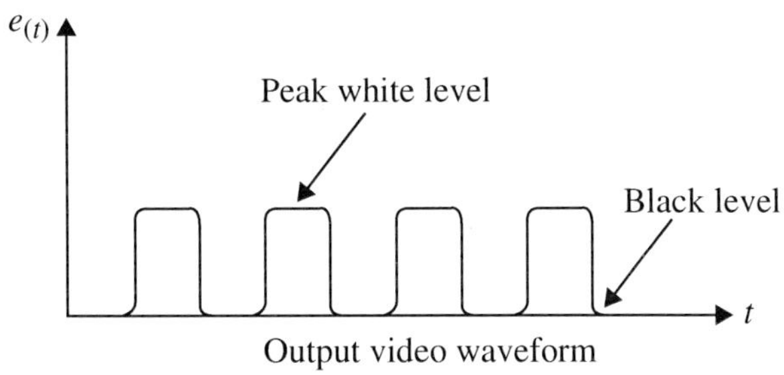

Figure 1.10 Video bandwidth.

$$\text{Horizontal resolution} = n_a \times \text{aspect ratio} \times K$$
$$= 585 \times \frac{4}{3} \times 0.69 = 533$$

so cyclic changes along one horizontal line is equal to ≈ 266.

1.9 VIDEO BANDWIDTH

To find the bandwidth of the video signal, the chess board pattern is taken. A video signal of the form square wave is switching continuously along the line between voltage levels corresponding to black and white it is shown in the Figure 1.11.

Of this period, 12μ sec are used for the blanking of the fly back retrace. Thus the 533 black and white alterations, *i.e.*, 266 cycles of complete square waves, are scanned along a horizontal raster line during the forward scan time of $(64 - 12) = 52\mu$ s. The period of this square wave is $52\mu s/266 =$ 5 MHz, giving the highest fundamental frequency of 5 MHz. This band width is adequate for the highest video frequency in the signal.

$$\text{Time for one horizontal trace} = 64\mu s - 12\mu s = 52\mu s$$
$$\text{Number of cyclic changes} = \frac{\text{horizontal resolution}}{2} = \frac{533}{2} = 266$$
$$\text{Time taken for one square wave} = \frac{\text{horizontal trace time}}{\text{no. of cyclic changes}}$$
$$t = \frac{52\mu s}{266} = 0.1954\ \mu s$$

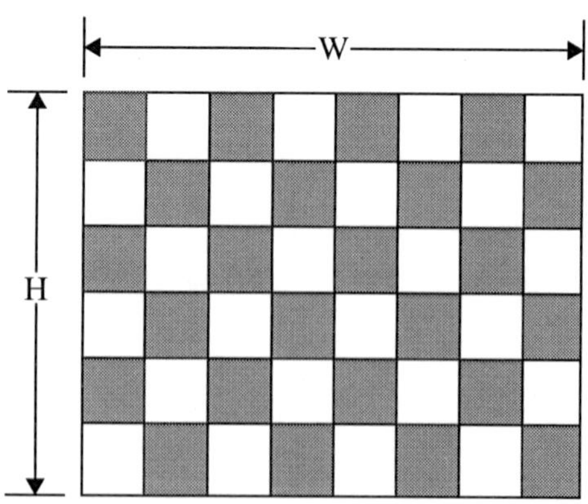

Figure 1.11 Chess board pattern for studying horizontal and vertical resolution.

$$\text{Maximum video frequency} = \frac{1}{t} = \frac{1}{0.1954\mu s} =\approx 5\,\text{MHz}.$$

1.10 CONSTRUCTION OF THE COMPOSITE VIDEO SIGNAL

The composite video signal is formed by the electrical signal corresponding to the picture information in the lines scanned in the TV camera pick-up tube and the sync signals introduced in it.

The three components of a composite video signal are illustrated in Figure 1.12 also illustrates how these three components are added to produce composite video signal. Figure 1.12 gives the successive values of voltage or current amplitude for the scanning of three horizontal lines in the image. The lower 75% is used for camera signal with the upper 25% for synchronizing pulses. In the camera signal the lowest amplitudes correspond to the whitest parts of the picture. While the darker parts of the picture have higher amplitude. In this way the signal is transmitted using negative polarity transmission. In the negative transmission white parts of the picture correspond to the low amplitude in the transmitted picture carrier signal. Higher amplitudes represent darker picture information until the black level is reached.

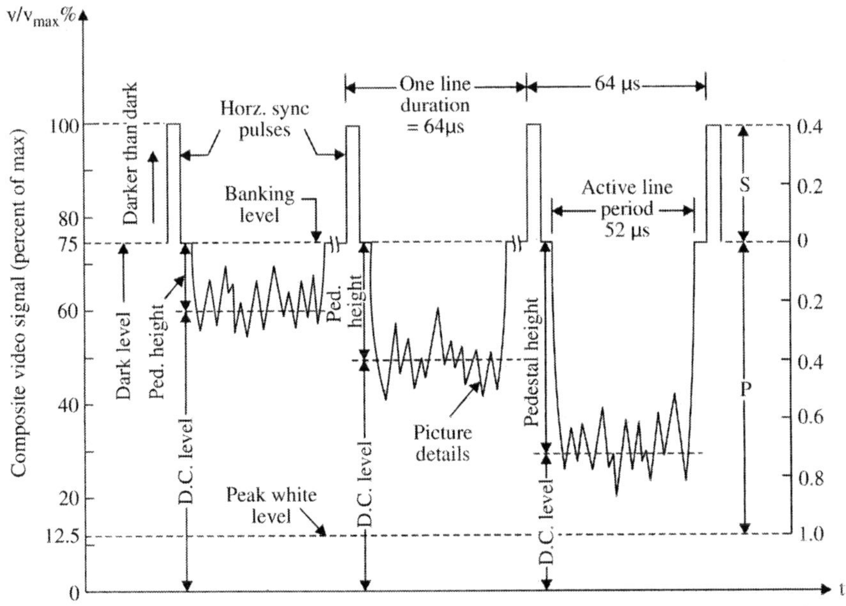

Figure 1.12 (a) Composite video signal.

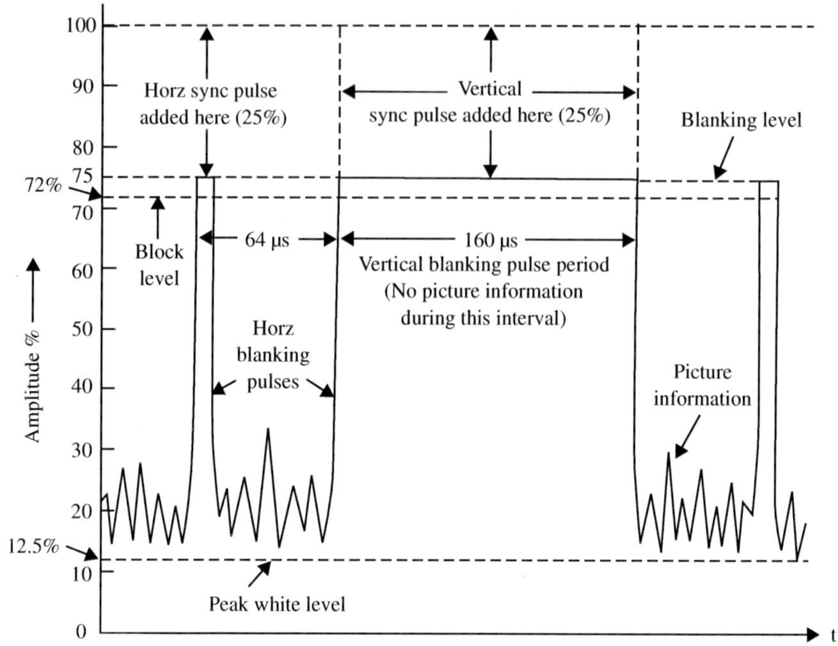

Figure 1.12 (b) Composite video signal for three lines.

1.10.1 Horizontal Sync Details

The horizontal blanking period and sync pulse details are illustrated in the Figure 1.12(b). The horizontal blanking period and sync pulse details for one horizontal line are shown in Figure 1.12(b).

$$\text{Total line period for one horizontal line} = 64\mu s$$

$$\text{Blanking period} = 12\mu s$$

During this blanking the synchronizing pulse are inserted. The blanking period consists of three sections they are:

(i) Front porch.
(ii) Line sync pulse.
(iii) Back porch.

1.10.1.1 Horizontal Line Period H

It is the total duration of one complete horizontal line period and it is given by:

$$H = \frac{1}{F_H} = \frac{1}{15{,}625} = 64\mu s.$$

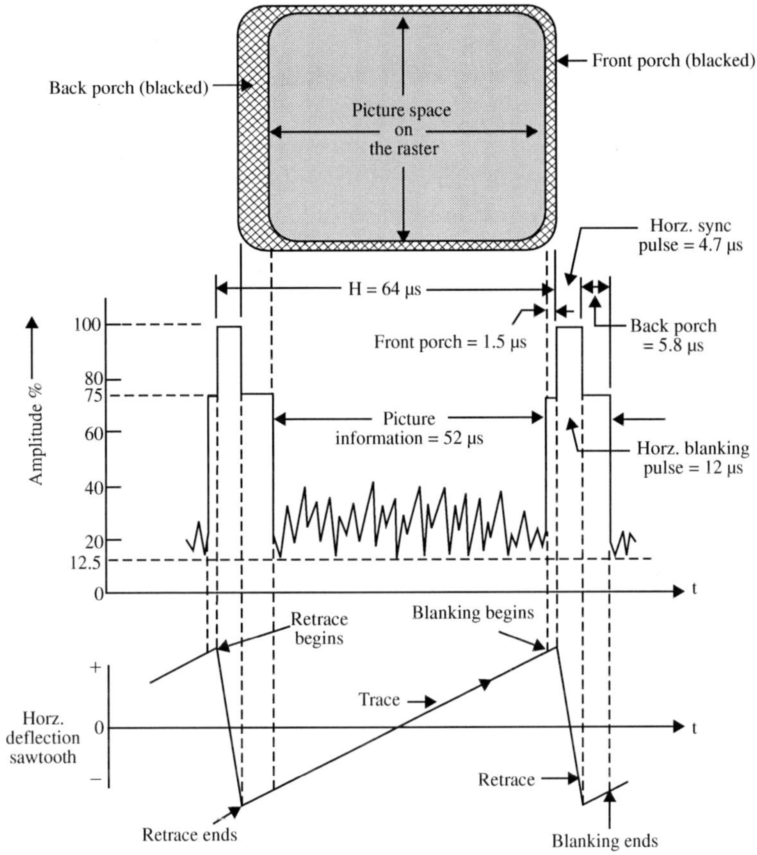

Figure 1.13 (a) Horizontal sync and (b) Blanking signal standards.

1.10.1.2 Horizontal Blanking Period

It is the part of each line during which line sync pulse is inserted. During this period, the beam fly back is initiated and completed and the beam cut off by the black level amplitude of video signal. This value is $0.19\,H$ and it is shown in Figure 1.13(b).

$$0.19H = 0.19 \times 62\mu\text{s} = 12\mu\text{s}$$

1.10.1.3 Horizontal Sync Pulse

It is the important short pulse sent from the transmitter to correct the horizontal scanning rate at the receiver when it deviates from the desired rate. This sync pulse is used to synchronize transmitter and receiver.

$$\text{Width} = 0.07 \times H = 0.07 \times 62\mu\text{s} = 4.7\mu\text{s}$$

1.10.1.4 Front Porch

The start of the sync does not coincide with the blanking but it follows after about 2% of the line period. This short period is called the front porch. Front porch is 2.5% width $H = 1.5\mu$ sec. This interval allows the receiver video circuit to settle down from picture video voltage level exists at the end of the picture line to the blanking level before the starts of sync pulse occurs. Thus sync circuits at the receiver are isolated from the influence to end of the line picture details.

1.10.1.5 Back Porch

This period of $5.8\mu s$ at the blanking level allows plenty of time for line fly back (retrace) to be completed. It also permits time for the horizontal time-base circuit to reverse direction of current for the initiation of the scanning of next time. The back porch also provides the necessary amplitude equal to the blanking level (reference level) and enables to preserve the DC content of the picture information at the transmitter.

At the receiver this level which is independent of the picture details is utilised in the automatic gain control (AGC) circuits to develop AGC voltage proportional to the signal strength picked up at the antenna.

1.10.1.6 Pedestal

In certain systems, the blanking level is blacker than black. The difference between the two levels is called the 'pedestal'.

1.10.1.7 Details of Horizontal Scanning

Periods	Time (μs)
Total line period (H)	64
Horizontal blanking	12 ± 0.3
Horizontal sync pulse	4.7 ± 0.2
Front porch	1.5 ± 0.3
Back porch	5.8 ± 0.3

1.10.2 Vertical Sync and Blanking Standards

At the end of the each odd field and even field a vertical sync pulse is added. The width of the vertical sync pulse has to be kept much larger than the horizontal sync pulse as shown in the Figure 1.14.

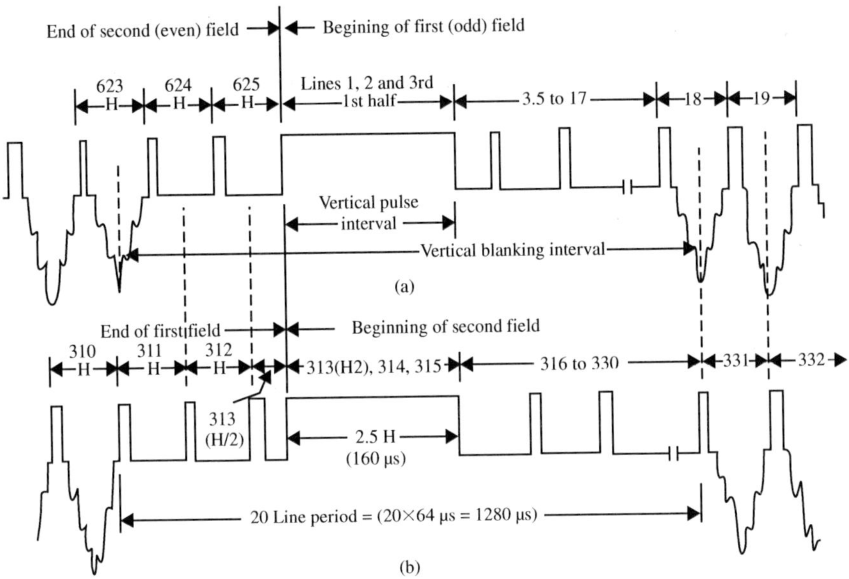

Figure 1.14 Composite video signal for horizontal and vertical sync pulses.

The standard specifies that the vertical sync period should be 2.5 to 3 times greater than the horizontal line period. If the width is less than this, it is difficult to distinguish between horizontal and vertical pulses at the receiver. If the width is greater than this (160μs) the transmitter requires larger power to operate.

1.10.2.1 Vertical Sync Pulse

In the 625 line system the vertical sync pulse has a duration of 2.5 line period (*i.e.*, $2.5 \times 62\mu s = 160\mu s$). At the end of the first field the vertical sync pulses is added during the period (312.5 to 315) similarly at the end of the second field another sync pulse is added during the line period of (1 to 2.5).

From the above, it is clear that one vertical sync pulse ends at the 1/2 line period and other at the full line period. This alignment of vertical sync pulses results in a relative misalignment of the horizontal sync pulses they do not appear one above the other but occur at half-line intervals with respect to each other.

1.10.2.2 Vertical Blanking Period

It is the period during which the picture information is completely suppressed and the flyback retrace of the field time base is initiated and completed. While the beam is cutoff by the blank level. This duration is equal to 20 line periods.

$$\text{Vertical blanking period} = 20H = 20 \times 62\mu\text{s} = 1280\mu\text{s}$$

1.10.2.3 Serrated Vertical Pulse

Horizontal sync pulses are available both during vertical trace and retrace period but there is no horizontal sync pulse available during 2.5 line vertical sync period. Due to this the horizontal sweep oscillator operating at (15,625 Hz) may come out of synchronization during each vertical sync period.

To maintain the synchronization five narrow slots of 4.7μs width are formed in vertical sync period at the interval of 32μs. The rising edges of these pulses are used to trigger the horizontal oscillator. This insertion of short pulses is known as serrated pulses. This is illustrated in Figure 1.14.

1.10.3 Equalizing

In TV transmission both vertical and horizontal sync pulses are transmitted along with video signal. In the receiver the horizontal sync pulses are separated by high pass filter. The vertical sync pulses are separated by

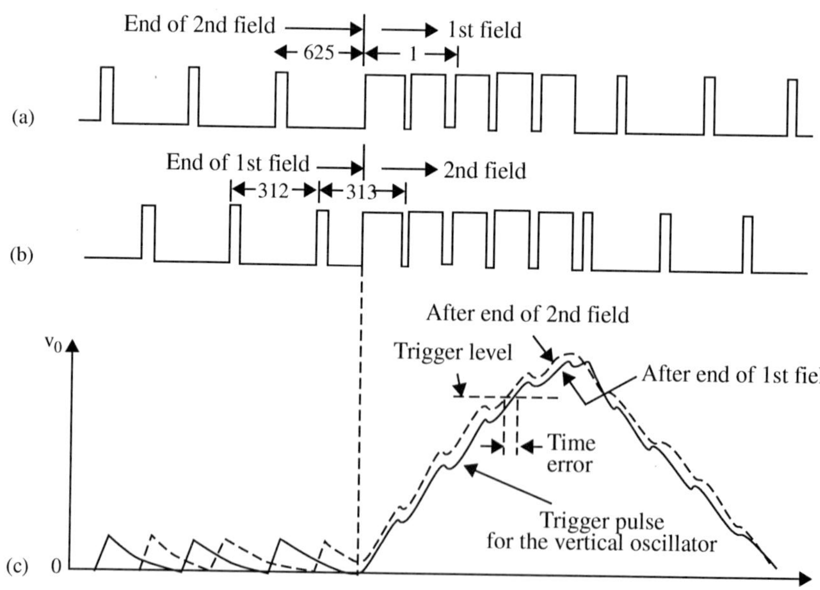

Figure 1.14 Field sync and blanking signal standards. (a) Odd field; (b) Even field; and (c) Serrated vertical sync detail.

integrator circuit or low pass filter. The integrated output signal is used to trigger the vertical oscillator and hence the vertical oscillator frequency is locked to 50 Hz as shown in the Figure 1.14(b).

The output waveform obtained at the integrator is shown in the Figure 1.14(c). At the beginning of the first field the separation between the line sync pulse (625) and vertical sync pulse is a full line period of 64μs. But at the beginning the separation between vertical and line sync pulse (313) is only half line period of 32μs.

Due to this half line period difference between odd and even field causes triggering timing error for vertical oscillator (Figure 1.15). To overcome this problem five equaling pulses of both the sides of the vertical sync pulses are added. These equalising pulses are used to provide identical wave shapes in the separated vertical sync signal for even and odd fields to obtain constant timing for good intra-lase. Both the pre-equalising pulses and post equalising pulses are of fixed pulse width of 2.3μs each.

The addition of these pulses shift the half line difference away from the beginning and end of vertical sync pulses pre-equalizing having 2.3μs duration result in the discharge of the capacitor essentially to zero voltage in both the fields post equalizing pulses are necessary for a fast discharge of capacitor to ensure triggering of the vertical oscillator at proper time (*i.e.*, exactly at an interval of $\frac{1}{50}^{\text{th}}$) of a second. If the discharge of the capacitor is slow then the trailing edge of the integrator may false trigger the oscillator.

1.11 VIDEO MODULATION

Amplitude modulation is used for the video signal. The composite video signal has distinct polarity for black and white levels in the picture. Hence two types of modulations are possible to represent the black and white levels they are: (i) Positive modulation and (ii) Negative modulation.

1.11.1 Positive Modulation

When the intensity of the picture increases, then the video signal amplitude increases, this is called the positive modulation. In this, the increase of brightness towards white level increases the carries amplitude. The peak white has 100% modulation. While the black and sync pulses have lower and minimum modulation.

1.11.2 Negative Modulation

When the intensity of the picture increases the video signal amplitude decreases. It is known as negative modulation. In this, the sync top levels

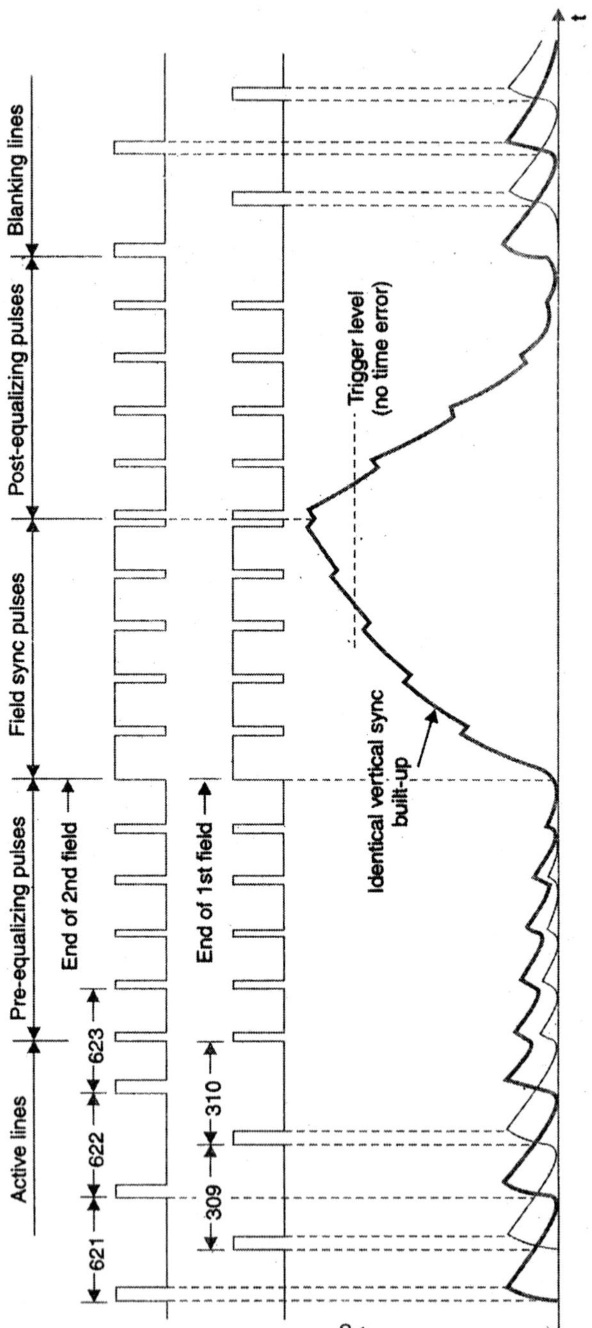

Figure 1.15 Equalizing pulses and integrator output.

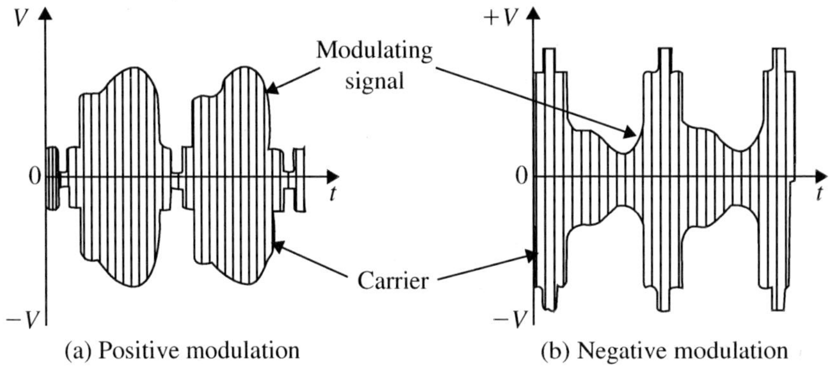

Figure 1.16

are at 100% modulation. Blanking level corresponds to black is at 75%, and increase in brightness towards white causes a decrease in carrier amplitude down to minimum of 10% at the white.

1.11.3 Merits of Negative Transmission

(i) Less disturbance due to noise pulses.

(ii) Less transmitting power is required to transmit because most of the picture information is white.

(iii) As a picture frame consist of more bright areas the white level which has 12.5% of signal amplitude is more. The white lever having this 12.5% of amplitude is appearing in large quantity in a picture frame. Hence it is enough to have less transmitter power for –ve transmitter.

(iv) Most of the AGC circuits in receivers measure the peak level of the incoming carrier signal and adjust the gain of RF and IF amplifiers.

According to in negative modulation system, the sync pulse level is at 100% peak level and it is used as a reference level for AGC *i.e.*, minimum

Positive Transmission	Negative Transmission
Video signal corresponding to white level has the amplitude of 75 to 100%.	Video signal corresponding to white level has only 12.5% of video signal amplitude.
Blank level corresponds to only 25% of video signal amplitude.	Blank level corresponds to 75% of video signal amplitude.
Sync level corresponds to 25% to minimum video signal amplitude.	Sync level corresponds to 75% of video signal amplitude.
Noise appears as white dots.	Noise appear as block dots.
Transmitter power is more.	Transmitter power is less.
Efficiency is less.	Efficiency is more.

white level amplitude less transmitter power. But in +ve transmission with higher white lever amplitude more transmitter power is needed. As −ve transmission needs less transmitter power the efficiency is more.

1.12 VESTIGIAL SIDEBAND TRANSMISSION

The video signal bandwidth of the 625 line system is 5 MHz. To reduce the channel bandwidth of video signal we are going to the vestigial sideband modulation. Because in double sideband AM.

We need 11.25 MHz channel bandwidth on both the side bands, the video signal frequency extend up to 5 MHz on both the side of picture carrier (PC) with attention slop 0.5 MHz, a 1 MHz is wasted on both the sidebands.

We have to provide a guard band 0.25 MHz between the channels totally channel bandwidth = 5.5 + 5.5 + 0.25 = 11.25 MHz.

As both sidebands have the same information so it is enough to send one sideband with carrier. But we cannot completely suppress one sideband using the practical filters. Also in the 10 W video frequency around the picture carrier, there are very fine details of the picture *i.e.*, the very fine details of the picture can be reproduced only by the 10 W frequency signals. hence we cannot completely attenuate the 10 W frequency signals on the suppressed LSB.

1.12.1 Transmission and Reception of Vestigial Sideband Signal Transmission

Since an extension bandwidth is required for the video signal it is desirable to use bandwidth saving technique. Both the two sidebands of the AM modulated carrier contains the signal information and one sideband may be suppressed altogether provided the carrier is present. This single sideband transmission technique reduces the bandwidth requirements to half *i.e.*, 5 MHz. However it is not possible to do this in the case of television signal.

In this transmission full upper side band is transmitted along with vestige of (small part) of lower side band in the lower side band 0–0.75 MHz as shown in the spectrum. The remaining frequencies on the LSB are completely suppressed hence the channel here by using VSB transmission we can say band saving is equal to 4.25 (*i.e.*, 11.25 − 7 = 4.25 MHz).

$$\begin{aligned} \text{BW} &= \text{attenuation slop} + \text{vestige part of lsb+usb+att.slop+guardband} \\ &= 0.5 + 0.75 + 5 + 0.5 + 0.25 = 7\,\text{MHz.} \end{aligned}$$

Practically it is not possible to design a filter which will cut the unwanted band while passing the carrier frequencies and low frequency components of the other sideband without objectionable phase distortion. As compromise, a part of only one sideband is suppressed. The transmitted signal consists of one complete sideband together with the carrier, and vestige of the partially suppressed sideband as shown in Figure 1.17(a).

1.12.2 Reception

In the VSB signal, the lower modulation frequencies from 0 to 0.75 MHz are present in both sidebands while the rest of the frequencies from 0.75 to 5 MHz are present only in one sideband. If all the channel frequencies are equally amplified in the RF and IF amplifiers of receiver and then

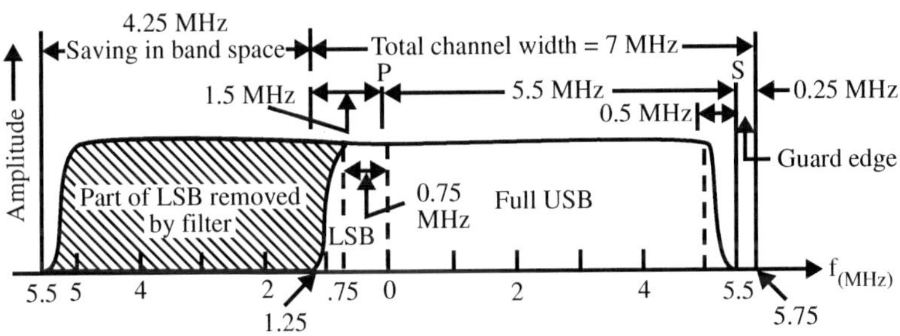

Figure 1.17 (a) VSB transmission.

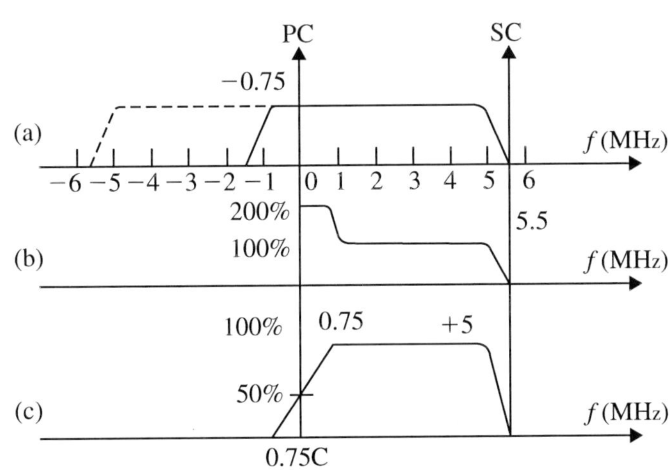

Figure 1.17 (b) VSB channel; (c) Video power content and (d) Receiver response.

it is fed to video detector, it will increase the amplitude, to a total of twice that amplitude output relative to the higher frequencies. The double sideband frequencies from 0.75 to 5 MHz that have only a single sideband component, as illustrated in Figure 1.17(c).

In order to make this eventuality, it is necessary to shape the picture IF response curve of the receiver, so that the frequency in the double sideband range are provided less amplification than the rest. It is done by placing the picture carrier half way down the slop of a response curve. The response curve slops linearly over the double sideband frequency range from −0.75 to +0.75 MHz with respect to picture carrier as shown in the Figure 1.17(d). It is observed that the width of the sloping edge is twice the width of the vestigial sideband and the sum of the output contribution from the two sideband together make the equal amplitude at all the frequencies.

1.12.3 Demerits of Vestigial Sideband Transmission

(i) A small portion of the transmitter power is wasted in the vestigial sideband filters which remove the remaining lower sideband.
(ii) In order to provide vestigial sideband compensation, the picture IF carrier is placed at 50 percent response of the video IF amplifier. This introduces a loss of about 6 dB in the signal to noise voltage ratio.
(iii) Some phase and amplitude distortion of the picture signal occurs, because of the filter design at transmitter.
(iv) The designing of the VSB filter is not easy. So, most of the transmitter power is wasted in the filters.
(v) In the receiver only 50% of gain is provided to picture carrier PC. Though the signal strength reduces and the signal to noise ratio (S/N ratio is also reduced).
(vi) The overall gain is reduced up to 6 dB for picture carrier.
(vii) Due to very small guard band the receiver tuning is very critical.

1.12.4 Merits of Vestigial Sideband Transmission

(i) Channel bandwidth reduces to 7 MHz.
(ii) The transmitter power is also reduced.
(iii) More number of channels can be transmitted.

1.13 INTER-CARRIER SOUND MODULATION

In television signal transmission, Amplitude Modulation (AM) is used for transmitting picture signal. Frequency Modulation (FM) is used for transmission of audio signal. The carrier frequency for AM and FM modulation have a difference of 5.5 MHz.

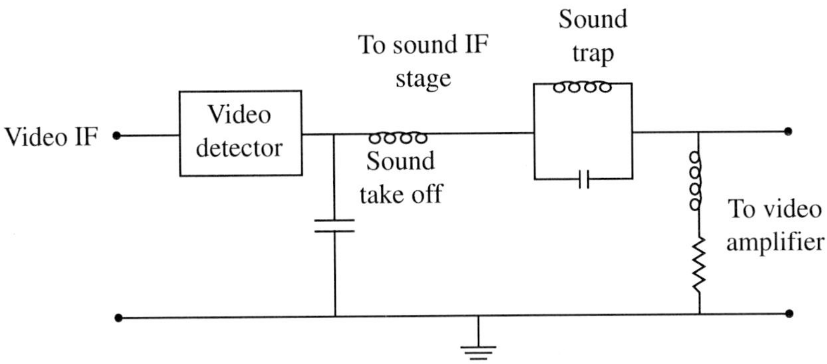

Figure 1.18 Inter-carrier sound modulation.

In the video detector circuit the strong picture IF carrier at 38.9 MHz acts as a local oscillator and heterodynes with the attenuated sound IF carrier at 33.4 MHz to produce a difference frequency of 5.5 MHz. The resulting at video detector the sound IF is made to beat with picture IF frequency to transfer the audio signal to new sound IF of 5.5 MHz. This is known as inter-carrier subsystem. As shown in Figure 1.18, a parallel tuned circuit commonly known as sound IF trap is inserted in the signal path. It is tuned to the centre frequency of 5.5 MHz.

The sound take-off circuit (trap) is used to pick up the sound IF signal (5.5 MHz). The sound trap circuit is used to block any sound signal from entering the picture tube.

These offer high impedance to the sound component of the detected signal and remove it effectively from the video signal path. A tuned secondary circuit delivers the inter-carrier IF to sound section of the receiver. In some TV receivers the inter-carrier sound signal is allowed one stage of amplification in the video amplifier and then separated through a trap circuit.

1.14 TELEVISION BROADCASTING CHANNELS

For television broadcasting, channels have been assigned are listed in Table 1.1.

Table 1.1

Lower VHF range	Band-1	41–68 MHz
Upper UHF range	Band-3	174–230 MHz
UHF range	Band-4	470–582 MHz
UHF range	Band-5	606–790 MHz

Table 1.2

	Channel	Frequency Range	Picture carrier (MHz)	Sound carrier (MHz)
Band I				
41–68 MHz	1	41–57	Not used for TV	Not used for TV
	2	47–54	48.25	53.75
	3	54–61	54.25	60.75
	4	61–68	62.25	67.75
Band III				
174–223 MHz	5	174–181	175.25	180.75
	6	181–188	182.25	187.75
	7	188–195	189.25	194.75
	8	195–202	196.25	201.75
	9	202–209	203.25	208.75
	10	209–209	210.25	208.75
Additional channel	11	216–223	217.25	222.75
	12	223–230	224.25	229.75

Band-2, 88–108 MHz is used for FM broad casting. The channel allocations in band I and band III are given in Table 1.2. In band I channel (1) is not used for TV broadcasting and it is assigned for other services.

1.15 TELEVISION CAMERA TUBES

A camera tube is an important unit that is used to convert the optical image into an electrical signal. The television camera tube serves the function of the eye of the television system. It must have the characteristic similar to the human eye. It must have the characteristic of sensitivity to the visible spectrum, a wide dynamic rage with respect to the light intensity and good resolution capability. Basically TV camera tubes works under the principle of photo conduction and photo emission.

TV camera tube consist of:

(i) An opto-electrical pick-up tube
(ii) A lens system
(iii) An signal processing unit
(iv) The pick-up tube converts the optical information into electrical information.
(v) Lens system or optical unit–focuses the picture on the camera plate. In colour camera tubes the received light is split into three primary colours.

(vi) Signal processing unit operation is same for both the colour and monochrome camera tubes but only the difference is
- One pick-up tube is needed in black and white camera.
- Three pick-up tubes is used for three colour information in colour camera tubes.

1.15.1 Characteristics of Camera Tube

The characteristics of camera tube are the following:

1.15.1.1 Light Transfer Characteristics

It is the relationship between the illumination on the target plate (x-axis) and the corresponding output current (y-axis). This is drawn on a log-log scale. The slope of these characteristics is called "gamma". It gives the relation between light falling on the camera tube and the electrical current produced.

1.15.1.2 Spectral Response

It is the ability of the camera tube to respond equally to all the colours. For black and television tube should have the same spectral response as that of the eye. But colour TV tubes have greater response to the respective primary colours.

1.15.1.3 Sensitivity

It is the ability of the camera tube to respond even for a very small illumination. The camera tube should be more sensitive. If dark current of the camera tube is high then the sensitivity will be reduced. Good camera tubes have high sensitivity.

1.15.1.4 Dark Current

This is the current produced when there is no light falling on the camera tube, a very small amount of charge carriers move across the junction of the target plate this current is called dark current. Due to this a very small amount of dark current (nano amps) flows in the camera tube output. But the dark current should be minimum otherwise it will affect the sensitivity. Good camera tubes have less dark current.

1.15.1.5 Lag Characteristics

It is the ability of the photo sensitive layer to follow faster changes in illumination of the camera tube. When a previous image appears on the screen along with the present image then this will cause a fog like appearance on the screen. This is due to very small lag of the inability of the camera tubes to illumination.

1.15.1.6 Resolving Power

It is found that the number of black and white lines of resolution in the picture is increased the signal. Current produced is not able to attain the full changes in black and white levels. The ability to reproduce the fine details of the image is called resolving power.

1.15.2 Camera Tube Types

Based on the principle of three photo electric effects used for converting variations of light intensity into electrical variations they are:

(i) Photo emission
(ii) Photo conduction
(iii) Photo voltalc

Based on the target plate the camera tube is classified into different types they are:

(i) Vidicon
(ii) Plumbicon
(iii) Saticon
(iv) Newicon
(v) Chalnicon
(vi) Silicon diode array vidicon

1.15.3 Vidicon

Vidicon camera tube was developed by RCA more than 50 years it is in use because of its small size and easy to operate. It works on the principle of photo conductivity. Here the resistance of the target material decreases when exposed to light.

1.15.3.1 Construction

The vidicon camera tube has a photoconductive target plate and an electron gun along with focusing and scanning coils. With an optical image focused on the target, it produces a charge image its is scanned by the electron beam

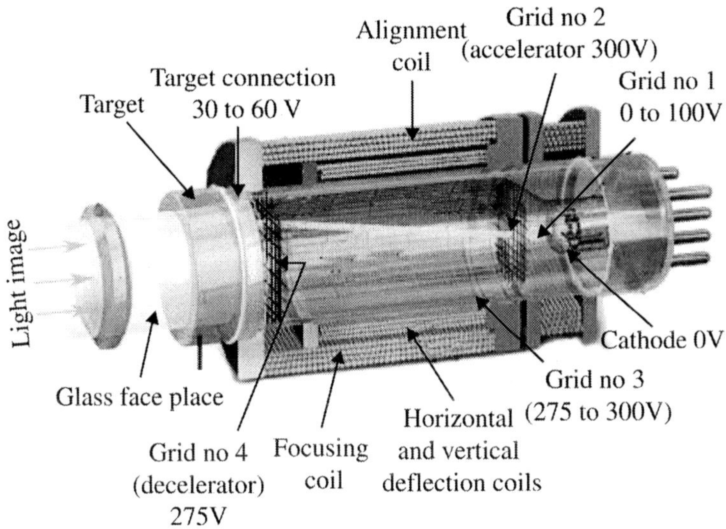

(a)

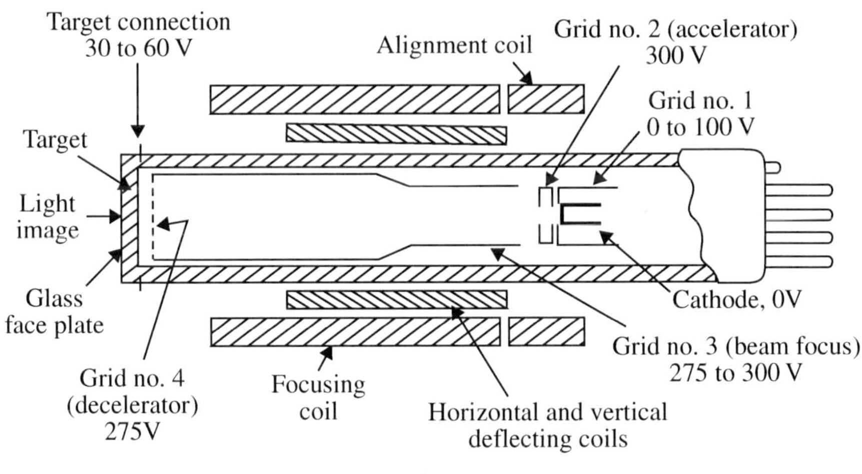

(b)

Figure 1.19 (a) and (b) Vidicon camera tube.

from the gun to develop the video signal voltage across a load resistor in the beam circuit. The vidicon camera tubes has the following the sections:

(i) Photo Conductive Target Plate: The target plate consists of two layers. The first layer is thin photo conductive layer made up of either gelinium or antimony compounds. This is deposited on a transparent conductive film, coated on the inner surface of the face plate. This conductive coating is known as signal electrode. A DC supply is connected through a

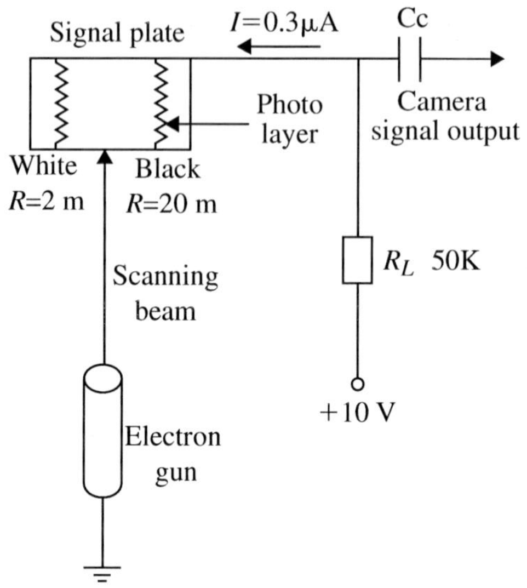

Figure 1.19 (c) Target construction.

load resister R_L, to the image side of the photo layer. The second layer is extremely thin coating of photo conductive material is a semiconductor. It has very high intrinsic resistance capable of changing with light and storing a small area of change. Target construction is shown in the Figure 1.19(c).

(ii) Scanning Section: The scanning section is used to move the electron beam horizontally (left to right) and vertically (top to bottom) to scan the entire target plate. An horizontal and deflection coils are used for this purpose. An alignment coil is provided to adjust the scanning beam position if necessary. Also a uniform decelerating field is provided by grid number 4. So that the electron beam approaches the target plate with low velocity to prevent any secondary emission.

(iii) Electron Gun Section: Heater, cathode, control grid, focusing coils and accelerating anode forms the electron gun section. Cathode is used to emit electrons and control the electrons coming out of the cathode. Focusing coils are used to focus the electrons beam to a sharp point on the target plate.

1.15.3.2 Working Principle

The working of vidicon can be explained by using (i) Charge image and (ii) Storage action.

(i) Charge Image: The photo layer has a thickness of about 0.0001 cm, and behaves like an insulator with a resistance of 20 M ohms. When there is no light falling on it. Whenever a light focused omit the photon energy enables more electrons to go to conduction band and this reduces resistivity. When bright light falls on the any area of the photo conductive coating resistance across the thickness of that portion get reduced to about 2 M ohms.

Whenever an image is falling on the target plate the photo layer obtains different resistance at each point depending on the light intensity. According to each point assumes different potential with respect to DC supply. An area with high illumination may attain a potential of about +39 V on the beam side. Similarly dark areas on account of high resistance, have a potential of about +35V. So a pattern of positive potential is formed the gun side of the photo layer. Because of this a charge image formed on the target plate that corresponds to the incident optical image.

(ii) Storage Action: Though light from the scene falls continuously on the target each element of the photo coating is scanned at intervals equal to the frame time. This result in storage action and net change in the resistance at any point on the photo conductive layer depends on the time which elapses between two successive scanning and intensity of incident light.

Since the storage time for all the points on the target plate is same the net change in resistance of all elementary areas is proportional to light intensity variations in the scene being televised.

1.15.3.3 Signal Current

As the electron beam scans the target plate, it encounters different positive potentials on the side of the photo layer that faces the gun sufficient electron beam is deposited on the photo layer surface to reduce the to reduce the positive potential of each element towards zero cathode potential of target plate.

The amplitude of current and output voltage taken across R_L are directly proportional to the incident light on the image plates. The electrons not deposited on the target plate are return back and are not used.

Large current would cause higher voltage drop across R_L, the output voltage is negative for while areas. Video signal developed across the load resistance R_L is very small in amplitude and it is amplified by conventional amplifiers before it leaves the camera unit.

1.15.3.4 Light Transfer Characteristics

A vidicon camera tube output characteristic is shown in Figure 1.20(a). Each curve represents the characteristics curve for a particular dark current.

Dark current is set by adjusting the target voltage sensitivity and dark current both increase as the target voltage is increased. The output for the bright light is 0.4 μA with a dark current of 0.02 μA.

(a) *Low illumination:* High target voltage $V_t = 40$ to 60 V for industrial use $\gamma = 0.8$ to 0.9.
(b) *Medium illumination:* $V_t = 25$ to 40 V for studio broadcast, $\gamma = 0.6$ to 0.8.
(c) *High illumination:* (Low target voltage) $V_t = 10$ to 20 V for tele-scene, $\gamma = 0.4$ to 0.6.

1.15.3.5 Image Lag

Due to time lag in the photo conductive layer, the output current is not able to follow the fast changes in the brightness on the target plate. This is called as image lag of vidicon. Due to this a smaller with a tail followed on this fast moving object. That is the white level of the picture is extended over the black level in the form of tail. This is shown in Figure 1.20(b).

Advantages

(a) Resolution is high.
(b) Long life and small size.
(c) By the varying the target voltage, it can be operated at different lengths of sensitivity.
(d) Gamma cancellation circuit is not necessary.

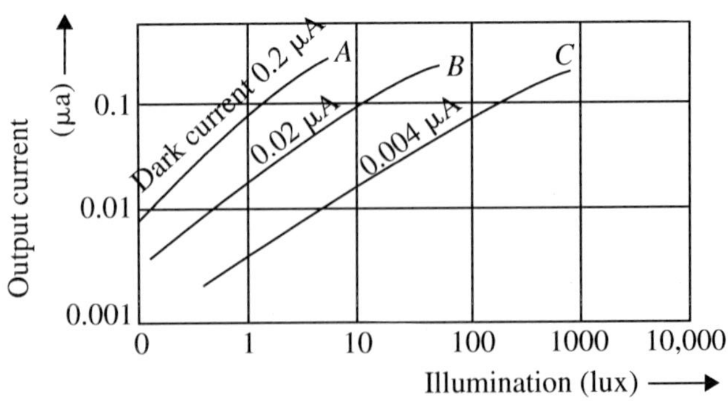

Figure 1.20 (a) Light transfer characteristic of vidicon.

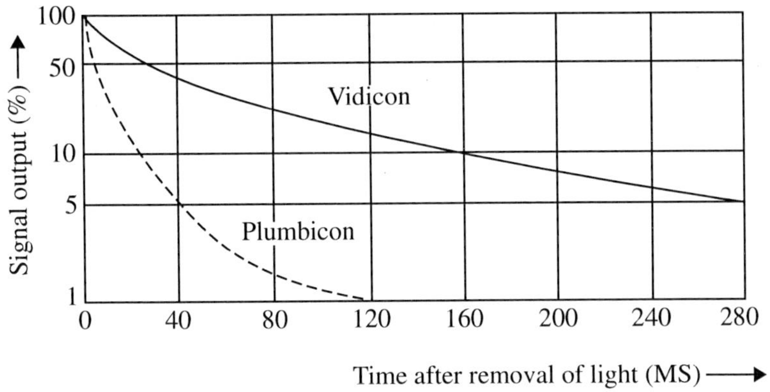

Figure 1.20 (b) Light transfer characteristics.

Disadvantages

(a) High dark current.
(b) Poor sensitivity.
(c) High image lag.

1.15.4 Plumbicon

Plumbicon is a small, light weight television camera tube introduced by Philips by 1963, employing lead oxide as the photo conductive layer it has overcome the many of the drawbacks of the vidicon tube. This has very fast response and produces high quality pictures even at low light levels. And it has low time lag and low dark current as compared with vidicon camera tube.

1.15.4.1 Construction

Plumbicon camera tube (Figure 1.21) consists of a target plate on the light falling side. It has a electron gun the other side which consists of a cathode, control grid and accelerating anode. Cathode is used to emit the electrons and control grid is used to control the electrons coming out from the cathode.

The focusing coil magnetic field focuses the electrons on to the target plate. The alignment coil produces a magnetic field which correctly aligns the electron beam on proper locations of the target plate. The accelerating anode prevents secondary emission from the target plate by reducing the velocity of the electron beam.

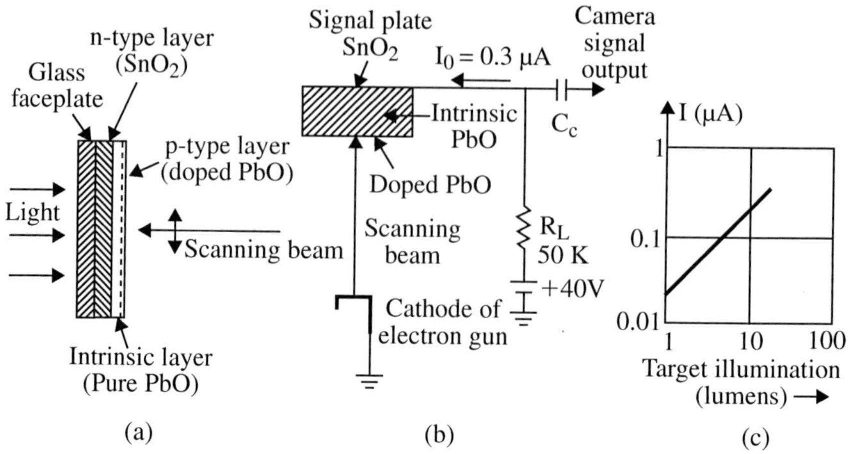

Figure 1.21 Plumbicon camera tube. (a) Target plate, (b) Target construction and (c) Light transfer characteristics.

1.15.4.2 Target Plate

Construction of target plate is shown in Figure 1.21(b). To form the target plate, pure lead oxide (PbO) is vapour deposited on the thin conducting film of tin oxide formed on the inside of the target plate. The inner surface of a glass plate of the camera tube is coated with a transparent conductive layer of SnO_2 (tin oxide). This region act as an n-type ($N+$) layer and forms the signal plate. On scanning the ($N+$) layer it deposit a photo conductive layer of (PbO) pure lead monoxide which is intrinsic.

The pure intrinsic (PbO) is doped to form a p-type semiconductor on the beam side of (PbO). Thus an intrinsic layer of (PbO) is sandwiched between thin n-type and p-type layers.

Hence these three layers form a P-I-N junction photo diode. The continuous array of pin diodes are formed and that are reverse biased by the target voltage set at (40–50 V). The overall thickness is about 15×10^{-6} m. In this layer only image is formed and the electron beam scans the image. Thus in the target plate every dot is a pin photo diode with a capacitor. Construction of electron gun section and scanning section are same as vidicon camera tube.

1.15.4.3 Working Principle

The conductive film of tin oxide is connected to the target supply of 40 V through an external load resistance R_L to develop the camera output signal, when there is no light, every element (pin diode) is under reverse bias.

So there is no conduction and no output current. A typical value of dark current is around 4 nA.

Whenever light is falling on the target plate, the resistance of the pin diode decreases due to photo excitation of semiconductor junction. Thus it forms a charge pattern on the target plate according to the image. Because of this the capacitor is discharged during the scanning of the electron beam, some electrons are moving through the target plate and so the output current is flowing. The current and output voltage taken across the R_L is directly proportional to incident light intensity on the target plate.

The amplitude of signal current depends upon the intensity of the areas on image pattern *i.e.*, corresponding to bright areas signal current amplitude is minimum. Thus in the camera tube output video signal is produced according to the scanned image.

1.15.4.4 Light Transfer Characteristic

The output current verses target illumination response (characteristic curve) of plumbicon is shown in the Figure 1.21(c), the transfer characteristic of plumbicon is a straight line with a gamma of about 0.9 the output is nearly proportional to the light input the sensitivity of plumbicon is better than that of vidicon. The resolution of plumbicon is somewhat poor due to the light scatter within the thick target.

1.15.4.5 Spectral Response

Figure 1.22 compares the sensitivity of the image orthicon, vidicon and plumbicon for light of different colours. The response of human eye centred about 5,500 A which is the wavelength for green. In the Figure 1.22 the plumbicon response is closest to this response. Here the slope of the response curve is higher than the vidicon tube, Due to the reduced recombination of photo generated holes and electrons. This type of tubes have high spectral response of plumbicon is very close to that of human eye.

Advantages

(a) Small dark current.
(b) Lag is low
(c) More sensitivity.
(d) Small in size.

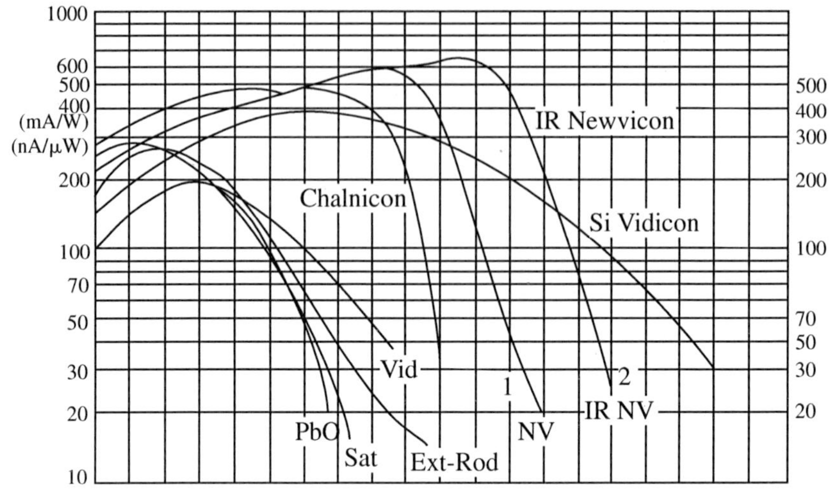

Figure 1.22 Spectral response.

Disadvantages

(a) High cost.
(b) It has poor spectral response for red colour.
(c) Poor resolution.

1.15.5 Silicon Diode Array Vidicon

Multi-diode vidicon is a another variation of vidicon camera tube but of the photovoltaic type. In which the target is not deposited on the glass face plate of the camera tube where the target plate is formed using a thin *n*-type silicon wafer instead of deposited layers on the glass plate, because of this an array of silicon photo diode is formed on the target plate. The advantages of the silicon diode array vidicon, low lag time, and high sensitivity for the visible light, which can be extended to the infrared section. The target construction is shown in the Figure 1.23.

One side of the substrate (*n*-type silicon) is de-oxidised to form a +ve of silicon di-oxide (SiO_2). It act as an insulating layer. By photomasking and etching process, an array of fine opening is made in the oxide layer to expose the *n*-type silicon islands, and a very thin layer of *p*-type silicon is formed. A fine layer of gold is deposited on each *p*-type opening to make a signal output.

The other side of the subscribe is given an anti-reflection coating. Due to this type of construction an array of photo diodes are formed. The resulting *p*-*n* photo diodes are about 8 μm in diameter. The size of the target plates is typically 0.003 cm thick, 1.5 cm square surface having an array of

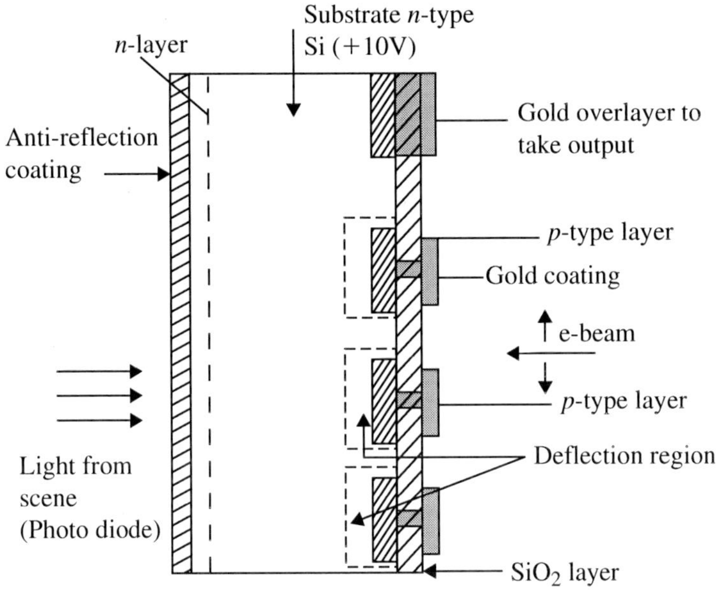

Figure 1.23 Silicon diode array vidicon.

540 × 540 photo diodes. This target plate is mounted in a vidicon type of camera tube.

1.15.5.1 Scanning and Operation

The photo diodes are reverse biased by applying +10V to the *n*-layer on the substrate. The substrate is illuminated by the light focused on to it from the image. This incidence photon of light generates the electron-hole pairs in which diffuse to the *p*-type regions by the depletion field of the diodes. The *n*-type of the substrate material has poor mobility for *p*-type carrier holes. Because of the +10V, the holes are moving towards the *p*-side of the depletion layer, establishing a positive charge on the gold over lay contact pad of the diodes. This reduces the reverse bias voltage of the diode.

When the electron beam scans the surface sequently, at deposits electron and returns the diode side of the target to the cathode potential. The increase in the target current thus caused by the incident electrons at the respective diodes represents the video signal. For a bright light, a peak signal current of about 7μA is produced.

Advantages

(a) It has low lag time.
(b) Low dark current.

(c) High sensitivity to visible light.
(d) Long life and lightly over loaded.
(d) Good resolution.

Disadvantages

(a) Speed of response is not good.
(b) Dark current is temperature dependent.

1.15.6 Newvicon

Newvicon camera tube has high resolution and very high sensitivity in the entire visible spectrum extending into the infrared region. It was developed by Matsushita Electric Corporation in 1974.

It uses a 'hetero-junction' photoconductive layer consisting of sub-layer of formed by a mixture of zinc selenide (ZnSe) and a sub-layer of zinc telluride (ZnTe) and cadmium telluride (CdTe).

The multi-layer target structure is shown in Figure 1.24. On the inside surface of the Nesa glass, a layer of *n*-type zinc selenide is deposited. Another layer of *p*-type cadmium telluride, a compound of zinc, cadmium and tellurium in amorphous form is deposited on this.

The ZnSe layer acts as a light transmission barrier to prevent crystallization of the amorphous CdTe layer. This followed by a final layer of antimony trisulphide, to suppress the secondary electron emission on the gun side surface. A *pn*-junction is formed at the boundary of zinc selenide and amorphous Zn-Cd-Te layer.

1.15.6.1 Target Operation

In newvicon camera tube, the target voltage setting is somewhat critical. Too high voltage setting results in appearance of spots and higher dark current. Two low voltage setting results in poor lag characteristic and high light burns. Newvicon tube does not permit Automatic Sensitivity Control (ASC) by means of adjustments in target voltage. Newvicon has the highest sensitivity of all the camera tubes, three to four times as high as that of plumbicon, with a very wide spectral response extending well into the infrared region as shown in Figure 1.22.

Depending on the intensity of light falling on the target plate, the positive charges are stored a on the capacitor. During scanning process the electron beam flowing through the target plate. To neutralize this positive charge so the current flow due to the electrons, produce voltage across R_L, connected to the signal plate.

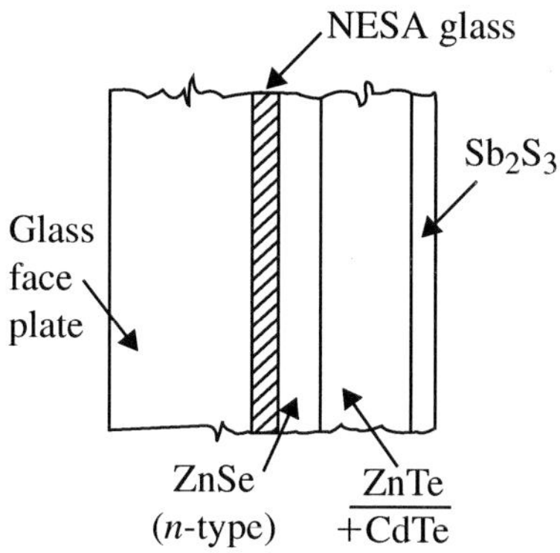

Figure 1.24 Newvicon target plate.

Because of its high target capacitance, the image lag of a newvicon tube is significantly higher than that of a plumbicon tube, the tube does not show photoconductive lag like vidicon, and hence under normal operating conditions, it is faster than vidicon by a factor of 2. Newvicon is resistant to image burns caused by high brightness and high contrast.

1.15.7 Chalnicon

Chalnicon also called Pasecon, was developed by Toshiba Electric Company in 1972. The target is of multi-layer construction with cadmium selenide as the photo-conducting layer, shown in Figure 1.25. The cadmium selenide layer is deposited by vapour deposition. Then a backing layer of cadmium selenium trioxide ($CdSeO_3$) is formed at a higher temperature. This is followed by arsenic trisulphide (As_2S_3) deposited on the scanned surface, on the gun side, to eliminate secondary electron emission.

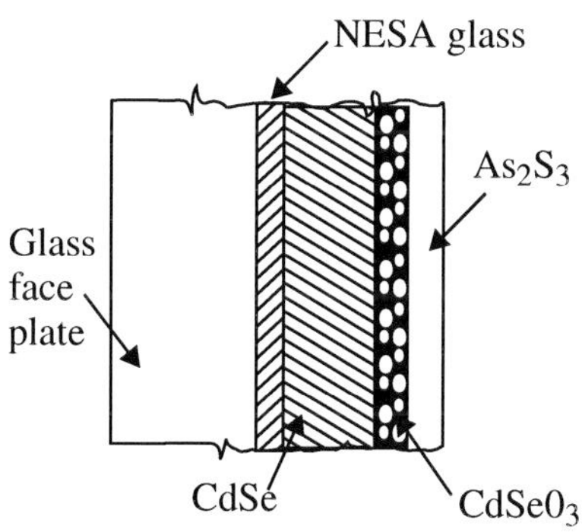

Figure 1.25 Chalnicon target plate.

Cadmium selenide gives an efficiency of almost 100%, an electron hole pair produced by every incident photon absorbed. Doping with cadmium telluride improves the red and infrared sensitivity. The addition of a porous layer of arsenic selenide (As_2Se_3) porous layer on the scanned side reduces the lag by reducing the capacitance.

1.15.7.1 Target Operation

Depending upon the temperature the target voltage are (20 to 50 V) is somewhat critical. The low voltage causes an image lag. The lag due to the junction capacitance and beam resistance is more at low lights. While excessive voltage produces a white spots.

Dark current is low, 1–2 nA. Spectral response is wide, extending well into the red. It is used in cameras meant for surveillance.

Depending on the intensity of the light facing on the target plate, positive charges are stored on the capacitor. During the scanning process the electron in the beam are flowing through the target to neutralize this positive charge. So across the load resistance R_L connected to the signal electrode have the video output.

Comparison of Camera Tube

Parameters	Image Orthicon	Vidicon	Plumbicon
Developed by year	USA, 1945	USA, 1951	Holland, 1963
Photo electric conversion	Photo emission	Photo conductive	Photo conductive
Target plate	Photo cathode	Sb_2S_3	PbO Sb_2S_3
Special features	High quality signal current maximum for black	Variable sensitivity long large	SPL target for red response
Sensitivity	High	Good	Better
S/N ratio	30 dB	50 dB	50 dB
Speed	No lag	Severe lag	Reduced lag
Resolution at 400 lines	30–50%	55%	40–50%
Dark current	30μA	20 nA	3 nA
Spectral response	Close to eye	Good	Poor in red
Gamma	1 near black 0.5 near high light	0.4–0.9	0.9–1
Cut off			650–850
Operation	Elaborate	Simple	Simple
Cost	Costlier	Cheaper costlier	Size
Size	Bulkey and Compact	Small and Compact	Medium
Length	15–20 inch	5–8 inch	8 inch
Diameter	3–4 inch	0.6–1.6 inches	1.2 inches
Life hours	1500–6000	5000–2000	2000–3000

1.16 SOLID STATE IMAGE SCANNERS

The solid state image sensor does not need an electron gun, scanning beam, high voltage or vacuum envelope of a conventional camera tube, since the entire image sensors assembly is contained in a flat silicon chip with an array of metal electrodes.

The operation of solid state image sensor is based on the functioning of charge coupled devices. The CCD act a shift register formed by a string of very closely spaced Metal Oxide Semi-conductor (MOS) capacitors. It can store a transfer analog charge signals either electrons or holes. The signal may be produced electrically or optically.

The construction details and mechanisms by which the charge transfer occurs are shown in the Figure 1.26. The electron consists of *p*-type substrate. One side of the substrate is oxidised to form a film of silicon di-oxide which is an insulator film by photolithographic process. The array metal electrodes are called gates.

This results in the creation of a very large number of tiny MOS capacitors on the entire surface of the chip. The application of small positive potentials to gate electrodes results in the formation of depletion region just below them. These are called potential wells. The depth of each well

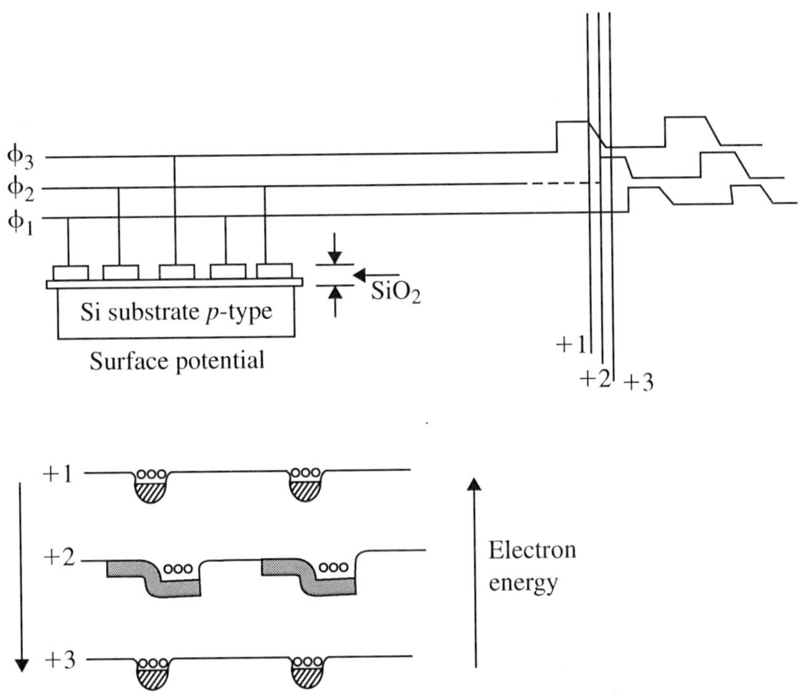

Figure 1.26 Solid state image scanners.

varies with a magnitude of applied potential. The electrode *s* operates in group of three with every third electrode connected to a common conductor. The spot under the centre of each triplet serves as one light sensitive element.

The operation of image sensor is explained as follows the image is focused on the silicon chip. The light causes electrons to be produced within the silicon. More charges are generated with more light. In each trio, the centre electrode is positive. The charge generated with more light. The charges generated by that element collect at the surface of the silicon under the centre electrodes. As a result the pattern of collected charges represents the image.

The charge of one element is transferred along the surface of the silicon by applying a more positive voltage to the adjacent electrode, while reducing the voltage on it. The accumulation of charge carries under the first potentials of two consecutive trios is shown in Figure 1.26.

1.17 MONOCHROME PICTURE TUBE

The picture tube is used to convert the video (electrical) signals into optical picture signals. A black and white picture tube is shown in Figure 1.27(a).

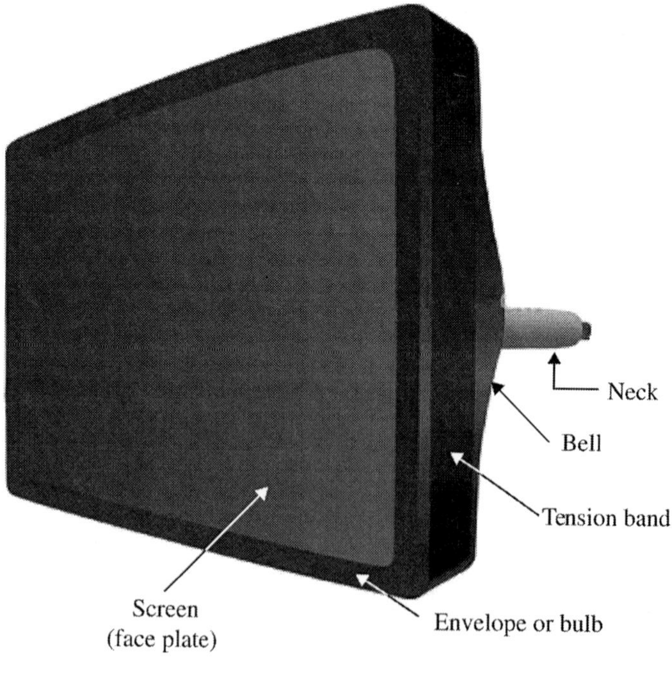

Figure 1.27 (a) Picture tube.

1.17.1 Construction

The constructional details of the picture tube can be clearly seen from its cross sectional view shown in Figure 1.27(b). Modern monochrome picture tubes employ electrostatic focusing and electromagnetic deflection. The deflection coils are mounted externally in a specially designed yoke that is fixed close to the neck of the tube. The coils which fed simultaneously with vertical and horizontal scanning current to deflection beam at a fast rate to produce the raster. The composite video signal is injected at the grid or cathode of the tube. It modulates the electron beam to produce brightness vertically on the screen. This results in reconstruction of the picture on the raster, but bit by bit as a function of time.

The main parts used in the picture tubes are:

(i) Electron gun.
(ii) Focusing anode.

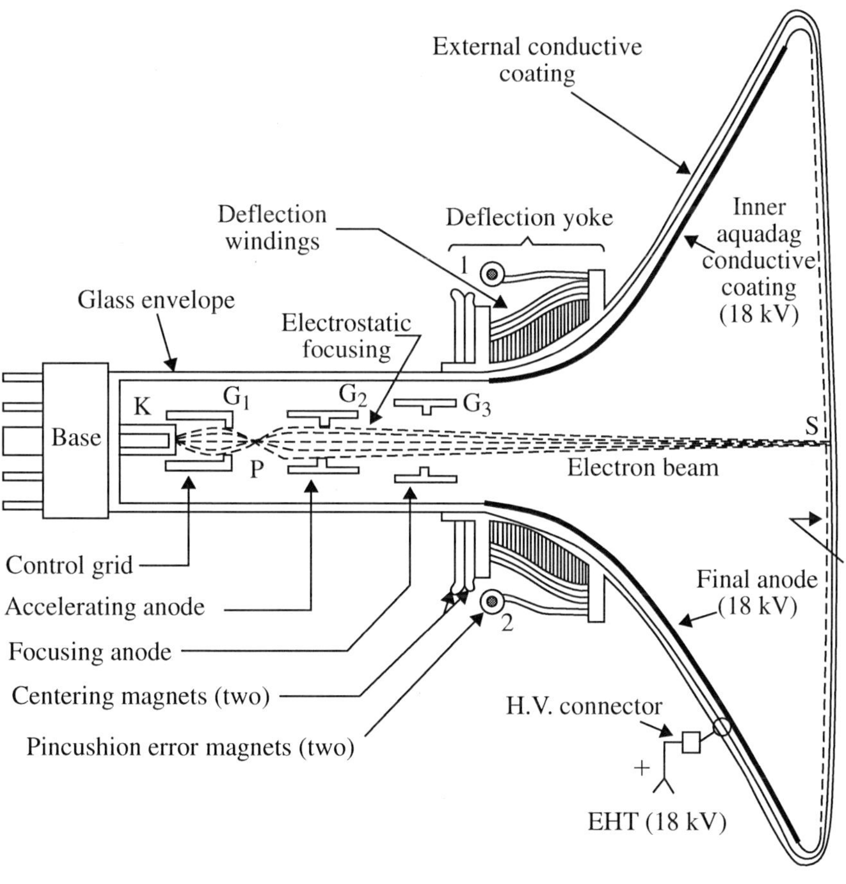

Figure 1.27 (b) Cross section view of picture tube.

(iii) Deflection coils.
(iv) Final anode.
(v) Phosphor screen.
(vi) External conductive (aquadag) coating.

1.17.1.1 Electron Gun

The various electrodes that constitutes the electron gun are shown in Figure 1.27(b). The electron gun unit consist of cathode, control grid (G1). Accelerating anode (G2) and focusing grid (G3). In this the cathode is in the form of cylinder made by nickel and thoriated tungsten or barium and strontium oxide is coated at its end. It is indirectly heated by a filament. Thoriated tungsten or barium and strontium oxides having low work function and so releases a sufficient number of electrons when heated.

Next to cathode (K) along the tube axis is the control grid (G1) almost, covers the cathode (K) cylinder but a small aperture in G1 allows the electrons to pass through negative potential is applied to the control grid (G1) which is used to control the space charge of electrons emitted from the cathode (K) surface.

Next to the control grid is screen grid (G2). It is maintained at 400 V and focusing grid (G3) is maintained at different positive potentials with respect to the cathode elements of the electron gun.

All the grids, cathode, and heater elements of electron gun are connected to a base pin. Through this base pin only the rated voltage for different grids are supplied.

1.17.1.2 Focusing Anode Section

Electrostatic focusing method is used to focus the electron beam. The positive potential at the screen grid (G2) extends through opening of the control grid right to the cathode surface. This field brings (converge) all the electrons into a small spot called as cross over point '*P*'. This is known as the electrostatic lens action. Convergence of the beam are shown in the Figure 1.27(b).

The second lens system that consists of screen grid (G2) and focus electrode (G3). Draws electrons from the cross over point and brings them to a focus at a viewing screen. The focus anode is larger diameter and is operated at high voltage than the first anode. Due to this arrangement, the electrons moving at different angle from the small point once again meet at another point on the axis of the tube. The voltages supplied to the grid is choosen, so that the convergence point is on the screen of the picture tube. This is shown in Figure 1.27(b).

1.17.1.3 Deflection Coil Section

To deflect the electron beam in horizontal and vertical direction we are using electromagnetic system. It consist of two pairs of horizontal and vertical deflection coils mounted on the picture tube as shown in Figure 1.27(c).

One pair of coil is placed above and below produces the horizontal deflection and another pair of coil is placed left and right produces vertical deflection.

The two pairs of coils are collectively called as a deflection yoke. This yoke is fixed outside and close to the neck of the tube just before it begins to flare out.

The sweep current is passed to horizontal and vertical deflection coils to deflect the electron beam from left to right and top to bottom. When the picture information is traced. The maximum angle which the beam can be deflected without hitting the side of the picture tube is called deflection angle. Each deflection coil gets its respective sweep input from the associated sweep circuit.

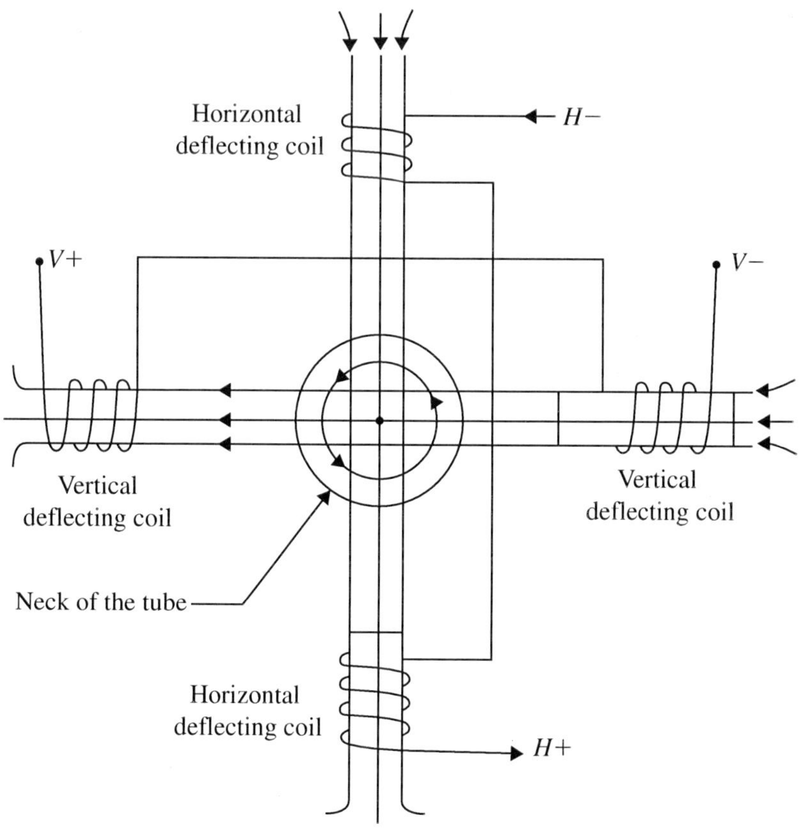

Figure 1.27 (c) Deflection yoke.

In the deflection yoke centering magnet and pincushion magnet are also provided for movements of electron beam at the corners.

Deflection Angle: It is the maximum angle through which the beam can be deflected without striking the side of the picture tube. Different values of deflection angles are 70°, 90°, 110° and 114° as shown in Figure 1.27(d).

The deflection angle of 110° means the electron beam is deflected 55° from the centre. The advantage of a large deflection angle is that for equal picture size, the picture tube is shorter and then installed in a small cabinet. The large deflection angle requires more power from the deflection circuit. Bigger picture tubes have larger axial length *e.g.*, 51 cm tube has a deflection angle of 114° while the 36 cm tube has deflection angle of 90° to conserve energy.

1.17.1.4 Final Anode Section

A final anode is included in the tube, to provide sufficient velocity and energy for the electron beam. Here a black graphic material coating called aquadag that is used as final anode. It is connected through a specially provided pin at the top side or side of the glass bell to a very high potential of over 15 kV. This aquadag coating is provided with a high voltage of 10 kV or more. The secondary electrons emitted from the screen are attracted by these aquadag coating.

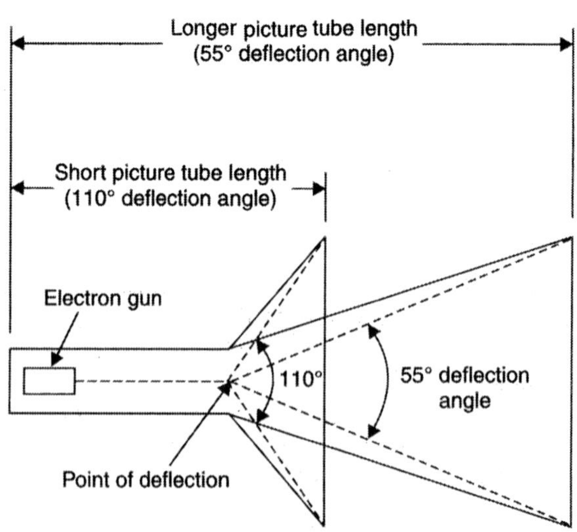

Figure 1.27 (d) Deflection angle of picture tube.

1.17.1.5 Phosphor Screen

The inner side of the picture tube is coated with phosphor chemicals are light metals such as zinc and cadmium in the form of the sulphate and phosphate compounds. The high velocity electrons of the beam on hitting the phosphor excite its atoms, with the result that corresponding spot fluorescence and emits light.

Persistence: The phosphorescent characteristics of the chemicals used on the phosphor screen. After glow on a light it remains on the screen for a short time after the electron beam move away from the screen. This after glow is called 'persistence'. It is to be noted that the persistence should be less than 1/25 second for picture tube screen. So that one frame does not persist into the next and it cause blurring of objects in motion.

1.17.1.6 External Conductive Coating

Aquadag is also coated on the outer surface of the glass bell. A spring clip is used to connect this coating with the chassis ground. This coating is used to filter the AC ripples in high voltage and to provide a perfect higher voltage.

1.17.2 Control Circuits of the Picture Tube

A typical external control circuit of a picture tube is shown in Figure 1.27(e).

1.17.2.1 Brightness Control (P1)

The video signal is DC coupled to the cathode (K) from the output of the video amplifier. Brightness control potentiometer (P1) controls the density of the electron beam which controls the brightness on the screen. Therefore the cathode has high DC potential, a less positive voltage is fed to (G1) to keep it at a suitable negative voltage with respect to cathode. By varying the (P1) potentiometer the viewer can adjust brightness as necessary.

1.17.2.2 Contrast Control (P3)

The contrast control (P3) controls the magnitude of video signal fed to cathode (K). It forms a part of the video amplifier and can be varied to adjust the contrast in the reproduced picture.

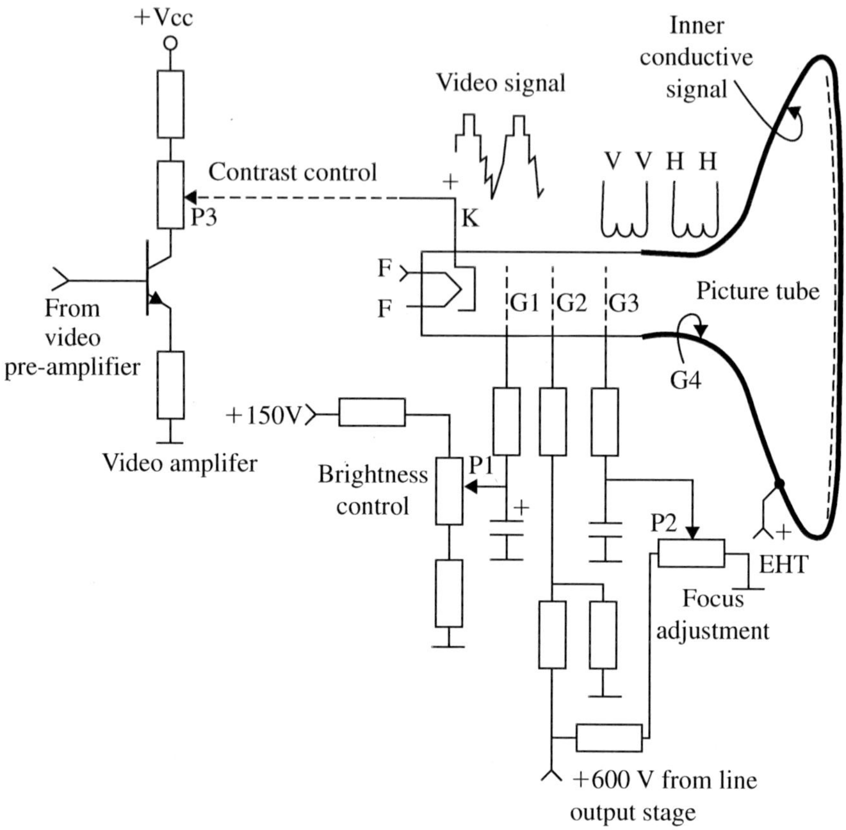

Figure 1.27 (e) Control circuit of picture tube.

1.17.2.3 Focusing Control

The screen grid (G2) is fed to fixed DC voltage. It is around 150 V for 51 cm and 100 V for 36 cm tube. The focus grid (G2) voltage is kept variable between 0 and 400 V through the potentiometer (P2). It is obtained from the line output circuit by rectifying and filtering the 15,625 Hz pulses.

Modern picture tube does not require the critical focus adjustment. Because (P2) is not located in the front panel of the cabinet.

1.17.2.4 Aluminised Screen

All the picture tubes have a very thin (0.1–1.4μm) coating of aluminum on the back surface of the screen phosphor. Figure 1.27(f) with the anode voltage of 10 kV or more, the electrons in the beam have enough velocity to pass through the aluminum and excite the phosphor. This aluminium layer is normally connected to high voltage anode coating.

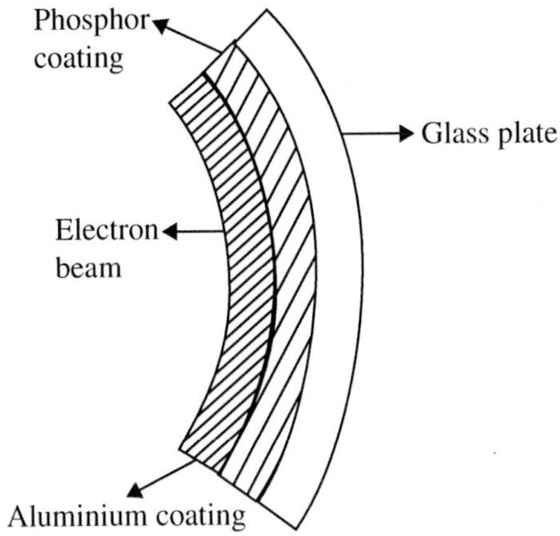

Figure 1.27 (f) Aluminised screen.

Advantages of Aluminum Layer

(a) Aluminum layer increases the light output from the phosphor by a factor of about 2.
(b) It protects the phosphor coating from damage due to ions by absorbing their energy.
(c) Since it is connected to high anode voltage, it helps in collecting the secondary electrons emitted at the screen.
(d) The aluminum coating also improves the contrast in the picture by preventing the back scattered light.

1.17.2.5 Screen Burns

A screen burn at the centre is a deflect due to the electrons. The dark spot burn can also be caused by intense beam of electrons immediately after the TV set is switched off, but the EHT voltage is present for a long time. Due to this it heat the cathode and it emits the electron even after the heater filament is switched off.

Because of the absence of deflection field, these electrons continuously strike the screen at centre for a long time. So the phosphor screen at the centre is burned and lost quickly. This problem is called 'Screen Burn'.

By proper construction and use of some special circuits like time constants of the G1, G2 and the time base supply line circuits must be large enough. So that sufficient beam current is maintained to discharge the EHT capacitor before deflection collapses after the set is switched off.

1.17.2.6 Ion Trap

In olden picture tubes a magnetic beam, bender commonly known as 'ion-trap' it was employed to deflect the heavy ions away from the screen. In modern picture tubes it is not necessary to provide an ion-trap because it has a thin metal coating on the screen. The aluminized screen coating makes the ion-trap unnecessary.

1.17.2.7 High Voltage Filter Capacitor

A grounded coating is provided on the outer surface of the picture tube. This provides shielding from stray fields and also acts as one plate of the capacitor, the other plate being the inner anode coating with the glass tube serving as an insulator between the two. Thus it forms a capacitor and filters out the high voltage supply.

The high voltage filter capacitor can hold a charge for a long time after the anode voltage is switched off and before opening the picture tube the capacitor must be discharged.

1.17.3 Transfer Characteristics of Picture Tube

Figure 1.27(g) shows the transfer characteristics of the picture tube. This curve is similar to the grid plate characteristics of vacuum tube.

The video signal is given to the control grid or cathode of the picture tube. It controls the brightness variations on the screen. To have an optimum average brightness, the grid of the picture tube itself with fixed bias. By

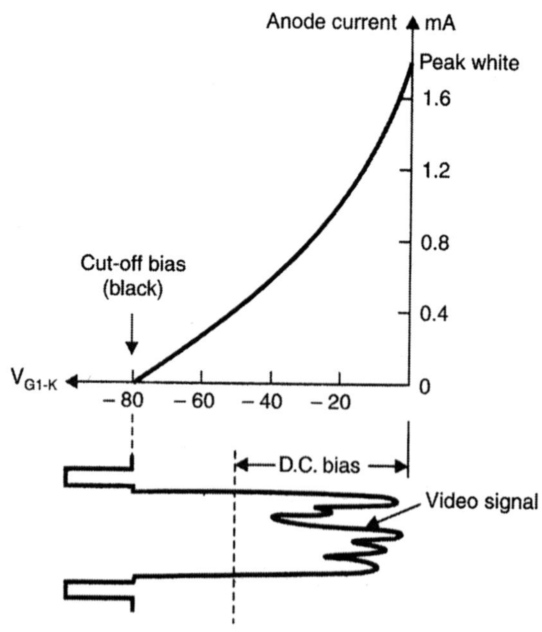

Figure 1.27 (g) Transfer characteristic of a picture tube.

varying the voltage on the control grid, the beam current and also the screen brightness is varied. This method is called intensity modulation or z-axis modulation.

The peak to peak amplitude of the AC video signal determines the contrast in the picture, between peak white with maximum beam. Current and black at cut-off. The contrast control in the video amplifier which controls the peak-peak amplitude of video signal.

1.18 MONOCHROME TUBE SPECIFICATIONS

Parameter	Type of picture tube		
	310C1P4	**470C1P4**	**610C1P4**
Size	310 nm	470 nm	610 nm
Deflection angle	110°	114°	110°
EHT voltage	10 kV	16 kV	18 kV
Focusing anode voltage VG4	100 V typical	0–400 V	0–400 V
1st anode voltage	300 V	400 V	400 V
Grid voltage	−33 to −77 V	−41 to −99 V	−41 to −99 V
Heater voltage	−12 V	6.3 V	6.3 V
Heater current	75 mA	300 mA	300 mA
Tube length	24 cm	29.5 cm	36.2 cm
Heat warm up time	14 sec	15 sec	14 sec

1.19 TELEVISION STANDARDS

Unlike commercial radio broad casting, the reception of TV signals a direct relation with a transmission of these signals. Scanning at the receiver is to be synchronised with the transmitter scanning. Also the receiver must be capable of performing equally well for all broadcasting stations. To achieve this various transmission standards have been fixed.

The various types of standards are listed below:

(i) Federal communication commission (FCC).
(ii) National television system committee (NTSC).
(iii) International radio consultative committee (IRCC).
(iv) Phase alternation with line (PAL).
(v) Sequential technique with memory (SECAM).

Table 1.3 gives the international television standards-CCIR system and FCC (NTSC system).

Table 1.3

Standards	**CCIR-A (England)**	**CCIR-B (India)**	**CCIR-C (Belgium)**	**CCIR-D**	**CCIR-E (French)**	**CCIR-F**	**NTSC (USA)**
Lines	405	625	625	625	819	819	525
Fields per second	50	50	50	50	50	50	60
Picture/sec	25	25	25	25	25	25	30
Lines/sec	10,125	15,625	15,625	15,625	20,475	20475	15,750
Line period	98.7 μsec	64 μsec	64 μsec	64 μsec	48.4 μsec	48.4 μsec	63.5 μsec
Video BW (MHz)	3	5	5	6	10	5	4.2
Gamma	0.4–0.5	0.5	0.5	0.5	0.6	0.5	0.45
VSB (MHz)	+0.75	−0.75	−.75	−0.75	−2	−0.75	−0.75
Sound carrier relative to picture (MHz)	−3.5	+5.5	+5.5	+6.5	+11.15	+5.5	+4.5
Video modulation AM	+	−	+	−	+	+	−
Sound modulation	AM	FM	AM	FM	AM	AM	FM
Sound carrier deviation		± 50 kHz		± 50 kHz			± 25 kHz
Channel BW (MHz)	5	7	7	8	14	7	6
Picture to sound power	4:1	5:1	4:1	5:1	4:1	4:1	10:1
Carrier Level % of Peaks							
Black	35%	75%	25%	75%	25%	25%	75%
Blanking	30%	75%	25%	75%	25%	25%	75%
White Peaks	10%	100%	100%	10%	100%	100%	12.5%
Picture IF		38.9 MHz					45.75 MHz
Sound IF		33.4 MHz					41.25 MHz
Colour Subcarrier		4.43 MHz					3.58 MHz
Colour System	PAL/SECAM	PAL/SECAM	PAL	SECAM	SECAM	SECAM	NTSC

1.20 CONSOLIDATED CCIR SYSTEM B STANDARDS

The main characteristics of CCIR system B for monochrome television, adopted in India are given below:

Video Characteristics	
No. of lines per picture	625
Interlace ratio	2:1
Field frequency	50 fields/sec
Picture frequency	25 fields/sec
Line frequency	15,625 lines/sec
Aspect ratio	4:3
Scanning sequence for lines	Left to right
Video bandwidth	5 MHz
Scanning sequence for fields	Top to bottom
Composite Video Signal Characteristics	
Line period H	64μsec
Line blanking period (HB)	18.5–19.2% of H, 11.8–12.3μsec
Front porch (FP)	2–2.8% of H, 1.3–1.8μsec
Line sync pulse (HS)	7–7.7% of H, 4.5–4.9μsec
Field period V	20μs
Field blanking period FB	[(18 to 22) $H + 12$]μsec 1164–1420μsec
Duration of pre-equalising pulse sequence	2.5H, 160μsec
Duration of vertical pulse sequence	2.5H, 160μsec
Duration of post equalising pulse sequence	2.5H, 160μsec
Interval between field sync pulses (serration)	7 to 7.7% of H, 4.5 to 4.9μsec
Duration of equalising pulse	3.4 to 3.75% of H, 2.2 to 2.4μsec
Build up time of field blanking pulse	6μsec
Radio Frequency Characteristics	
Channel bandwidth	7 MHz
Sound carrier relative to picture carrier	+5.5 MHz
Sound carrier relative to the nearest edge of channel	−0.25 MHz
Vestigial sideband lower	−0.75 MHz
Minimum attenuation of vestige beyond −1.25 MHz	20 dB
Type and polarity of video modulation	A5, AM negative
Sync level as percentage of peak carrier	100%
Blanking level as percentage of peak carrier	72.5–77.5%
Difference between black level and blanking level as percentage of peak carrier	0 to 2%
Peak white level as percentage of peak carrier	10–12.5%
Sound modulation type	F3, FM
Carrier deviation	50 kHz at 100% modulation
Ratio of radiated powers of vision and sound	5:1

[Contd...]

[Contd. . .]

Standard	CCIR-B	NTSC
Channel width	7 MHz	6 MHz
Horizontal sync, blanking and sync frequency	15,625 Hz	15,750 Hz
Vertical scan blanking and sync frequency	50 Hz	60 Hz
Number of frames transmitted per second	25	30
Number of fields transmitted per second	50	60
Number of lines per frame	625	525
Type of scanning	Interlaced	Interlaced
Picture IF	38.9 MHz	45.75 MHz
Sound IF	33.4 MHz	41.25 MHz
Second sound IF	5.5 MHz	4.5 MHz
Colour subcarrier	4.43 MHz	3.58 MHz

KEY TERMS AND CONCEPTS

Camera tube	Converts optical image into electrical signal.
Picture tube	Converts electrical signal into picture image signal.
Persistence of eye	It is storage capability of eye.
Contrast ratio	The ratio of the differential brightness of a small grey object to its surrounding background brightness level.
Scanning	It is the process to convert the each picture element into electrical signal.
Aspect ratio	It is the ratio of width to height of picture frame, for television, it is standardized at 4:3.
Flicker	It will give more disturbance to the viewer when the picture is made alternately bright and dark.
Interlaced scanning	In a picture frame total no. of lines are divided into two groups called 'fields'. Each field is scanned alternatively. This way of scanning is called 'interlaced scanning'.
Resolution	The ability of image to reproduce the fine details of picture image is called 'resolution'.

[Contd. . .]

[Contd...]

Sync pulse	In order to have synchronization between transmitter and receiver. A horizontal sync pulse and vertical sync pulse is added during the end of each vertical and horizontal trace.
Blanking pulse	To make the retrace period invisible the blanking pulses are added end of trace period and retrace period.
VSB	The method of transmitting a full upper sideband and vestigial of lower side band is called 'VSB'.
Photo emission	In photo emission electrons are emitted, more light develops more electrons.
Photo conductivity	In photoconduction, the resistance or conductance is changed; more light reduces the resistance.
Electrostatic focusing	The term focusing used for making a 'narrow beam' and when a voltage has a steady value, its field is called 'electrostatic', which means it does not vary with respect to time.
Deflection coils	In order to deflect the electron beam in horizontal and vertical direction deflections coils are used.
Composite video signal	Composite video signal consist of a camera signal, blanking pulses and synchronizing pulses.
Pedestal	The difference between black and blanking level is called 'pedestal'.
Pedestal height	The distance between DC level and blanking level is called 'pedestal height'.
Front porch	It is a brief cushioning period of 1.5μs length. It is inserted between the end of a line and leading edge of the sync pulse.
Line sync pulse	It is having a period of 4.7μs. These pulses are separated at the receiver and after differentiation used for triggering the horizontal deflection oscillation.
Back porch	Back porch is having a period of 5.8μs. it gives the necessary time for the retrace (fly back) to be completed.

[Contd...]

[Contd...]

Luminance	It is the average luminous flux incident on to a surface.
Viewing distance	The best viewing distance of a television which is normally about 3 to 8 times the picture height.
Inter-carrier sound system	At the video detector stage, the audio IF is beat with video IF with new sound or audio IF of 5.5 MHz (38.9 MHz–33.4 MHz) frequency is obtained. This method of sound signal production is called 'inter-carrier sound system'.

REVIEW QUESTIONS

Part A

1. What are the various blocks present in the TV transmitter?
2. What are the various blocks present in the TV receiver?
3. Why scanning is necessary for the TV Transmitter?
4. Define horizontal scanning.
5. Define vertical scanning.
6. What is flicker? How this can be eliminated?
7. What is the need for interlaced scanning?
8. What is equalizing pulse?
9. What is serrated pulse?
10. What is the need for pre and post equalizing pulses?
11. What is the function of synchronizing pulse?
12. What is the function of the blanking pulse?
13. Define resolution.
14. What is the importance of aspect ratio?
15. Define aspect ratio.
16. How much bandwidth is needed for synchronous pulse?
17. State the reason between the resolution and bandwidth.
18. What is composite video signal? List the components present in it?
19. What is the function of DC components?
20. What is mean by positive modulation?
21. What is mean by negative modulations?
22. Compare the difference between the positive modulation and negative modulation.
23. What are the factors that affect the bandwidth of the channel?
24. List the bandwidth required for VHF and UHF, TV channel in TV broadcast.
25. What are the different TV standards available?

26. What do you mean by image lag?
27. What is the basic principle of camera tube? List its types.
28. What are the important characteristics of vidicon and plumbicon?
29. What is mean by photo emission and photo conduction?
30. What is the principle of monochrome principle picture tube?

Part B

1. Draw and explain the operation and construction of vidicon camera tube?
2. Draw and explain the working principle and the construction of plumbicon camera tube.
3. With a neat diagram explain the multidiode vidicon.
4. Compare different camera tubes.
5. With a neat diagram explain the construction and working principle of monochrome picture tube.
6. Explain the following terms: (a) Phosphor screen; (b) screen burn; (c) Aluminized screen and its advantages and (d) high voltage capacitance.
7. With a neat diagram explain the construction and working principle of Vidicon camera tube.
8. With a neat diagram explain the construction and working principle of Plumbicon camera tube.
9. With a neat diagram explain the operation of multidiode diode tube.
10. Compare the characteristics of different camera tubes.
11. Explain the various Types of Scanning Process and its need.
12. What is meant by flicker and how it is eliminated? Explain in detail interlaced scanning with suitable diagram.
13. Explain in detail about composite video signal with suitable diagram.
14. Explain in detail About VSB transmission of video signal state its merits and demerits.
15. Explain the following: (a) TV Broadcast Channel Allocation and (b) Different TV Standards.
16. Explain in detail about VSB Transmission of video signal state its merits and demerits.

QUESTIONS AND ANSWERS

1. **Mention the major function of the camera tube?**
   The major function of the camera tube is to convert an optical information of the scene into electrical form and it may be called the 'eye of the TV system'. It must possess the characteristics similar to that of the human eye.

2. **Define visual acuity.**
   Visual acuity can be defined as the ability of human eye to resolve fine details in a picture. It determines the number of scanning lines.
3. **Define brightness gradation.**
   Signal transmission path should have proper transfer characteristics to preserve the same brightness gradation as eye perceives while viewing the scene directly. If the non linearities occur the necessary correction has to be done in transmitter side and receiver side.
4. **State the factors that affect the gradation of reproduced picture.**
   The factors that affect the gradation of reproduced picture are brightness, contrast and viewing distance.
5. **What do you refer by persistence of eye?**
   The persistence of eye refers to the storage capability of the human eye. The sensation produced by the incident light on the nerves of the eye's retina does not cease immediately. It persist for 1/16 th of a second.
6. **Define contrast.**
   Contrast is defined as the intensity difference between the black and white parts of the picture above and below the average brightness level.
7. **Define contrast ratio.**
   The ratio of maximum to minimum brightness relative to the original picture is called the 'contrast ratio'.
8. **What is gamma of the picture?**
   Ratio of brightness variations in the reproduced picture on the screen of the picture tube, to the brightness variations in the original scene is known as 'gamma of the picture'.
9. **Define luminous intensity.**
   It is the measure of light intensity responsible for stimulating visual sensation.
10. **Define aspect ratio.**
    Aspect ratio can be defined as the ratio of width to height of the picture frame. For television, it is standardized as 4:3.

$$\text{Aspect ratio} = \frac{\text{width}}{\text{height}} = \frac{4}{3} = 4:3$$

11. **State the reason for having the aspect ratio and larger width and smaller height.**
    Reasons for having the aspect ratio and larger width and smaller height are:
    - The aspect ratio for motion picture is 4:3; by using this ratio we can telecast the motion picture without wasting any portion.
    - Most of the objects are moving in horizontal direction.
    - Human eye can be able to view the moment easily in horizontal than in vertical plane.

12. Define luminous flux.
Luminous flux can be defined as the radiated luminous power or power of visible light expressed in terms of its effect on the average or normal human eye.

13. Define luminance.
Luminance can be defined as the quantity of light intensity emitted per square centimeter of an illuminated area.

14. What do you understand by illuminance?
Luminance is the average luminous flux incident on to a surface.

15. What do you mean by resolving power or resolution?
It is the ability of the image reproducing system to represent the fine structure of an object.

16. Mention some important characteristics of human eye.
Visual acuity, persistence of vision, brightness and colour sensation are some of the important characteristics of human eye.

17. Define image continuity.
The persistence of eye is 1/16 per second. In order to have an illusion of continuity in motion pictures 24 frames are transmitted per second.

18. State the effect of interlaced scanning on bandwidth.
By employing the interlaced scanning, the basic concept of interchangeability of time and bandwidth is not violated, because more time is allowed for transmission and these results in decrease of bandwidth needed for each channel.

19. State the effect of field frequency on bandwidth?
With the increase in field frequency the time available for each fields decrease and these results in proportionate decrease of the active line period. Hence the bandwidth increases in direct proportion to the increase in the field frequency.

20. What is the bandwidth required for transmission of synchronizing pulses?
The equalizing pulses have a pulse width of $2.3\mu s$ with an allowed rise time of $0.2\mu s$. The highest frequency which must lie in the pass band of the effective transmission of these pulses is given by:

$$\begin{aligned}\text{Highest frequency} &= \frac{1}{2} * \text{allowed rise time}\\ &= \frac{10^6}{2 * 0.2}\\ &= 2.5\,\text{MHz}\end{aligned}$$

21. Why scanning is necessary in television system?
The whole picture is divided into number of small elements called 'pixels'. This pixels or picture elements should be converted into electrical

signals. But it is very difficult to convert all elements simultaneously. This problem can be solved by a method known as 'scanning'.

22. Define scanning process.

The optical information is converted into electrical; form is carried out by element by element in a sequential manner to cover the entire picture. The scene is scanned both in the horizontal and vertical directions simultaneously in a rabid rate. As a result sufficient number of complete picture of frames per second is obtained to give the illusion of continuous motion.

23. What do you understand by flicker?

The result of 24 pictures per second in motion pictures and that of scanning 25 frames per second in television pictures are enough to make an illusion of continuity. But, they are not rapid enough to permit the brightness of one picture or frame to blend smoothly in the next through the time when the screen is blanked between successive frames. This develops in a definite flicker of light that is very irritating to the observer when the screen is made alternately bright and dark.

24. How will you solve the flickering problem?

The flickering problem is solved in motion pictures by showing each picture twice. Hence 50 views of the scene are shown per second although they are still the same 25 pictures frames per second. As a result of the increased blanking rate, flicker is removed. In TV, interlaced scanning is used to avoid flicker.

25. What do you mean by interlaced scanning?

To reduce the flicker, the vertical scanning is done 50 times per second. In interlaced scanning the 625 lines are grouped into two fields. They are called 'even field' and 'odd field'. Each field contains 312.5 lines. Each field is scanned alternately. This method of scanning the alternate lines is called 'interlaced scanning'.

26. Give the reason for the choice of interlaced scan format.

- For an illusion of continuity in motion picture it should have a minimum picture refresh rate is 24/per second.
- To have a smooth blending in motion pictures. By showing the each picture frame twice ($25 * 2 = 50$ views/second).

27. Define resolution.

The ability of the image reproducing system, to resolve the fine details of the picture, distinctly in both horizontal and vertical direction is called resolution of the system.

28. Define vertical resolution.

The scanning system is capable of resolving picture details in the vertical direction it is referred to as 'vertical resolution'.

In 625 line system,

$$\text{The vertical resolution } Vr = Na * K$$
$$= 585 * 0.69$$
$$= 400 \text{ lines}$$

29. What is horizontal resolution?
The capability of the system to resolve maximum number of picture elements along the scanning lines determines horizontal resolution. This can be evaluated by considering a vertical bar pattern.

$$\text{The number of alternate black \& white bars} = Na * \text{aspect ratio}$$
$$= 585 * \frac{4}{3}$$
$$= 780$$

30. Why AM is chosen for picture signal transmission?
If FM is adopted for picture transmission, the changing beat frequency between the multiple paths delayed with respect to each other would develop a bar interference in the image with a shimmering effect as the bars continuously changes as the beat frequency changes therefore, no study picture is produced. Apart from that, circuit complexity and BW requirements are much less in AM than FM. Hence AM is preferred to FM for broadcasting the picture signal.

31. List the contents of a composite video signal.
Composite video signal consists of a camera signal, horizontal blanking pulses, vertical blanking pulses, horizontal synchronizing pulse, vertical synchronizing pulse, colour burst and equalizing pulse.

32. What do you mean by pedestal?
The difference between the black level and blanking level is known as the 'Pedestal'. Normally these two levels are being very close to merge with each other.

33. Define peak-white level.
The peak-white level is defined as the level of the video signal when the Picture detail being transmitted corresponds to the maximum whiteness to be handled.

34. Define pedestal height.
Pedestal height is the distance between the pedestal level and the average value axis of the video signal. It is used to find the average brightness level of the video signal when the signal loses its DC level.

35. What is the main function of the blanking pulses?
The composite video signal consist of blanking pulses to make the retrace lines invisible by increasing the signal amplitude little above the black level of 75 per cent during the time scanning the circuits develop retrace.

36. What are pre equalizing and post equalizing pulses?
To rectify the drawback which occurs on account of half-line discrepancy five narrow pulses are added on either side of the vertical sync pulse. These are called 'pre-equalizing' and 'post-equalizing' pulses.

37. What is serrated vertical pulse?
To maintain synchronization in the horizontal oscillator, a five narrow pulses of 4.7μs width are formed in vertical sync period at the interval of 32μs. The leading edges of these pulses are actually used to trigger the horizontal oscillator and are called 'serrated pulses'.

38. What is the necessity of pre and post equalizing pulses?
- Pre-equalizing pulses are of 2.3μs duration. It is used to discharge the capacitor to zero voltage in both the fields.
- Post-equalizing pulse is necessary for a fast discharge of the capacitor to ensure triggering of vertical oscillation at proper time.

39. Differentiate monochrome and colour camera tube.
In black and white camera only one pickup tube is needed but three such tubes or its equivalent is necessary in color cameras to develop separate signals for red, green and blue information present in the scene.

40. What is meant by positive modulation?
When the intensity of the picture increases then the video signal amplitude increases. This is called 'positive modulation'. Sync level starts from zero per cent; white level corresponds to 100%.

41. What is meant by negative modulation? State its advantages.
When the intensity of the picture increases then the video signal amplitude decreases. This is called 'negative modulation'. Hence the sync level corresponds to 75% to 100%. White level corresponds to zero level.
Advantages of negative transmission:
- Less disturbance due to noise.
- It is very difficult to develop AGC voltage at the receiver side.
- Less power is required because most of the video signals are white.

42. Compare positive modulation and negative modulation?

Positive modulation	**Negative modulation**
White level of video signal corresponds to 100% of total magnitude.	White level of video signal corresponds to 12.5% of total magnitude.
Black level of video signal corresponds to 25% of the total magnitude.	Black level of video signal corresponds to 75% of the total magnitude.
Less efficiency and high distortion	More efficiency and low distortion.
More power required	Less power required.

43. What do you mean by VSB transmission?

To reduce the band width of the video signal we are using the VSB modulation. The method of transmitting a full upper side band and a vestigial of lower side band transmission is called 'VSB modulation'. In VSB the channel bandwidth is reduced to 11.25 MHz to 7 MHz and the bandwidth is saving of up to 4.25 MHz.

44. What are the merits and demerits of VSB transmission?

Merits

- Bandwidth of the channel is reduced.
- More numbers of channels can be transmitted.
- Transmitted power is reduced.

Demerits

- Small power is wasted in VSB filters.
- Phase and amplitude modulation distortion may occur because of the design of filters at transmitter.
- Tuning is difficult at the receiver side.

45. List the bandwidth required for VHF and UHF, TV Channel.

VHF & UHF frequencies are used for TV broadcasting. Channel allocation for TV broadcasting.

Lower VHF range	Band-1	41–68 MHz
Upper UHF range	Band-3	174–230 MHz
UHF range	Band-4	470–582 MHz
UHF range	Band-5	606–790 MHz

Band II is not used for FM broadcasting.

46. What do you mean by image lag?

It is the inability of the photo sensitive layer to follow faster changes in illumination of camera tube.

47. What are the different types of a camera tube?

Based on the target plate we can classify the following types of camera tubes:

- Vidicon
- Plumbicon
- Saticon
- Newvicon
- Chalinicon
- Silicon diode array

48. State any four comparison of different camera tubes.

Device	**Vidicon**	**Plumbicon**
Photo electric conversion	Photo conductive	Photo conductive
Material used in target plate	Antimony	Tinoxide SnO_2 or selenium compound
Sensitivity	Good	Better
Spectral response	Good	Poor in red
Speed	Severe lag	Reduced lag

Device	**Saticon**	**Newvicon**
Photo electric conversion	Photo conductive	Photo conductive
Material used in target plate	Selenium+ arsenic + tellurium. + lead monoxide + pure PbO	Tinoxide + ZnSe+ZnTe +CdTe+Sb_2S_3
Sensitivity	Good	Very high
Spectral response	Good	Good
Speed	High lag	High

49. What is meant by photo emission and photo conduction?

In photo emission the target plate of electron beam deposit some charge on the target plate, which is proportional to the light intensity variation in the picture being televised. The conductivity of the target plate increases with the intensity of light falling on it.

CHAPTER 2

MONOCHROME TELEVISION TRANSMITTER AND RECEIVER

KEY TOPICS

- ✓ TV transmitter and TV signal propagation
- ✓ TV transmission antennas and monochrome TV receiver
- ✓ RF tuner, UHF, VHF tuner cancellation of phase errors
- ✓ Digital tuning techniques–AFT-IF subsystems
- ✓ AGC – Noise cancellation
- ✓ Video and sound inter-carrier detection and vision IF subsystem, video amplifiers requirements and configurations
- ✓ DC re-insertion, Sync separation, typical sync processing circuits
- ✓ Deflection oscillators, frame deflection circuits
- ✓ Line deflection circuit - EHT generation
- ✓ Receiver antennas

LEARNING OBJECTIVES

After reading this chapter, the learner should be able to:

- ♦ Radio waves are electromagnetic waves, which are radiated from transmitting antennas, travel through space to distance places, where they are picked up by receiving antennas.
- ♦ Depending up on the wavelength, there are three distinctive methods by which propagation takes places. They are:

  (1) Ground wave or surface wave propagation.
  (2) Sky wave propagation, and
  (3) Space wave propagation.
- ♦ The height of antenna must be high as possible in order to increase the line of sight distance. Generally horizontal polarization is preferred.
- ♦ RF tuners blocks the interference antenna pick up signals in the IF ranges and prevents them from entering the receiver and mixes with the local oscillator and then to the IF amplifier.
- ♦ The purpose VHF tuner unit is to amplify both picture and sound signals picked up by the antenna and to convert the carrier frequencies and their associated side bands into intermediate frequencies.
- ♦ The various sections of the IF subsystem are divided into two parts.

  (1) discrete circuits used in earlier receivers and
  (2) specially designed integrated circuits used by at present modern receivers.

- The frequency of horizontal sync pulse is 15,625 Hz so the noise pulse can easily pass through the high pass filter. This causes out of synchronization. AGC circuit is used to control the gain of RF and IF amplifiers.
- To produce a picture with contrast, about 80 to 100 volts at the picture tube. But the output of video detector is only about 2 volt peak to peak. So a video amplifier is used. DC reinsertion is needed to preserve the component and maintain the true representation of the picture.
- The height of antenna must be high as possible in order to increase the line of sight distance, generally horizontal polarization is preferred because SNR is favorable when antennas placed high above surface of earth.

2.1 TV TRANSMITTER PRINCIPLES

A television transmitter transmits both audio and video signals using a channel bandwidth of 7 MHz two types of transmitters are employed in within a main transmitter. They are:

(i) Picture signal transmitter (visual transmitter)
(ii) Sound signal transmitter (aural transmitter).

- The picture signal is amplitude modulated and sound signal is frequency modulated before transmission. Negative modulation is preferred because the effect of noise on the picture is less. In order to reduce the channel bandwidth vestigial side band transmission is a standard practice.
- Initially a camera is used to convert the optical signal into electrical signal. After adding sync pulses and blanking pulses we get composite video signal. Which is to be amplitude modulated.
- The sound signal is associated with the picture scene is obtained through one or more microphones. It is used to convert the sound signal into electrical signals. After pre-emphasis a proper amplification the sound signal is frequencies modulated and then send to the aural transmitter.
- After processing the two modulated signals are combined using a diplexer unit. Then it is fed into the common antenna.

To modulate these signals, two types of modulations are used:

(a) High level modulation.
(b) Low level intermediate frequency modulation.

2.1.1 High Level Modulation

The high level modulation is preferred for large power output, in this both audio and video signals are amplified to the require level before modulation.

Naturally it has high power level, so grid bias modulation is employed for the signal.

If this type of modulation is used, it is very difficult to demodulate at the receiver.

2.1.1.1 Disadvantages of High Level Modulation

(a) The video modulation is needed to supply a high video voltage, in order of several hundred volts to the load, that are highly capacitive. These loads need large current at the video frequencies.
(b) The load available to the driver stage is variable, since grid modulation is employed.
(c) To provide vestigial side band properties, VSB filters are employed it requires high power at the operating channel. The linearity and high video driving powers will introduce design problems.

2.1.2 Low Level Modulation

In low level modulation, the signal is modulated before amplification. The power level of this transmitter is less than 10W so, it is easy to demodulate at the receiver. So low level modulation is more commonly used.

2.1.2.1 Advantages of Low Level Modulation

(a) As the modulation is done at low power level and at IF (Intermediate Frequency), solid state device can be utilized for the construction of a transmitter which becomes more reliable and efficient.
(b) The VSB filters can be used immediately after the modulation stage lumped components can be employed for its design to shape both the lower and higher side band response.
(c) In the UHF and VHF bands, the video transmitters are identical this is cost effective.
(d) The stages next to the VSB filters can be permanently tuned for a wide band linear operation. This results simplified tuning procedure.

2.1.3 Factors that Affect TV Transmission

(i) TV uses line of sight transmission. Therefore the transmission range is limited to very small distance. Also the transmitted power goes on reducing with increasing distance. The signal strength should be sufficiently large to reduce the effect of noise and produce good quality picture in TV transmission.
(ii) The height of the antenna is also an important factor. By increasing the height of the antenna the transmitted power will reduces. So the

transmitting antenna should be, mounted on the centre point. The point should be high as possible.

(iii) In the service area the population density is not uniform around the transmitting station. The area containing high population can be given more transmitting power. Here the directional antennas can be used to distribute the power and avoid wasting of power.

(iv) The audio and video signal are independently modulated by two different carriers and transmitted simultaneously. At the receiver the audio and video powers are different, because difference in attenuation. Also receiver needs low audio power than video power to create good quality sound.

(v) The adjacent channel should be kept apart from a space as possible. So that there is no adjacent channel interference.

2.1.4 Sound Signal Modulation

The frequency modulation is used for sound signal transmission. The bandwidth required for sound signal transmission is 150 kHz this bandwidth is less when compared to the entire television signal channel bandwidth of 7 MHz thus the FM signal does not increase the TV channel bandwidth. So FM sound signal transmission is widely used in TV signal.

2.1.4.1 Advantages of FM Sound Signal Transmission

(a) The increase in channel bandwidth of 7 MHz is insignificant even though more bandwidth is needed for FM.

(b) The effect of noise interference is minimum in FM. Since the amplitude of modulated carrier remains constant. The noise can be easily removed by placing a amplitude limiter at the receiver end.

(c) Transmission efficiency is high in FM. Because the amplitude of FM is independent of modulation. So the low level modulation also can be used in FM.

(d) The effect of adjacent channel interference is reduced by providing a guard band between two channels.

(e) Co-channel interference effect is minimum in FM. The co-channel interference means signal from other transmitting station which is operating at same carrier frequency. These signals are relatively weak in amplitude but tend to interference with the signals. The amplitude limiter is placed in FM receiver which removes these interference signals and thus reduces the co-channel interference.

2.1.5 Need for AM Modulation for Picture Signal

(i) The bandwidth of the video signal is from 0 (DC) to 5 MHz the amplitude modulation generates two side bands. For an complete video information can be obtained by transmitting full upper side band and a part of lower side band. This is called 'VSB'. These can be accommodated with overall channel bandwidth of 7 MHz.

(ii) But FM is not used for picture signal transmission because it contains more number of sidebands. So it will increase the channel bandwidth. The increased channel bandwidth is not affordable since no of channels transmitted in particular frequency band may reduce, so FM is not suitable for picture signal transmission.

2.1.6 IF Modulation System

Modern TV transmitters employ IF modulation at low level these systems are simple in design, easy to operate and superior in performance. The block diagram of low level IF modulation is shown in Figure 2.1(a).

In these low level IF modulation, the video signal is amplitude modulated at the standard IF frequency of 38.9 MHz and the audio signal is frequency modulated at the standard IF frequency of 33.4 MHz.

To select a particular channel IF carriers are increased to the required frequency level by heterodyning them with a signal from a crystal oscillator. The VSB is given at IF level. Both heterodyned signals are given to the respective linear RF power amplifiers and then given to the diplexer unit for combining the audio and video signal.

2.1.6.1 Visual Exciter

The visual exciter is used to process the video signals. This is shown in Figure 2.1(b). Visual exciter consist of following sections:

(a) A video processing unit.
(b) A video modulator which is a diode bridge modulator or diode balanced modulator commonly employed for low level modulation.
(c) Phase compensator or delay equalizer.
(d) Frequency up convertor.
(e) VSB filter.
(f) Linear and power amplifier.

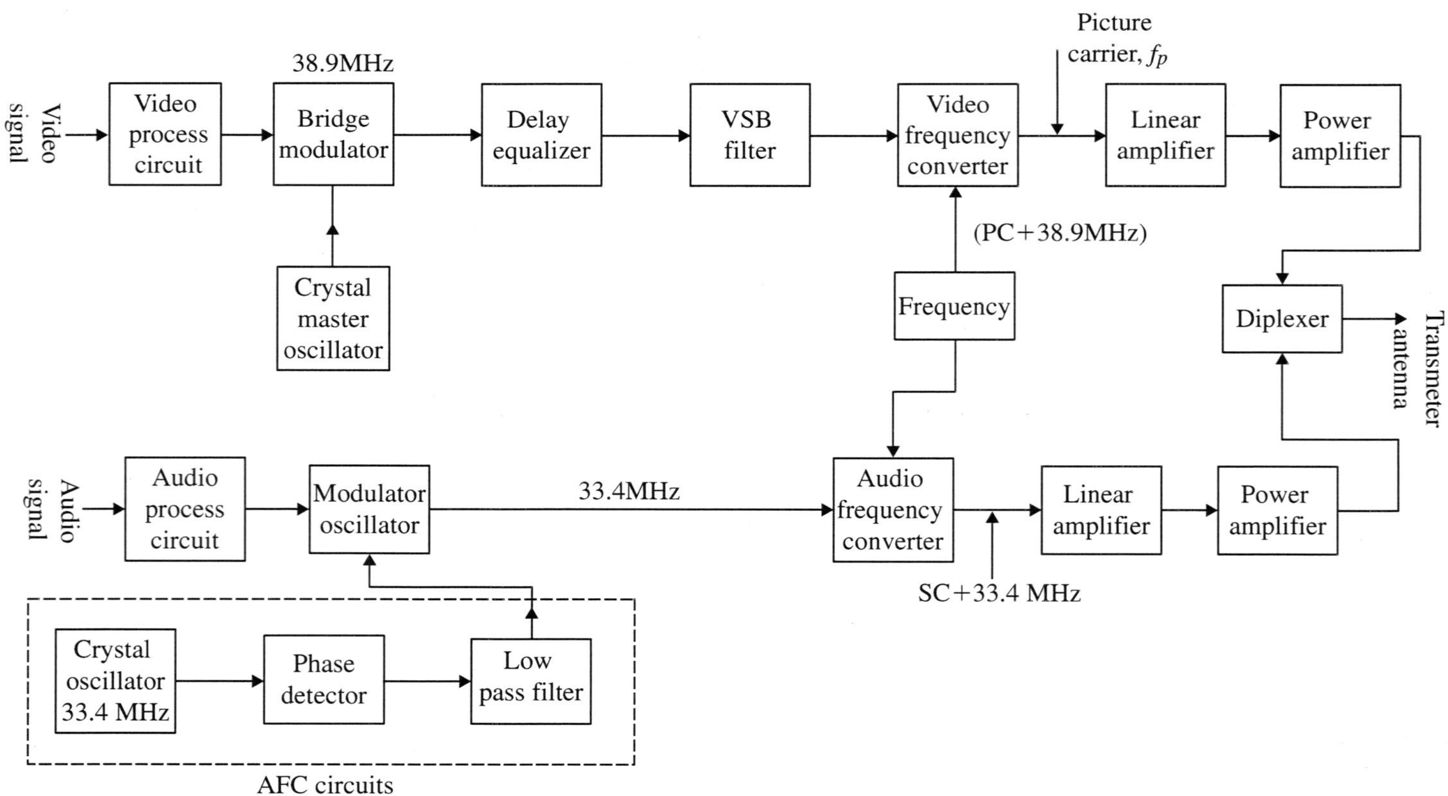

Figure 2.1 (a) The block diagram of an IF modulated TV transmitter.

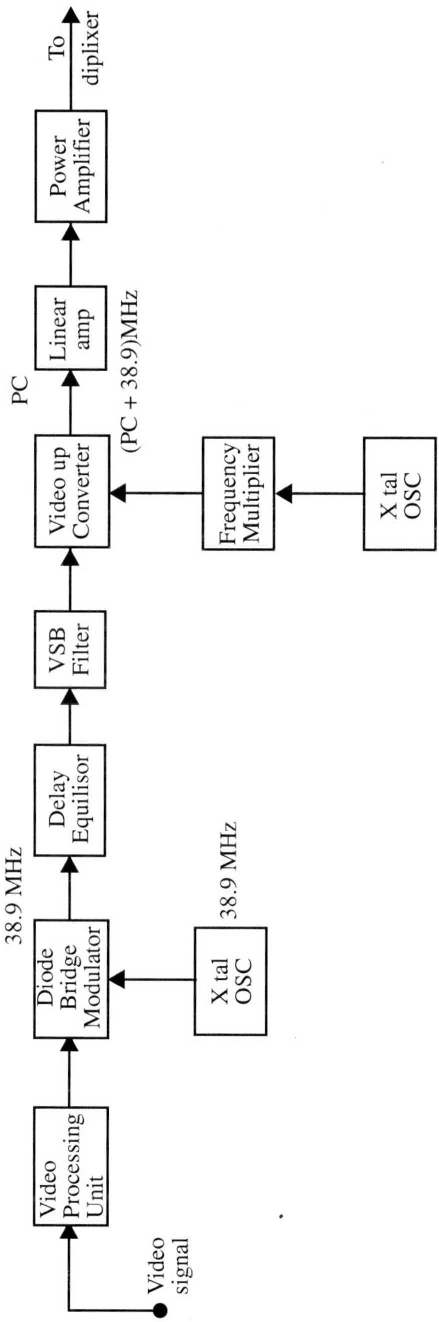

Figure 2.1 (b) Represents the block diagram of visual exciter unit.

2.1.6.2 Video Processing Unit

In this video processing unit a signal input 1 V P-P is given to a stabilising amplifier and a sync pulse generator. These units reduce the effect of noise and hum, sync pulse compression and other distortions. These units have control for sync level gain, black setup, picture sync ratio, sync stretching, linear correction, and helps in bringing the video signal into a standard form.

After amplification the sync pulse, blanking pulse are added to get the composite video signal.

2.1.6.3 Video Modulator

A diode modulator or a diode balanced modulator is commonly used for low level modulation as shown in Figure 2.1(c).

In diode bridge modulator, the diodes are switched ON and OFF by the carrier voltage $E_c \cos \omega_c t$ by the positive and negative cycles.

During the positive half cycle of carrier signal, R_L is short circuited. So, output across R_L is zero. During the negative half cycle, R_L is in series with R_S so the output is proportional to the potential divider ratio formed by R_S and R_L.

The instantaneous amplitude of the output voltage is:

$$\begin{aligned} E_o &= \frac{R_L}{R_S} + R_L(E_O + E_m \sin \omega_m t) \cos \omega_c t \\ &= A_o(1 + m \sin \omega_m t) \cos \omega_c t \end{aligned}$$

where

$$A_o = E_o\left(\frac{R_L}{R_S}\right) + R_L$$

and

$$m = \frac{E_m}{E_o}.$$

From the above equation the video signal is amplitude modulated by the carrier signal.

2.1.6.4 VSB Filter or Delay Equalizer

The VSB filter employs lumped components for its design and consists of four sections of low pass filters network which attenuate the frequency beyond 1.25 MHz above the carrier frequency by about 20 dB or more. The attenuate sideband becomes the lower sideband by up converting the modulated signal. To adjust phase distortion, the phase compensator is built

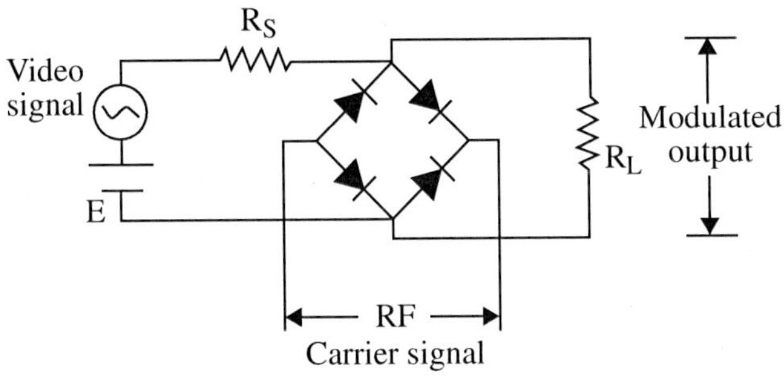

Figure 2.1 (c) Represents the block diagram of visual modulator.

in with the VSB filter. The VSB filter and delay equalizers are normally designed in a combined unit.

2.1.6.5 *Frequency Up Convertor*

The frequency up converter has two inputs. One is from VSB filter output and other is from the crystal oscillator.

The frequency produced by the crystal oscillator is equal to the picture carrier plus the video IF of 38.9 MHz (P.C +38.9 MHz).

These two frequency inputs are heterodyned in the frequency up converter stage. The band pass filter is used at the output of up converter stage. It is tuned to F_c, only VSB modulated picture carrier signal passes through the band pass filter. Because of mixing process, lower and upper sidebands are interchanged.

2.1.6.6 *Linear and Power Amplifier*

Power amplifiers are wide band amplifier is used to amplify the modulated channel carrier signal from the up converted frequencies. The power levels of this amplifier are very high and at the same time its bandwidth is high.

Using the linear and power amplifier, the video signal is amplified to the required level before fed into the diplexer unit.

2.1.6.7 *Audio or Aural Exciter*

Aural exciter is used to process the audio signals as shown in Figure 2.1(d). This section consist of:

(a) Audio processing unit
(b) Audio modulator
(c) Audio up converter
(d) Linear and power amplifiers

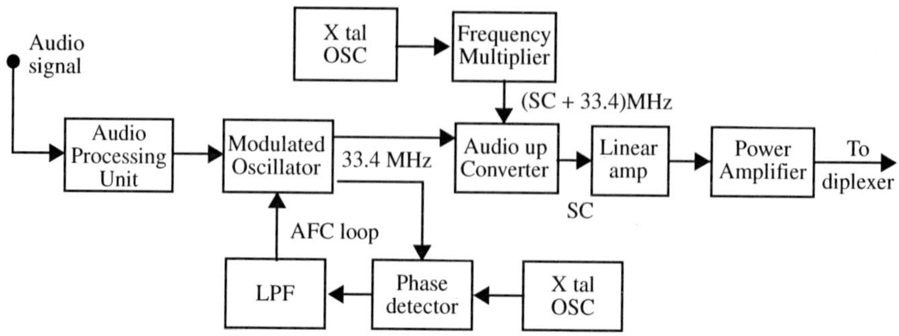

Figure 2.1 (d) Aural exciter.

(a) Audio Processing Unit: In this section, pre-emphasis circuit and amplifier are used to improve the signal to noise ratio and amplify the audio signal to the required level.

(b) Audio Modulator: Audio modulator has two inputs one is from the video processing unit and the other is from crystal oscillator.

These two frequencies are given to the varactor diode in parallel with LC tank circuit. The output of this audio modulator is sound IF of 33.4 MHz.

In between the crystal oscillator and audio modulator, one low pass filter and phase detector are used.

If there is any change in the sound IF, the phase detector produces an error voltage. This error voltage is given to the varactor diode to correct the sound IF. Here the audio signal is frequency modulated.

(c) Audio Up Converter: The output of the audio modulator is fed to the audio up converter as one input. Another input is taken from the crystal oscillator and frequency multiplier circuit. The output of this crystal oscillator produces (sound IF + sound carrier). These two frequencies are heterodyned to produce the required audio signal output.

2.1.6.8 Diplexer

For cost efficient, the output of both the video signal and the audio signal are combined in a single unit is called Diplexer, The visual (picture or video) and aural (sound) signals are mixed with help of diplexer to feed a common broad cast transmitting antenna system. The diplexer combines visual and aural signals with sufficient isolation between them. These signals are fed to common transmitting antenna.

2.1.7 CIN Diplexers

CIN-constant impedance notch. The block diagram for CIN diplexer is shown in Figure 2.2. CIN diplexer combines the output powers from video and audio transmitter are combined in a unit called diplexer and given to a common broadcast transmitting antenna system. It consists of two units of 3 dB directional couplers, two notch cavities connected by co-axial lines and dummy load matched termination. It offers constant input impedance at the output terminals.

2.1.7.1 Properties of 3 dB Coupler

(a) When the power is given to port 1, zero power comes at port 3 while at port 2 and 4 each, half the power appears. The port 4 power lags by 90° the power at port 2. When port 2 and 4 are given equal powers of the same frequency maintaining the phase lag of 90° between port 4 and 2, the zero power comes at 3 and combined power (p2, p4) appears at port 1.

(b) When port 2 and 4 are terminated by equal impedance in phase and port 3 is terminated by the characteristic independence Zo.

(c) When ports 2 and 4 are terminated by Zo, impedance at port 1 is also irrespective of the impedance of port 3.

2.1.7.2 Aural Notch Cavity

Aural notch cavity is a co-axial resonant cavity. It is tuned to form a series resonant circuit at the sound carries frequency *i.e.*, it looks like a short

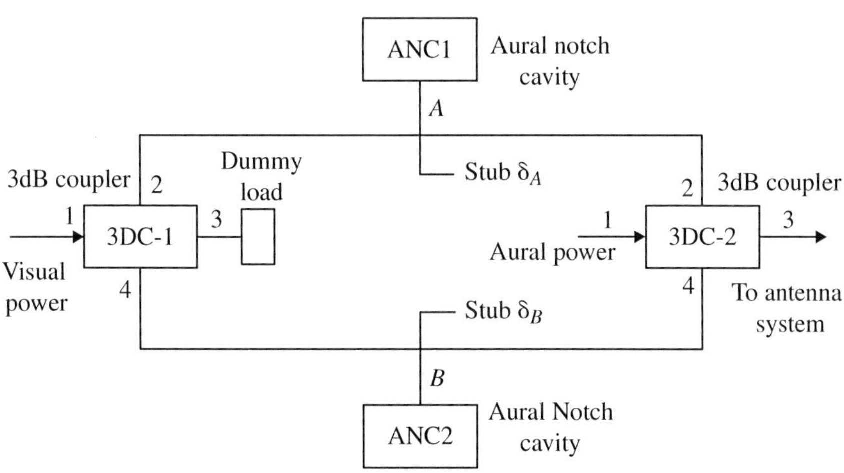

Figure 2.2 CIN diplexer.

circuit to the audio frequency. The length of the associated short circuit stub is chosen to form a non-resonant circuit at the picture carrier frequency. Hence it offers very high impedance in video band on the transmission line. At the same time, it does not affect picture signals but short circuits the aural frequencies.

2.1.7.3 Working of CIN Diplexer

This unit consists of:

(a) Two 3 dB directional coupler (DC)
(b) Two aural notch cavities (ANC).

These two 3 dB directional couplers are joined using co-axial transmission line and ANC's are connected in parallel with the co-axial line.

- DC1 and DC2 have 4 ports. Port 1 is the input port and other three are output ports.
- Visual power is fed into port 1 of DC1. This signal equally passed through port 2 and port 4 of DC1 with 90° phase shift. Port 3 is connected with dummy load to avoid leakages.
- These two separated signals are passed through co axial line to port 2 and port 4 of DC2. These two are added up and taken from port 3 of DC2.
- Audio power is fed into the port 1 of DC2, this signal divided equally into port 2 and port 4 of DC2 with 90° phase shift. These are passed to ANC1 and ANC2 and reflected to come back to port 2 and 4 of DC2.
- The reflected signal are added and taken from the port 3 of DC2 after this processing, the audio and video signals fed to the common antenna without mixing.

2.2 TV SIGNAL PROPAGATION

Radio waves are electromagnetic waves, which are radiated from transmitting antennas, travel through space to distant places, where they are picked up by receiving antennas.

Depending upon the wavelength, there are three distinct methods by which propagation takes places. They are:

(i) Ground wave or surface wave propagation.
(ii) Sky wave propagation.
(iii) Space wave propagation.

2.2.1 Ground Wave or Surface Waves Propagation

Vertically polarized electromagnetic waves are radiated at zero or small angles with ground. They are guided by the conducting surface of the ground along which they are propagated, such waves are known as ground or surface waves. As the ground waves travel along the surface of the earth, their attenuation is proportional to frequency. The attenuation is reasonably low below 1500 kHz. Therefore, all medium wave broadcast and long wave telegraph and telephone communication is carried out by ground wave propagation.

2.2.2 Sky Wave Propagation

Similarly sky wave propagation which is dependent on reflection from the ionosphere is not suitable because at frequencies beyond 30 MHz, it becomes impossible to deflect the radiated beam back to the earth.

In ground wave propagation, a frequency above 1600 kHz does not serve any useful purpose as the signal gets very much attenuated within a short distance of its transmission. Therefore, most radio communication in short wave bands up to 30 MHz is carried out by sky waves. When these waves are transmitted high up in the sky, they travel in the straight line until the ionosphere is reached. This region begins about 120 km above the surface of the earth. The region consists of large concentrations of charge gaseous ions, free electrons and neutral molecules. The ions and free electrons cause to band all passing electromagnetic waves.

2.2.3 Space Wave Propagation

Propagation of radio waves above about 40 MHz is not possible through either sky wave or surface wave propagation. Therefore, the only alternative for transmission in the VHF and UHF bands, despite large attenuation is by radio waves which travel in a straight line from transmitter to receiver. This called 'space wave propagation'.

2.2.4 Line of Sight Transmission

Assume the surface of the earth surface to be flat and different rays of wave propagation can reach the receiver from transmitter as shown in Figure 2.3(a). Where h_t is height of the transmitting antenna. h_r is height of the receiving antenna. d is distance between the two antenna. $A - B$ is direct wave and $A - C - B$ is reflected wave both contribute to the field strength at the receiving antenna.

Assuming the earth surface to be perfectly reflecting, the total field strength E, due to both direct and reflected waves then the large valve of

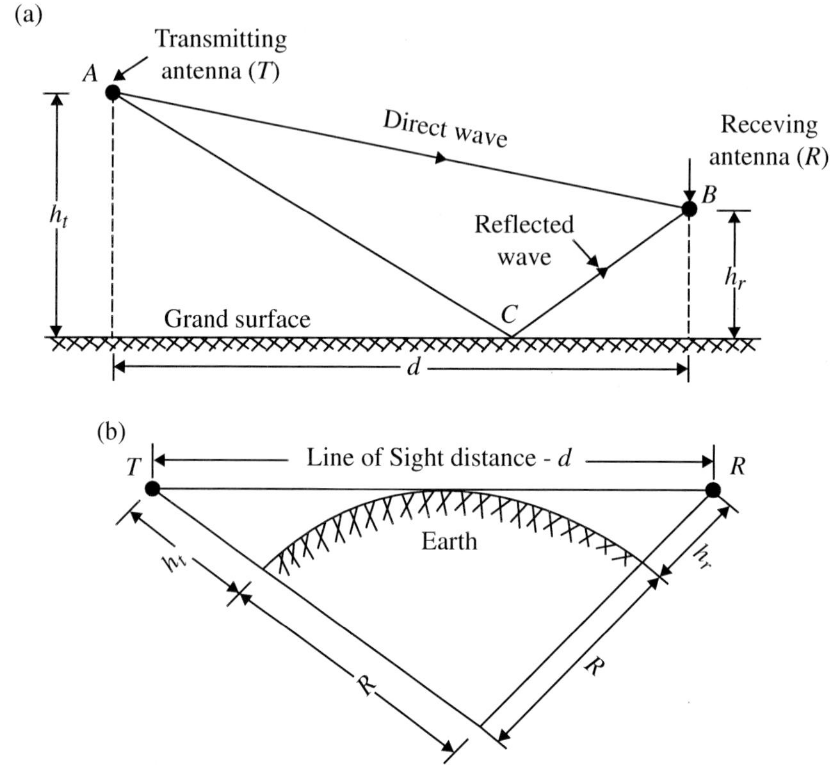

Figure 2.3 (a) Space wave propagation. (b) Computation of line of sight distance.

'd' can be expressed as:

$$E = \frac{2E_0}{d} \sin\left(\frac{2\pi h_r h_t}{d}\right)$$

$$E^* \simeq \frac{4\pi F h_t h_r}{d^2} \times E_0$$

where E_0 is the field strength at unit distance from the transmitters. F is the frequency of the transmitted signal.

The field strength varies inversely as the square of the distance between the two antennas but is directly proportional to their heights.

2.2.5 Effect of Earth's Curvature

Earth's curvature limits the maximum distance between transmitting and receiving antenna. This is shown in Figure 2.3(b) the maximum line of sight distance '*d*' between two antennas can be easily calculated. Neglecting (h_t^2) and (h_r^2), being small compared to R, the line of sight distance is:

$$d \simeq 4.22\left(\sqrt{h_t} + \sqrt{h_r}\right)\ \text{km}$$

where h_t and h_r are expressed in metres.

Electromagnetic waves bent slightly to surface of earth and increases the line of sight distance to small amount. Ground coverage will increase in height of both transmitting and receiving antenna. Due to this reason the television transmitting and receiving antenna are placed as high as possible and local noise disturbance pick-up is reduced by placing the antennas at high attitudes.

2.2.6 Range of Transmission

A simple calculation for a transmitting antenna height of 225 m above ground level the radio horizon is 60 km. If the receiving antenna height is 16 m above the ground level the total distance is increased by 76 km. Greater distance between antennas may be obtained by locating them on very tall buildings or hill rocks. Links longer than 120 km are not used for TV transmission because of limitation of radiated power.

Depending on the service area may extend for:

(i) 120 km for the channels in the VHF band.
(ii) 60 km for the channels in the UHF band.

2.2.6.1 Booster Stations

The areas of shadowed by mountains or are too far away from the transmitter for good reception of signals, some booster stations are used. It is located in a place where it can receive and broadcast the programme to receiver in adjoining areas.

2.2.7 Transmitter Power

Satisfactory level of the signal exist when the signal strength is a function of power radiated, transmitting and receiving antennas heights and the terrain above which propagation occurs. Peak signal to R.M.S noise ratio of 45 dB is generally considered adequate to produce good quality pictures.

Field strength is indicated by amount of signal received by a receiving antenna at height of 10 m from ground level. It is calculated in microvolt per metre of antenna dipole length. The field strength for very good reception in thickly populated and built up area is 2500 μv/metre for channels 2 to 4 (47 to 68 MHz) and 3550 μv/m for channels 5 to 11 (174 to 233 MHz) for channels in UHF band field strength of about 5000 μv/m becomes necessary. This is so because of the lower sensitivity of the receiver for higher channels.

The picture transmitter that produces the AM video signal has a peak power rating of 0.5, 1, 5, 10, 20 or 50 kW. The Effective Radiated Power (ERP) is much greater because it includes the antenna gain. The second transmitter that produces the FM sound can have a power output of 0.5 to 1.5 times the picture signal power.

2.3 INTERFERENCE IN TV SIGNALS

In addition to thermal and man-made noise, the carrier signal must complete with various other forms of interfering signals originating from other TV stations, radio transmitters, and industrial radiating device and TV receivers.

When the interference signal has a frequency that lies within a channel to which TV receiver is tuned, the extent of interference depends only on relative field strength of desired signal and interfering signal. If the interfering signal frequency spectrum lies outside the desired channel, selectivity of the receiver aids in rejecting the interference.

2.3.1 Co-Channel Interference

Two stations operating at the same frequency, if located closely will interfere with each other. This phenomenon which is common in fringe areas is called 'co-channel interference'.

As two signal strength in any area almost equidistant from two co-channel station become equal, a phenomenon known as "venetian blind" interference occurs. This takes the form of horizontal black and white bars. These bars tend to move up or down on the screen. The horizontal bars

Figure 2.4 Venetian blined interference caused by beat frequency between picture carries on the same channel.

are a visible indication of beat frequency between two interfering carriers. Figure 2.4 shows bar pattern appear on the screen.

Co-channel interference was a series problem in early days of TV transmission, when channel allocation was confined to VHF band only this necessity repetition of channels at distance too far from each other. Nowadays large number of channels in UHF band are available hence such a problem does not exist.

The sharing of numbers of channel numbers is carefully planned so that within the 'service area' of any station signals from the distant station under normal condition of repetition are so weak as that cannot to be perceived. The use of highly directional antenna is very helpful in eliminating co-channel interference.

2.3.2 Adjacent Channel Interference

Adjacent channel interference occurs as a result of beats between any two of the frequencies or between a carrier and any side bands as shown in Figure 2.5.

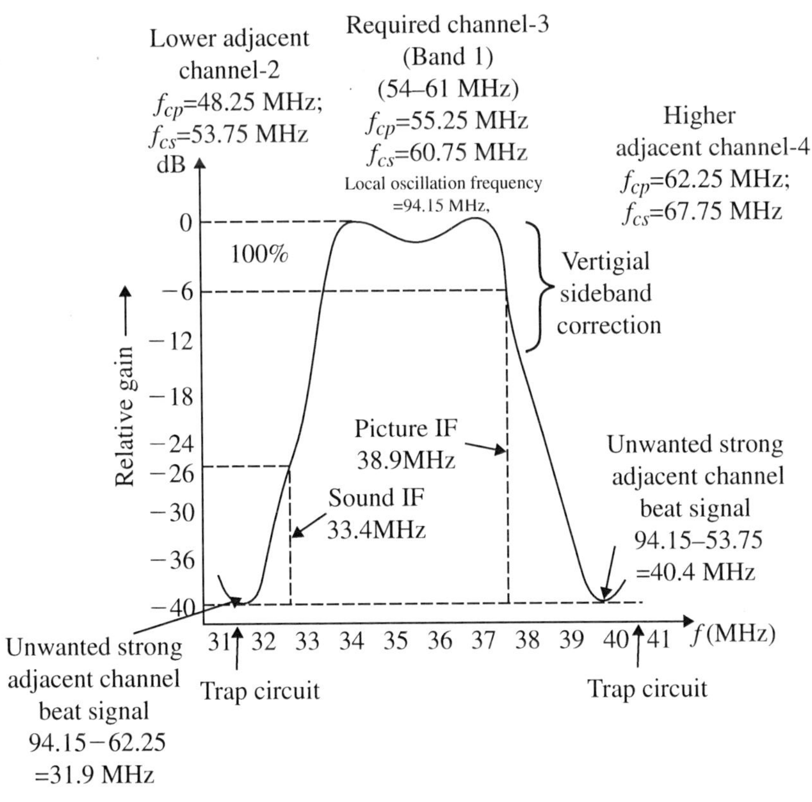

Figure 2.5 Adjacent channel interference.

A course dot structure is generated on the screen if picture carrier of the received channel beat frequency with sound carrier frequency of lower adjacent channel. The next most prominent source of interference is the one produced by picture sideband components of upper adjacent channel. The beat frequency between adjacent picture carriers is 7 MHz. However the picture sideband of the upper adjacent channel may beat with desired channel carrier and produce an interfering image.

To prevent adjacent channel interference, several sharply tuned band eliminator filter (trap circuits) are provided in IF section of the receiver. In addition to this, the guard band between two adjacent channels also minimizes the intensity of any adjacent channel interference. A space of about 150 km between adjacent channel stations is enough to eliminate such an interference is normally allowed.

2.3.3 Ghost Interference

Ghost interference arises as a result of discrete reflection of the signals from the surface of buildings, bridges and hills, towers *etc.* Figure 2.6(a) shows path of direct and reflected electromagnetic waves from the transmitter to receiver. Since reflected path is longer than the direct path, the reflected signal takes a longer time to arrive at the receiver.

2.3.3.1 Multi-path Transmission

The direct signal is usually stronger and assumes control of the synchronizing circuitry and so the picture due to the reflected signal that arrives late, appears displaced to the right. Such displaced pictures are known as 'trailing ghost' pictures. In rare cases, the direct signal becomes we are receiver synchronization is now controlled by the reflected signal. Then the ghost picture, now caused by direct signal appears displaced to the left and is known as 'leading ghost picture'. Figures 2.6(a) and (b) shows the formation of trailing and leading ghost pictures on the receiver screen.

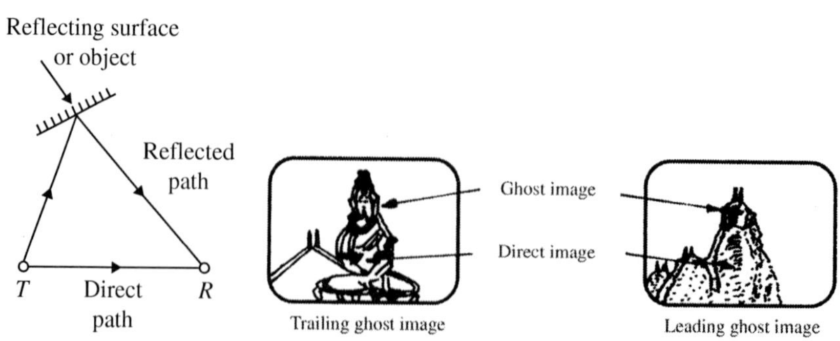

Figure 2.6 (a) Geometry of multi-path transmission and (b) Ghost interface.

2.3.3.2 Selecting AM for Picture Signal Transmission

At the UHF and VHF carrier frequencies there is displacement in time between the direct and reflected signal. The direction which develops due to interference between multiple signals is more objectionable in FM than AM since frequency varies in FM continuously.

If AM is used, the multiple signal path can almost generate a ghost image which is steady. Apart from that, circuit complexity and bandwidth requirements are less in AM than FM, therefore AM is preferred for broadcasting the picture signal.

2.4 TV TRANSMISSION ANTENNAS

The basic antenna for TV transmitter and receiver in the VHF band is the half wave dipole this is the type called resonant antenna which means that a specific length is used for particular frequency television signals are transmitted by space wave propagation and so the height of antenna must be high as possible in order to increase the line of sight distance. Generally horizontal polarization is preferred because SNR is favourable when antennas is placed high above surface of earth. The factors involved in this are, type of polarization used and type of antenna used. The selection of these factors depends on following requirements:

(i) To cover the rcquired geographical area.
(ii) The intensity of signal in the coverage area should be above the minimum level and should produce good quality picture.
(iii) The signal attenuation should be minimum over the transmission path.
(iv) The effect of noise interference should be minimum.

2.4.1 Turnstile Array Antenna

To obtain an omni directional radiation pattern in the horizontal plane, for equal TV signal radiation in all directions, an arrangement known as 'turnstile array' is often used. In this type of antenna two crossed dipole of length $\lambda/2$ are used in turnstile array arrangement as shown in Figure 2.7(a). Each dipole has a figure of eight pattern in horizontal plane, but crossed with each other.

The resultant field in all direction is equal to the square root of sum of the squares of fields radiated by each conductor in that direction. Figure 2.8 shows several turnstiles stacked one above other for vertical directives. To increase the directives in vertical plane turnstile antenna are stacked one over the other for vertical directivity. The distance between two antenna in vertical plane is also $\lambda/2$.

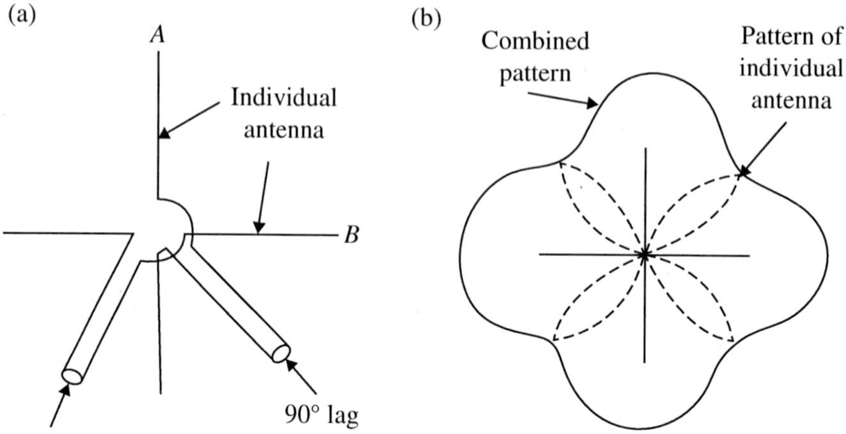

Figure 2.7 (a) Configuration and (b) Directional pattern of turnstile antenna.

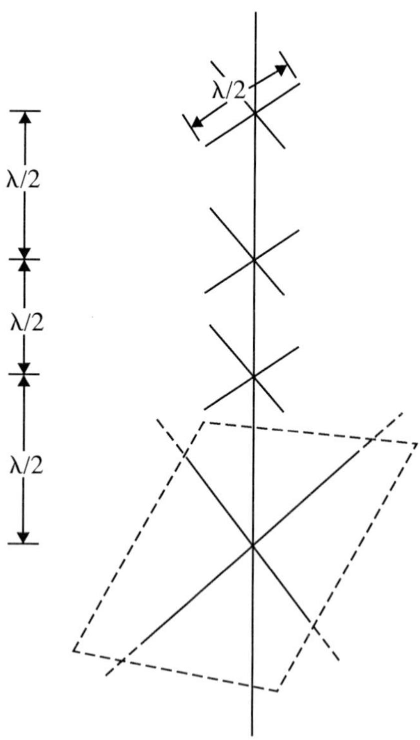

Figure 2.8 Stacked turnstile array.

2.4.2 Dipole Panel Antenna System

Another antenna system that is often used for band I and III transmitters consists of dipole panels mounted on the four sides at the top of the antenna tower.

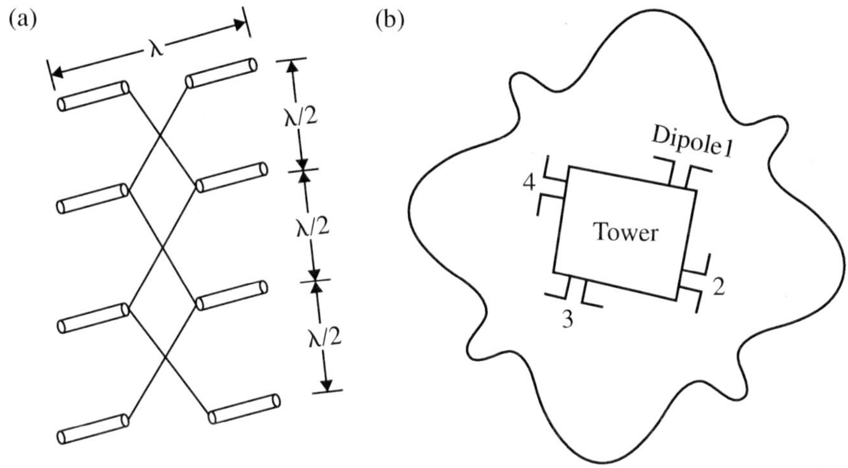

Figure 2.9 (a) Dipole antenna system and (b) Radiation pattern for tower mount dipole antenna.

As shown in Figure 2.9 each panel consists of an array of full wave dipoles mounted in front of reflectors for obtaining unidirectional pattern the four panels mounted on the four sides of the tower are so fed that the current in each lags behinds the previous by 90°. This is achieved by varying the field cable length by $\lambda/4$ to the two alternate panels and by reversal of polarity of current.

2.4.3 Combining Network

The AM picture signal and FM sound signal from the corresponding transmitters are fed to the common antenna through balancing unit called 'diplexer'. Antenna combining system as shown in Figure 2.10 is a bridge configuration in which first two arms are formed by two radiators of the turnstile antenna and other two arms consists of two capacitive reactance. Under balanced condition, video and sound signal through radiate by same antenna, do not interfere with the functioning of transmitter other than their own.

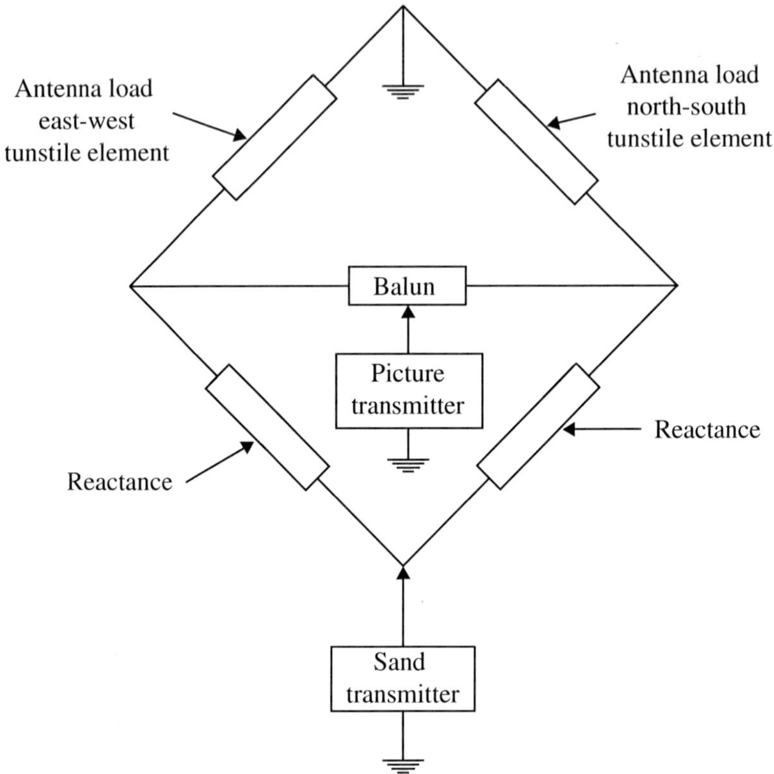

Figure 2.10 Equivalent circuit of diplexer.

2.4.4 Yagi Antenna

This is common antenna is used for TV receivers. The gain of the antenna is around 7 dB and its radiation pattern is very much directive in one direction (normally in receiving direction). Yagi antenna has three elements, reflector, director and dipole. Reflectors and directors are called 'parasitic elements'. They are used to increase the directives of the antenna in the direction required.

2.5 MONOCHROME TV RECEIVER

A block diagram of monochrome TV receiver is shown in the Figure 2.11 the important section in the receiver are:

(i) Antenna section.
(ii) RF tuner section.
(iii) Video signal and AGC section.
(iv) Audio signal section.

(v) Sync separation.
(vi) Horizontal and vertical oscillator section.
(vii) EHT section.

Let we see the functions of each block in the monochrome TV receiver.

2.5.1 Antenna Section

To receive the TV signal, antennas are mounted at the top building. Tuned antennas are used to receive good signal strength. The RF signal collected by the antenna, most commonly folded dipole is used by TV receivers for good reception. A folded dipole with one reflector and more directors, called Yagi-Uda antenna has a high gain antenna. On the antenna itself booster amplifier is mounted. It is mounted very close to the collecting element. That is dipole, booster amplifier is nothing but RF broad band amplifier. It increases the gain of the signal and also improves signal to noise ratio. If the signal level is sufficiently high, then booster amplifier is not required. A half wavelength antenna is mostly used for UHF band channels

2.5.2 RF Tuner Section

The broad band RF amplifier, mixer and local oscillator are mounted in a single unit is called 'RF tuner'. This is shown in Figure 2.12. The output of this unit is a IF signal. RF tuner is in the front end of the receiver. The incoming RF signal of a particular channel is amplified in the RF amplifier. The signal is in range of few milli-volts, therefore noise can superimpose on the required signal. Therefore the complete tuner section (RF amplifier, mixer and local oscillator) is completely shielded. There is a local oscillator whose frequency is always greater than the incoming signal frequency f_s.

- f_s-incoming channel frequency (picture and sound carrier).
- f_0-local carrier frequency is equal to $f_s + IF$.

In the mixer stage, a heterodyne process is done to produce sum frequency, difference frequency and harmonics the output of the mixer is $f_0 - f_s$. The local oscillator is adjusted above the channel frequency such that the subtraction of the two frequency produce picture IF and sound IF. For all the signals we should get the picture IF (38.9 MHz)and sound IF (33.4 MHz) at the output of mixer.

The standard intermediate frequencies for the 625-B system are:

$$\text{Picture IF} = 38.9\,\text{MHz}$$

$$\text{Sound IF} = 33.4\,\text{MHz}$$

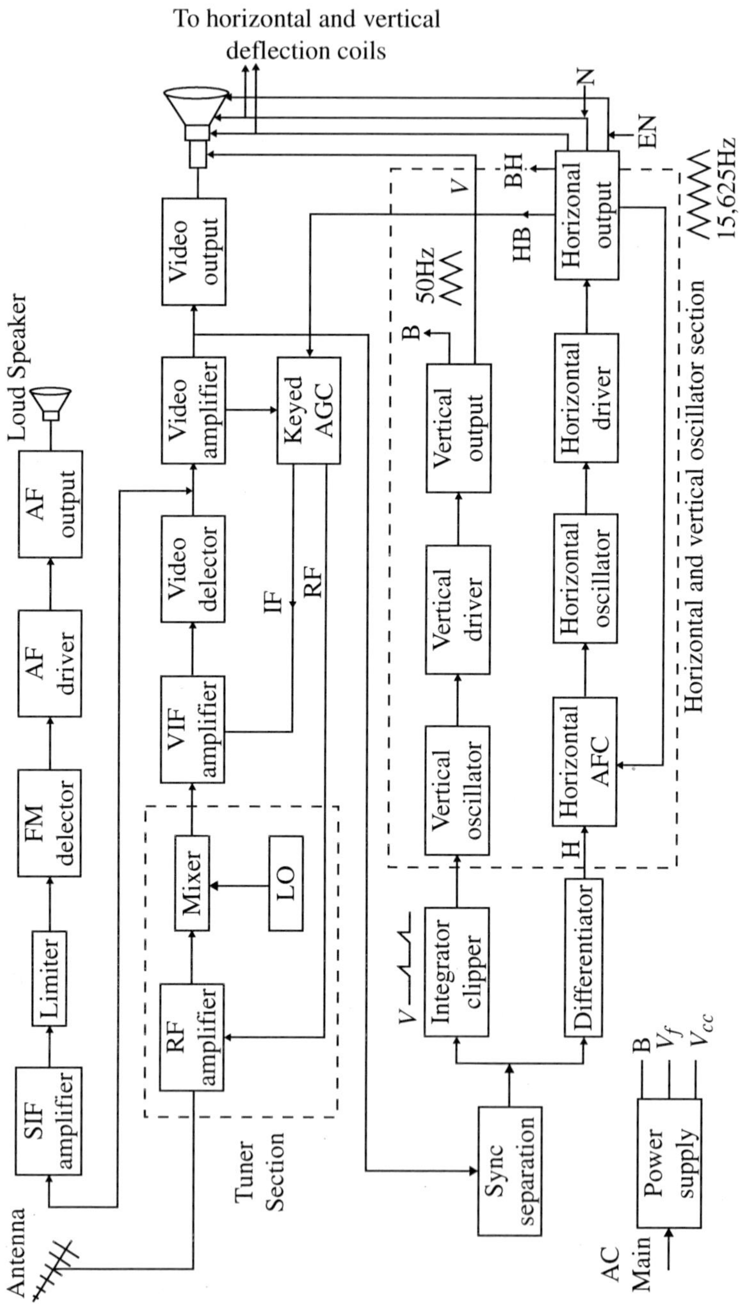

Figure 2.11 Block diagram of monochrome TV receiver.

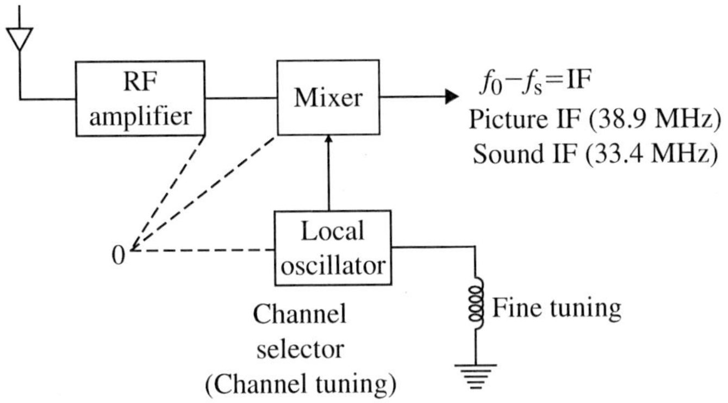

Figure 2.12 RF tuner section.

For each channel selection, the local oscillator produces a frequency equal to channel picture frequency +38.9 MHz or channel audio frequency +32.4 MHz.

2.5.3 Video Section

This section consist of:

(i) IF amplifier.
(ii) Video detector.
(iii) Video amplifier.
(iv) Picture tube.

The signal strength at the output of the mixer is very low. So, before giving it to the detector, two or three stages of amplification is needed.

After amplification it is fed into video detector. This detector is used to separate the video signal from the composite video signal. the separated video signal has also low amplitude. We need to amplify the signal before giving it to picture tube.

The picture tube needs 30 to 100 volts peak to peak amplitude for producing good picture.

2.5.3.1 Automatic Gain Control (AGC)

Signals received from long distance transmitters are weak and so maximum gain of amplifier is necessary, however signals received from near by transmitter are strong and reduced gain of amplifier is needed. But it is not possible to change the gain of amplifier while selecting different channels.

The gain of the video IF amplifier should be adjustable by AGC over a wide range to accommodate the input signal variation at the antenna from

50μ V to 0.5 V. Hence an AGC is used to automatically control the gain of RF and IF stages and provide a constant voltage.

2.5.4 Sound Section

Sound section consists of audio IF amplifier, FM demodulator, audio amplifier and loud speaker. Using sound trap circuit the audio signal is separated from video detector section. The sound carrier trap is the narrow band pass filter. It passes only 5.5 MHz signal. The remaining signal is suppressed. This FM signal is further amplified by IF amplifier. The sound IF amplifier amplifies the sound signal (5.5 MHz) and after amplitude limiting it is demodulated. The demodulated original audio signal is amplified by audio amplifier and power amplifier stages. This amplified audio signal is then fed to the loud speaker to reproduce sound. Volume control is added in this section.

2.5.5 Sync Separation

The composite video signal at the output of video amplifier is given to the sync separator the sync separator separates the horizontal and vertical sync pulses. These are then given to low pass (integrator) and high pass filter (differentiator) to generate a triggering pulse for horizontal and vertical oscillators.

2.5.6 Horizontal Oscillator

The horizontal sync pulses are separated by (differentiator) and then fed to horizontal AFC circuit. The AFC circuits maintain the H-sync frequency at 15,625 Hz. The horizontal oscillator generates a linear ramp signal at 15,625 Hz. The AFC generates triggering pulses for the horizontal oscillator. These pulses are applied to horizontal oscillator which produce saw tooth curves. These are amplified by horizontal amplifier and fed to pair of horizontal-deflection coil (H-DC). The horizontal deflection circuit works in feedback mode.

2.5.6.1 Vertical Oscillator

Vertical section consists of vertical oscillator, output amplifier and vertical deflection coil. The vertical oscillator is synchronized by trigger pulses from the integrator. The vertical oscillator operates at 50 Hz and generates a linear ramp signal. The vertical output amplifier is the driver circuit for vertical deflection coils. Then saw tooth signal is given to deflection coil, after proper amplification by vertical amplifier.

2.5.7 EHT Section

The horizontal output generates B++ supply and very high voltage for final accelerating anode of the picture tube *i.e.*, called EHT. During the retrace period of the horizontal scanning high voltage pulse of 6 to 9 kV are produced across the primary winding of the horizontal output transformer. These voltages are stepped up by an auto transformer to about 15 to 18 kV. This voltage is then rectified into a high voltage rectifying diode. Then the ADC voltage obtained is filtered and then applied to the final anode of the picture tube.

2.6 RF TUNER

The broad band RF amplifier, mixer and local oscillator are mounted in a single unit is called RF tuner. This is shown in Figure 2.13. The output of this unit is a IF signal. RF tuner is in the front end of the receiver. The incoming RF signal of a particular channel is amplified in the RF amplifier. The signal is in range of few milli-volts, therefore noise can superimpose on the required signal. Therefore the complete tuner section (RF amplifier, mixer and local oscillator) is completely shielded. There is a local oscillator whose frequency is always greater than the incoming signal frequency f_s.

- f_s-incoming channel frequency (picture and sound carrier).
- f_0-local carrier frequency is equal to $f_s + IF$.

In the mixer stage, a heterodyne process is done to produce sum frequency, difference frequency and harmonics the output of the mixer is $f_0 - f_s$. The local oscillator is adjusted above the channel frequency such that the subtraction of the two frequency produce picture IF and sound IF. For all the signals we should get the picture IF (38.9 MHz)and sound IF (33.4 MHz) at the output of mixer.

For example, for channel 5 tuning:

$$\text{Picture carrier} = 175.25\ \text{MHz}$$

$$\text{Sound carrier} = 180.75\ \text{MHz}$$

Now the local oscillator should be tuned to the frequency:

$$\text{Picture carrier } f_0 = (175.25\ \text{MHz} + 38.9\ \text{MHz})$$

$$\text{Sound carrier } f_0 = (180.75\ \text{MHz} + 33.4\ \text{MHz})$$

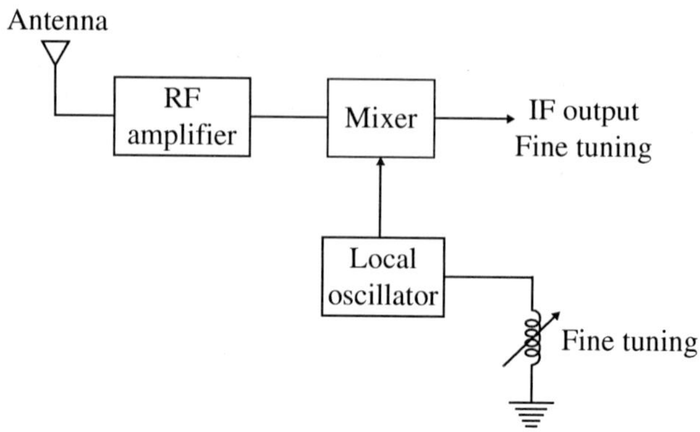

Figure 2.13 Block diagram RF tuner.

for each channel the oscillator frequency may change, but the picture and sound IF should be constant.

Mixer output frequency = L.O frequency – RF signal frequency
= IF range of frequency

to match the impedance of the antenna and receiver a balancing unit called "balun transformer" is used in between these two. Because, the transmission line has 300Ω impedance, but the receiver has only 75Ω impedance.

2.6.1 Functions of RF Tuner

The functions of RF tuners are as follows:

(i) It selects the channel to be received by switching pre-tuned circuits in the RF stage and the local oscillator.
(ii) It converts the RF into IF by mixing it with the local oscillator frequency and feed the signal to the video IF amplifier.
(iii) The RF tuner isolates the local oscillator signal from the antenna. The RF amplifier acts as a buffer, preventing radiation and interference to other receivers
(iv) RF tuners blocks the interference antenna pick up signals in the IF ranges and prevents them from entering the receivers and mixes with the local oscillator and then to the IF amplifier.
(v) It reject the image frequencies by means of RF selective circuits.

2.6.2 Types of Tuner

The VHF tuner may be single channel or multi-channel to cover overall band I and band III channels. The multi-channel tuner may employ different types of coil switching:

(i) In coil turret types switching the coils for various channels are mounted on the periphery of turret drum. The rotation of the turret type changes the coil in contact with the circuits.
(ii) In wafer switch type, switching is made with incremental coils or printed circuit coils for the channels.

2.6.2.1 *Turret Tuner or Drum Type Tunner*

The coils for various channels are mounted on a slotted drum type structure. Each channel is selected by rotating the channel selector. Each strip has coil for RF amplifier, mixer and local oscillator for single channel. That is why this type of tuner are called 'turret' or 'drum type tuners'. The turret tuner structure is shown in Figures 2.14(a) and (b) shows the strips on the drum.

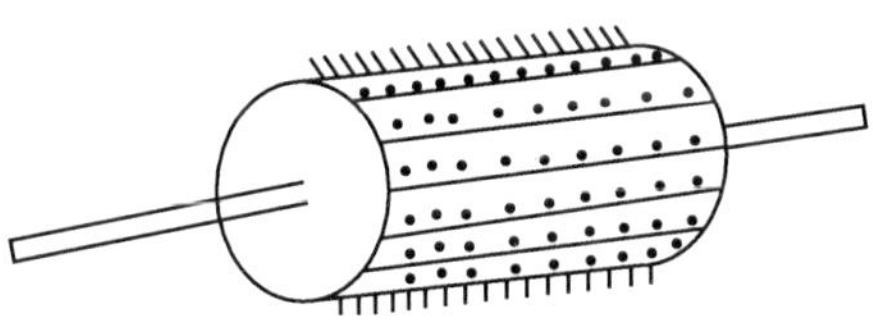

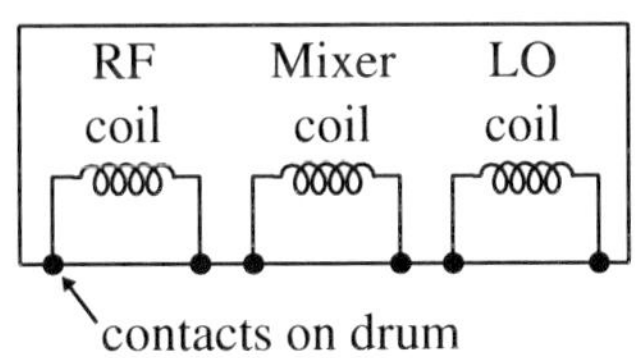

Figure 2.14 (a) Turret drum and (b) Strips on drum.

2.6.2.2 *Wafer or Incremented Tuner*

This tuner employs a wafer switch type configuration. Number of wafer switches is used to select the proper RF, mixer and oscillator coil. The coils are mounted around the outer rim of the switch. Few turns of wire are needed for the selection of lower channels and number of turns decreases the higher selection of higher channels.

This is also called incremental tuner because the change of channel is accomplished by progressive shorting of total inductance. The wafer arrangement is shown in Figure 2.14(c).

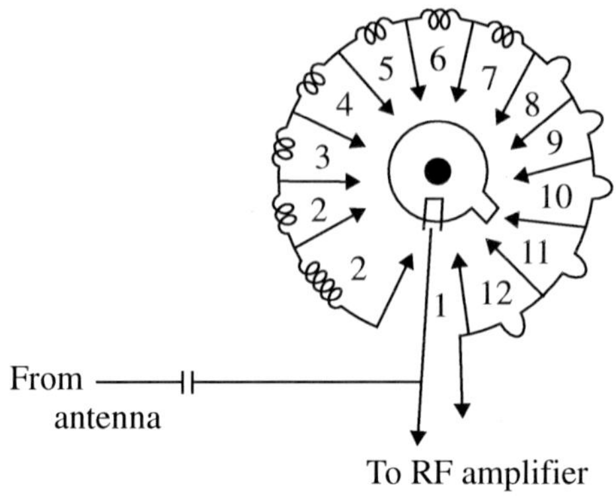

Figure 2.14 (c) Arrangement of wafer.

2.7 VHF TUNER

The VHF tuner consists of the following sections:

(i) Balun transformer.
(ii) High pass filter and trap circuits.
(iii) RF amplifiers.
(iv) Mixer stage, and
(v) Local oscillators.

This is shown in Figure 2.15.

2.7.1 Balun Transformer

The tuner must have an input impedance equal to the characteristic impendence of the antenna. So that the matching provides a maximum power transfer and avoid reflections on the line. So the Balun transformer is used for matching the impedance of the antenna (or) ribbon wire (300Ω) with the input independence of amplifier (75Ω). So the impendence matching is done by Balun transformer. This is shown in Figure 2.16.

2.7.2 HPF and Trap Circuit

The trap circuit are used are used to remove the unwanted spurious signals from entering the RF amplifier. Unwanted spurious signals which lie in the IF range of 32 to 42 MHz are rejected by two trap circuits connected on the either side of the HP filter. The high pass filter with 45 MHz cut off frequency is used remove the channel signal (41–47 MHz) below the channel 2. These HPF sections are essential because the RF amplifiers has large bandwidth and low 'Q', so RF circuits cannot be able to attenuate the

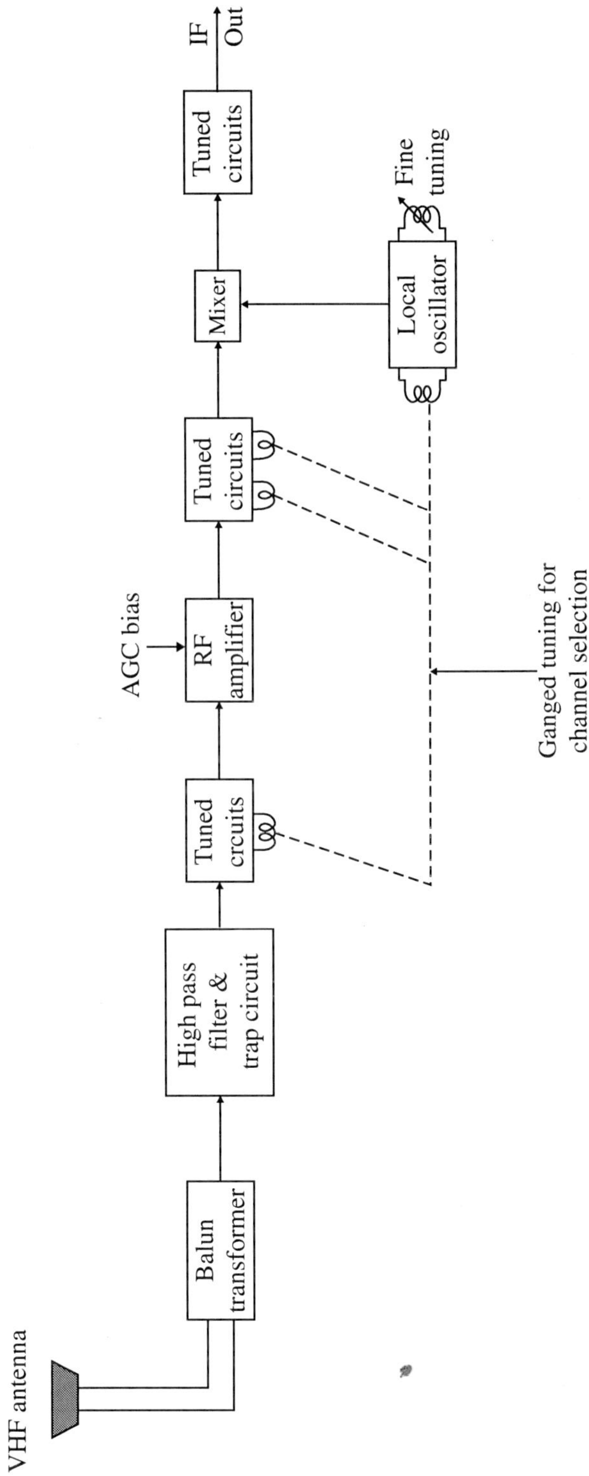

Figure 2.15 Block diagram of VHF tuner.

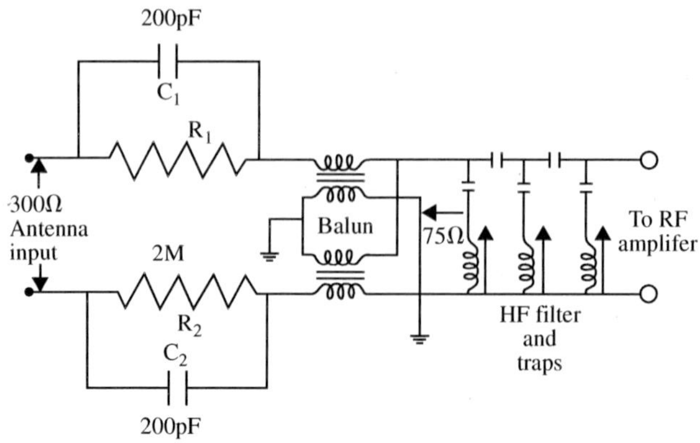

Figure 2.16 Balun circuit.

adjacent channel frequencies. And it is very difficult to reject such signals once they enter the mixer and IF stages.

2.7.3 RF Amplifier

The RF amplifier is designed to provide sufficient gain to weak signals. It has high signal to noise ratio. In addition to this RF amplifier is also used for the following purposes:

(i) It prevents radiations from the local oscillator.
(ii) It rejects image frequency signals.
(iii) It is best suited for selection of AGC.
(iv) It prevents radiation from the local oscillator.
(v) By rejecting un wanted signals, they provide good selectivity.

To maintain a good signal to noise ratio for weak signals, the application of AGC is generally delayed to the RF amplifier.

Double tuned circuits are used for coupling the RF amplifier to the mixer stage.

2.7.4 Mixer

The function of this circuit is used to convert RF signal frequency from different channels into IF band of frequencies. This is achieved by mixing local oscillator and RF signal. The output of the mixer is fed into IF amplifier for further amplification.

IF-sound carrier → 33.4 MHz

IF-colour carrier → 34.47 MHz

For example

$$\text{Mixer output frequency} = \text{L.O frequency} - \text{RF signal frequency}$$
$$= \text{IF range of frequency}$$

For channel 4 the oscillator frequency is:

$$(62.25\,\text{MHz} + 38.9\,\text{MHz}) = 101.5\,\text{MHz}$$

For channel 5 the oscillator frequency is:

$$(176.25\,\text{MHz} + 38.9\,\text{MHz}) = 215.15\,\text{MHz}$$

2.7.5 Local Oscillator

The local oscillator is used to generate the unmodulated sinusoidal voltage. The output is given to mixer stage for heterodyning process. For different channel selection, we have difference frequency is equal to sum of channel selection we have different local oscillator frequency. Usually the oscillator frequency is equal to sum of channel carrier frequency and IF range of frequency. The fine tuning control allows the oscillator frequency to be varied over a narrow range.

2.8 UHF TUNER

The block diagram of UHF-VHF tuner is shown in the Figure 2.17.

All channel receivers employ a UHF-VHF tuner are combined in a single unit and are assembled in a single compact shield. Separate antennas are used for receiving the UHF and VHF band frequency signals. The desired channel input is selected by a change over switch, which selects the input between the UHF and VHF antenna outputs.

At the input of the UHF tuner, a high pass network is provided, which prevents the passage of any VHF channel signal and also provides an impedance match between the antenna and the input of RF amplifier.

To match the impedance of the antenna and receiver a balancing unit called "balun transformer" is used in between these two. Because, the transmission line has 300Ω impedance, but the receiver has only 75Ω impedance.

The UHF tuner does not employ RF amplifier because of its high cost and complexity in design. The UHF antenna signal is coupled to the input tuned circuit and the diode mixer via a matching circuit. The local oscillator output is coupled to diode mixer that gives out the IF.

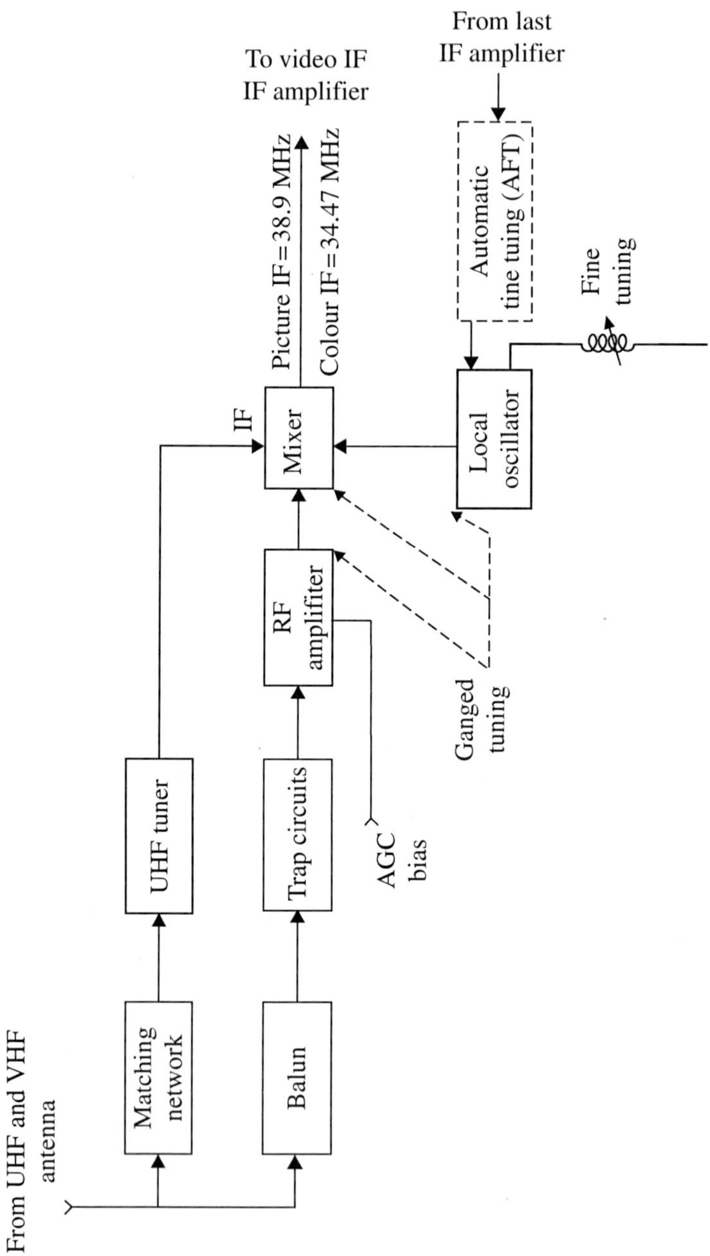

Figure 2.17 UHF-VHF tuner.

2.8.1 VHF Tuner

The block diagram of VHF tuner is shown in the Figure 2.17 when we tune the tuner in TV receiver to VHF band. Then the VHF antenna pickup the channel signals after the impendence matching the signals are given to the RF amplifier. The RF amplifier amplifies the selected frequency. Then in the mixer these selected channel signals is superheterodyne (mixed) to produce the required IF outputs. In these case the required IF output are selected by tuning the mixer output to difference frequency signals ($f_0 - f_s$).

2.9 DIGITAL TUNING TECHNIQUES

The transition from discrete to integrated circuits and availability of micro computers has totally transformed the method of channel selection in television receivers. In the new methods of channel selection, a frequency synthesizer is employed. It generates a number of crystal stable frequencies using only one crystal. These stable reference sources are used to lock the tuner oscillator frequency for each channel, at its correct value by a Phase Locked Loop (PLL) control system.

2.9.1 Phase Locked Loop Control

The PLL control compares two signals and develops a control voltage, proportional to the frequency or phase different between them. For channel selection the reference signal is the frequency obtained on dividing the stable crystal oscillator frequency and the signal to be controlled is the oscillator frequency. This is also divided by a suitable factor before feeding do the comparator in the PLL circuit.

2.9.1.1 PLL Operation

The basic operation of a PLL control as used for channel selection can be explained with the help of block diagram shown in Figure 2.18 assume that crystal accuracy of the output of 20 kHz tuned oscillator circuit is required. Precision crystal at 20 kHz is not easily available at low cost. Thus it is desirable to use a 20 MHz crystal to stabilize a 20 kHz oscillator. Such oscillators are always Voltage Controlled (VCO) configurations as shown in Figure 2.18 the output of 20 MHz crystal stabilised oscillator feds into a divide by 1,000 digital frequency divider.

$$\frac{20\text{ MHz}}{1000} = 20\text{ kHz}$$

The output signal of the divider is gives to the phase comparator. The phase comparator compares phase output of the divider with the reference

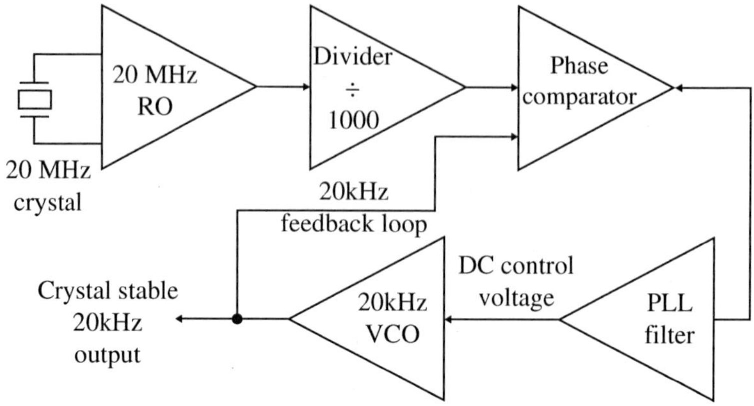

Figure 2.18 Block diagram of phase lock loop.

signal generated from the crystal stable 20 kHz frequency. If the two signals are sychronized, no correction is required. However if the VCO direct-off frequency, then the phase comparator develops the correction voltage proportional to the phase difference between the two input signals. The output from the PLL comparator is filtered by a PLL filter and develops a DC control voltage to the voltage control oscillator. This DC control voltage forces the oscillator to lock with the crystal stable frequency of 20 kHz.

2.9.2 Channel Selection by Frequency Synthesis

Frequency synthesis is an electronic system which can generate a number of stable frequencies while using only one crystal in an electronic tuner, frequency synthesis is used to generate crystal stable frequencies for the tuner oscillator in all VHF and UHF channels.

The set-up includes a highly stable crystal reference oscillator, a phase locked loop (PLL) and digital circuits controlled by a microcomputer (μC) the μC provides intelligent decision making and control for the system. It also eliminates the use of fine tuning control.

2.9.3 Phase Lock Loop

The stable frequency derived from crystal oscillator is reduced to the required level through programmable divider. Similarly the local oscillator frequency is reduced to 1 kHz and these two frequency signals are given to the phase comparator.

The phase comparator compares these two frequencies and produces an error voltage corresponds to the difference frequency. This error voltage is given to varactor tuning diode to correct local oscillator frequency for particular channel selection.

When the error voltage comes to zero, the PLL will lock in that particular frequency.

2.9.4 Channel Selection

The microcomputer (μC) controls the channel selection, tuning and other receiver functions by continuously scanning various inputs to IF (Intermediate frequency) for new commands (Figure 2.19) it also receives input from sensing circuits that determine whether a station signal is present or not and if so, how accurately the channel is tuned.

When new channel is selected by appropriate entry on the keyboard, the μC provides correct binary oscillator dividers. A step signal also goes to the sound IC to mute sound during channel selection the led drives are also fed with necessary binary input.

To mute the sound during the channel selection, a step input signal is given to the sound section IC. To display the selected channel number, LED drivers are also fed with necessary input from microcontroller. The numbers corresponds to each channel are stored in the memory of micro controller. Other function performed by this microcontroller is based on frequency synthesizer and are:

(i) Channel mute.
(ii) Digital LED display.
(iii) Remote control.
(iv) Manual tuning.

2.9.5 Tuner Oscillator Frequency Range

The crystal controlled reference frequency is chosen between 3 to 5 MHz but the tuner oscillator frequency differs for each channel. The programmable divider is used for proper frequency division between the local oscillator in the tuner and comparator.

The pre-scalar is a fixed divide-by-175 counter followed by another divided-by-2 counter. After this it is given to a programmable divider. This set up gives the required frequency division. So in reference oscillator section we have a total binary division of 3500 ($175 \times 10 \times 2 = 3500$) that leads to 1 kHz reference frequency.

For proper operation of PLL, the local oscillator input to discriminator also should have the value 1 kHz. From Figure 2.19 we find that in the local oscillator network we have divider-by-175, divide-by-2 and divide by $-n$

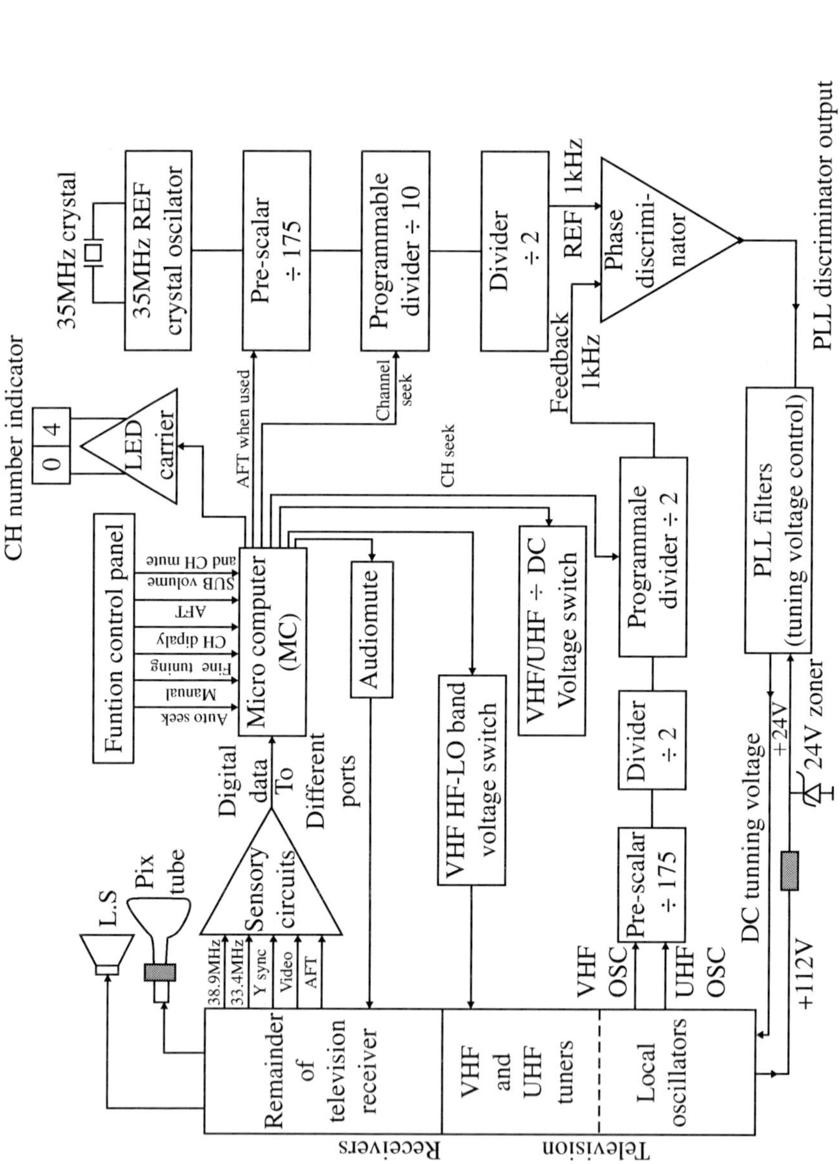

Figure 2.19 Simplified block diagram of microcomputer controlled electronic channel selection tuning system.

divider.

$$\frac{\text{Local oscillator frequency}}{175 \times 2 \times N} = 1 \text{ kHz } (1{,}000 \text{ Hz})$$

Thus

$$n = \frac{\text{local oscillator frequency}}{175 \times 2 \times 1000}$$

Consider lower VHF channel is 87.15 MHz. Similarly for channel 3 and 4, if it is 94.15 and 101.15 MHz respectively. Therefore N for these channels can be found as:

$$N2 = \frac{87.15 \times 10^6}{175 \times 2 \times 1{,}000} = 249$$

$$N3 = \frac{94.15 \times 10^6}{175 \times 2 \times 1{,}000} = 269$$

$$N4 = \frac{101.15 \times 10^6}{175 \times 2 \times 1{,}000} = 289$$

The numbers corresponding to each channel are stored in memory unit of the micro-controller. The main function of this setup is tuning and channel selection. A part from this, other functions performed by such system are:

(i) Auto seek programming.
(ii) Manual programming.
(iii) Digital LED display for the channel selection.
(iv) Operation with or without remote control.
(v) Audio mute during channel selection, and
(vi) AFT control.

Note: Value of N increases by 20 for each higher channel hence programmable divider ratio to be changed in reference oscillator chain of dividers.

2.10 AUTOMATIC FREQUENCY TUNING

The local oscillator frequency in the RF tuners is set to provide exact IF frequencies. However, despite many remedial measures to improve the stability of oscillator circuit. The proper tuning of a colour receiver is critical because the colour subcarrier must be properly positioned on the IF (Intermediate frequency) response curve. This is in Figure 2.20(a) illustrated consider that due to faulty tuning the local oscillator frequency changes to

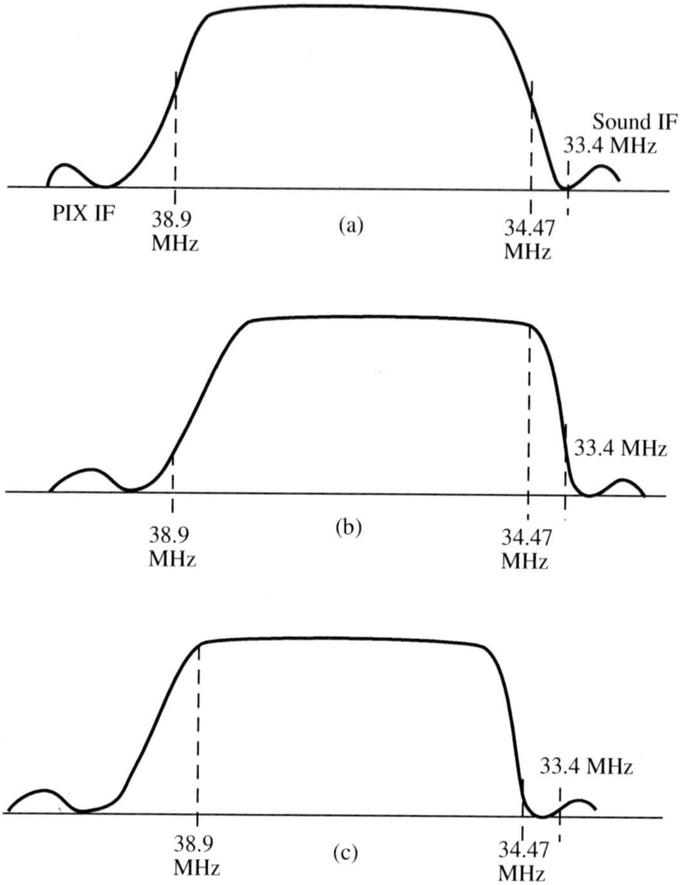

Figure 2.20 Effect of changes in the tuning on he location of picture, colour sub-carrier and sound IF frequency, on the IF response (a) correct and local oscillator, (b) high, and (c) low.

become high by 1 MHz. This as shown in Figure 2.20(b) will shift the picture, subcarrier and sound IF frequency by the same amount causing low frequency video side band to become weaker and chrominance sideband stronger.

If the local oscillator frequency changes to become less by 1 MHz, the opposite will happen as shown in Figure 2.20(c). As a result of decrease in colour level and in particular loss of colour burst, only a black and white picture will be seen on the receiver screen. Hence tuning is necessary for high quality colour TV reception. Improper fine tuning can result in loss of colour, weak colour, wrong colour and colour interference.

Thus automatic fine tuning circuit must be incorporated in colour receivers. Receivers employing synthesised frequency tuning may not need such a control but it always provided to take care of any channel frequency offset and in accurate cable frequencies.

2.10.1 AFT Circuits

Modern receivers employ an integrated circuit for AFC. It is either a separated IC or part of the IF subsystem. such a circuit is shown in Figure 2.21 note that the 38.9 MHz input is tuned circuit, the discriminator transformers and decoupling IC-RC networks are the only component that are external to the IC. The input as earlier is from the 3rd IF amplifier and DC control output feeds across the tuner varactor.

2.11 IF SUB SYSTEM

The IF section consist of three sections they are:

(i) IF section.
(ii) AGC circuit.
(iii) Video detector.

2.11.1 IF Section

The main function of this section is to amplify modulated IF (Intermediate frequency) signal over its entire bandwidth with an input of about 0.5 mV signal from the mixer to deliver about 4V into the video detector. This needs an overall gain of about 8,000 the gain should be adjustable Automatic Gain Control (AGC), over a wide range to accommodate input signal variations at the antenna from 50 μV to 0.5 V, to deliver about 4V peak to peak at the input of video detector.

2.11.1.1 Design Requirements

(a) The combined gain of the IF amplifiers should be about 8000 to deliver about 3 to 5 V peak to peak at the input of video detector.
(b) For the faithful reproduction of picture, sound and colour information, the stage must provide necessary bandwidth and selectivity.

2.11.1.2 Vestigial Side Band Correction

The important function of IF section is to equalize the amplitude of sideband components, because of vestigial side band correction. This is shown in Figure 2.22(d).

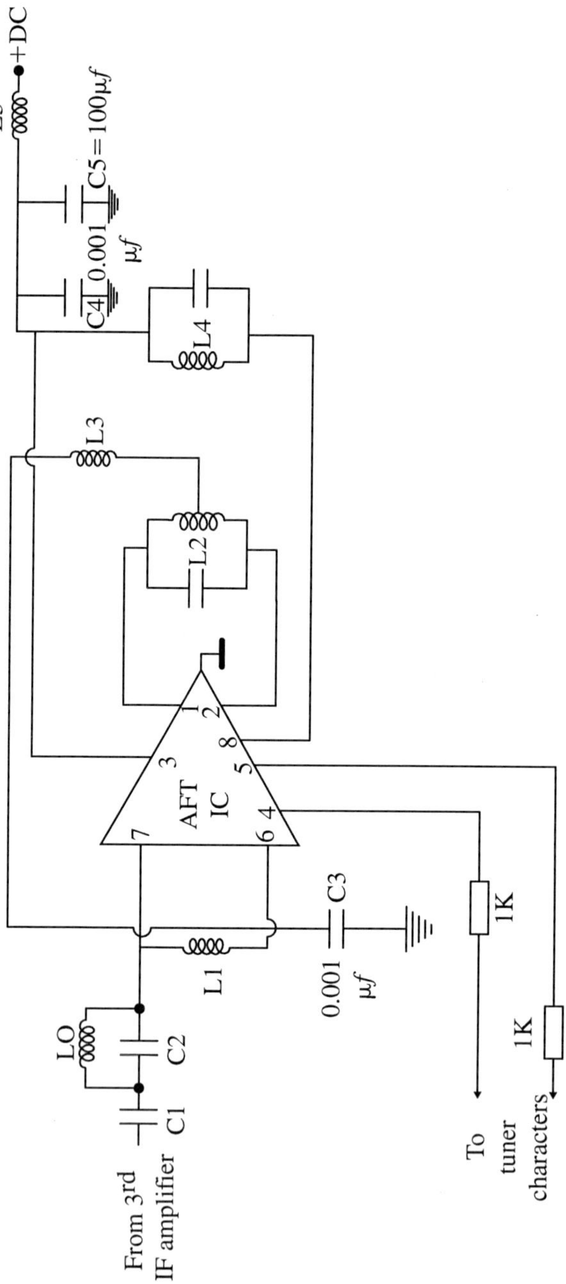

Figure 2.21 AFT integrated circuit.

These functions are:

(a) The upper edge of the overall response curve should have the required slope for vestigial side band correction. This is very useful to provide uniform signal at video detector over entire range.
(b) For vestigial side band correction the picture carrier frequency gain must be 6 dB down (50%) on the IF frequency response curve. It is also necessary to shape the IF response curve around the picture IF frequency.
(c) To reduce the gain at the sound IF to 5% (−26 dB) of the maximum gain, the lower edge of the IF response should be designed to have a steep slope. This is very useful for successful functioning of the inter-carrier system.
(d) In colour receivers, the lower skirt of the IF response should have a slope, so as to place the colour IF at about −5 dB level.
(e) To suppress adjacent channel interference, rejection filters must include in both monochrome and colour receivers.

The factors which influence the choice of intermediate frequencies in TV receivers are image rejection ratio.

Consider a desired input signal at 100 MHz and a local oscillator frequency is set at 110 MHz is shown in Figure 2.22(b). If the IF frequency is fixed at 10 MHz, similarly if another station is operating at 120 MHz, and the local oscillator frequency is set at 110 MHz and it develops an output of 10 MHz IF. Note that in case I the undesired signal which get received is spaced at a gap of twice the IF frequency and it is known as image signal:

2.11.1.3 Image Rejection Ratio

The unwanted signal whose frequency are equal to sum of desired signal frequency and twice the IF frequency are called 'image signals'. That is

$$\begin{pmatrix}\text{Image}\\ \text{frequency}\end{pmatrix} = \begin{pmatrix}\text{desired signal}\\ \text{frequency}\end{pmatrix} + \begin{pmatrix}\text{2 IF}\\ \text{frequency}\end{pmatrix}$$

Image rejection ratio is defined as the output due to the desired station divided by output due to the image signal. An RF amplifier is placed before the mixer which helps in reducing the image signal.

(a) At low IF frequencies (10 MHz), the image signal frequency is 120 MHz and it pass through the pass band of RF amplifier and the signals get attenuated. This is shown in Figure 2.22(a).
(b) But if the IF frequency is kept high as possible as shown in Figure 2.22(b), the image frequency is 80 MHz and desired signal frequency is 100 MHz and it has no chance of passing through the RF amplifier.

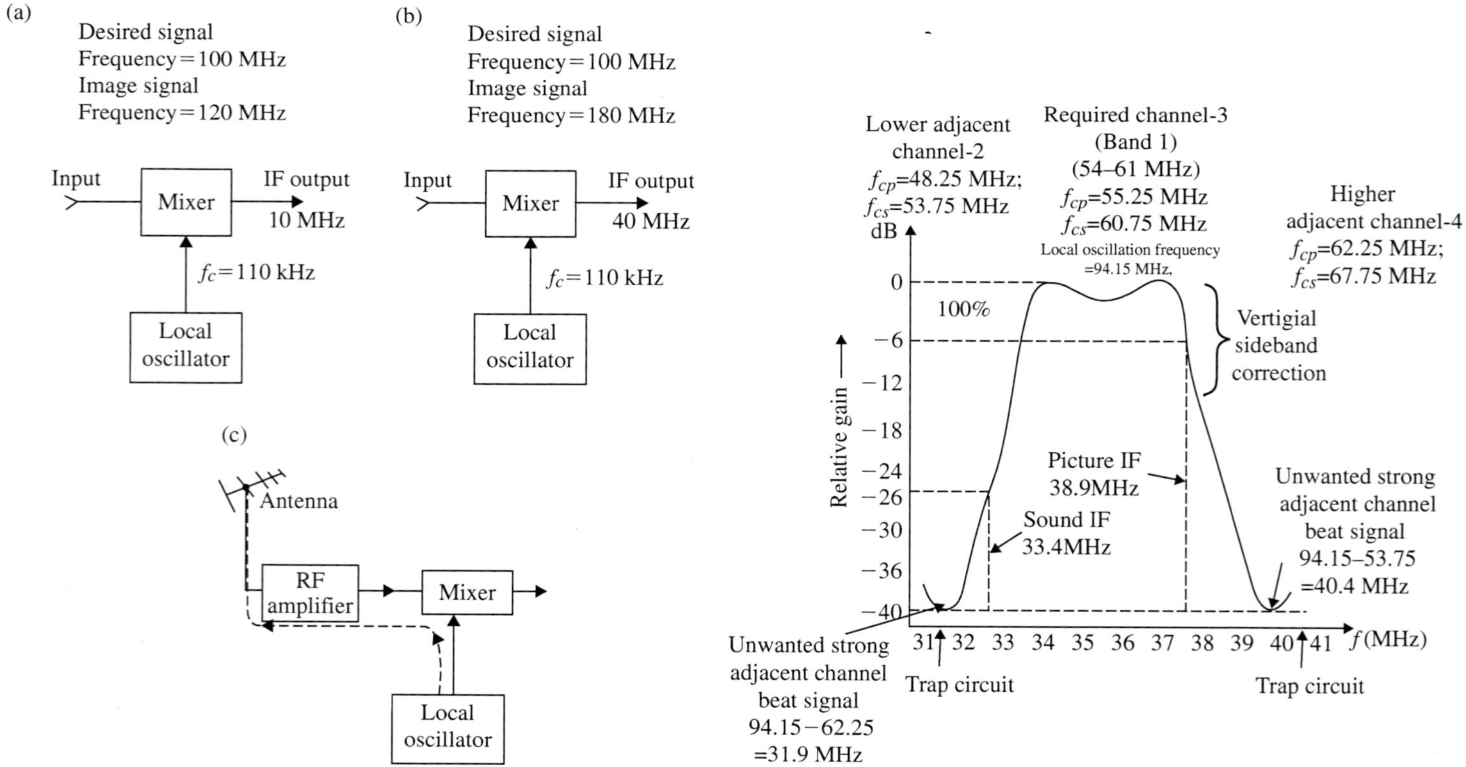

Figure 2.22 (a) Low IF range, (b) High IF range, (c) Local oscillator radiation and (d) Vesitigal side band correction.

Thus the use of the RF amplifier helps in reducing interference due to image signals and a higher IF frequency results in a very high image rejection ratio.

Frequency Range	Choice of Intermediate Frequency
For lower VHF: 41–68 MHz	Choice of IF is 28 MHz
For UHF band: 470–582 MHz	Choice of IF is 56 MHz. Because of direct pick up problem in this UHF band, so 56 MHz cannot be used, therefore the IF frequency must be less than 41 MHz

2.11.1.4 Pick Up to Local Oscillator Radiation from TV Receiver

If the output from the local oscillator of TV receiver gets coupled to the antenna it will get radiated and may cause interference in another receiver. This is shown in Figure 2.22(c). The advantage of frequency there is a separation between the resonant circuit of local oscillator and RF amplifier circuit. Thus lesser signal is coupled from the local oscillator through the RF amplifier to antenna circuit and interference due to local oscillator radiation is reduced.

2.11.1.5 Image Frequency Should Not Lie in the FM Band

The FM band frequencies lies in the range of (8.8 MHz to 110 MHz). The choice of Intermediate frequency (IF) frequency choosen close to 40 MHz, the image frequencies of the lower VHF band fall between 121 to 148 MHz and does not cause any interference in the FM band. Higher TV channels are much above the FM band.

2.11.1.6 Interference or Direct Pickup from Bands Assigned for Other Services

Amateur and industrial applications frequency band lies between 21 to 27 MHz. If the Intermediate frequency (IF) frequency is chosen above 40 MHz, even the second harmonics of this band will not cause any serious direct pick up problems.

2.11.1.7 Merits of IF Frequency

Amplifiers with large gain at relatively low frequencies are very easy to build. Some 45 years back, the TV manufactures used IF frequency as low as 12 MHz. This was due to poor quality of the components marketed at that time but now days, the active devices and IC which can perform well at high frequency are now easily available.

The merits of having high frequency are thus obvious and his has lead to the choice of IF frequencies close to 40 MHz. The IF frequencies recommended by:

Area	Sound IF in MHz	Picture IF in MHz
India and other countries	33.45	38.9
USA	41.25	45.75
British 625 system	33.5	39.5

2.11.2 Need For AGC Circuit

Depending on the distance transmitter and receiver, cable losses *etc.*, the strength of the incoming signal varies. Hence we have to use AGC circuit to control the gain of the RF and IF amplifiers. In receivers the AGC circuit produces a DC bias voltage according to the strength (peak value) of the incoming signals. The AGC produce a DC bias voltage to reduce the amplifier gain for strong signals. The AGC bias voltage should increase the gain of IF and RF amplifiers when the incoming signals are weak. So in receiver automatic gain control is used to control the gain of RF and IF amplifiers.

2.11.2.1 Requirements of AGC

(a) The AGC automatically adjust the gain of RF stage and IF stages of the receiver in order to maintain a constant level of signal at the video receiver.
(b) The AGC circuit should produce a DC bias voltage from the incoming video signal *i.e.*, provision should be made in the circuit to convert video signal AC to a DC bias.
(c) The AGC circuit should must act very fast to reduce the gain of the amplifier for strong signals.
(d) When incoming signals are weak, the AGC should apply a delayed DC bias to the RF amplifier.
(e) According to the amplitude of the incoming signals, the AGC circuit should produce a varying DC bias voltage.

2.11.2.2 Merits of AGC

(a) Intensity and contrast of the picture, once set with manual controls remain almost constant despite changes in the input signal strength from the antenna.

(b) Contrast in the reproduced picture does not change much when receiver is switched from one channel to another.
(c) Amplitude and cross modulation distortion on strong signals is avoided due to reduction in gain.
(d) AGC also permits increase in gain for weak signals this is achieved by delaying the application of AGC to the RF amplifier.
(e) Flutter in the picture tube due to passing aero planes and other fading effect is reduced.
(f) Separation of sync pulses becomes easy.
(g) Sound signal is also controlled by AGC.

2.11.2.3 Keyed AGC

Consider the block diagram of keyed AGC is shown in Figure 2.23(a). The AGC rectifier is keyed ON by the fly back pulses from the horizontal output

(a)

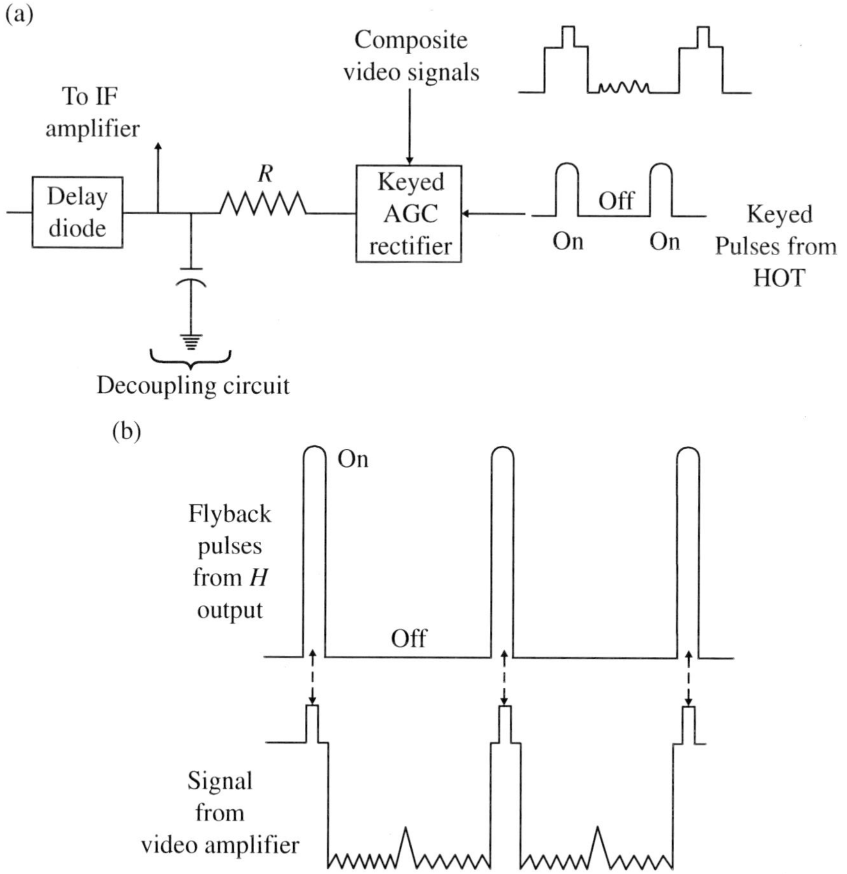

Figure 2.23 (a) Block diagram of keyed AGC and (b) Keyed AGC output waveform.

transformer because the AGC rectifier is not necessary to turn ON for the full composite video signal period.

Hence the rectifier is turned ON for a shorter duration using the fly back pulses (these pulses are generated in the HOT during horizontal retrace period). The rectifier produces a DC bias at a much faster rate. The DC bias is passed through decoupling circuit through the IF amplifier to control the gain through the diode delay the DC bias and then applied to RF amplifier to control the gain for weak signals.

Here, filter (R & C) is used to filter the DC voltage. Due to keyed AGC, the noise pulses in the signal have little effect on the AGC bias voltage.

Transistorised Keyed AGC

Basic keyed AGC circuit designed to develop a positive AGC voltage is shown in the Figure 2.23(c).

The p-n-p transistor Q1 is biased to cut-off under no signal conditions. When the negative going retrace pulses are applied to the collector terminal, the base-emitter junction gets forward biased when the video signal at the base of transistor Q1 approaches a minimum value.

The keying pulses are fed into the collector of the keyed transistor via diode D1. When transistor Q1 conducts, current flows through R1, R2, Q1, D1, winding on H.O.T and C1 there by charging it with the polarity marked on it. The voltage across capacitor C1 is the AGC output voltage.

The diode D1 prevents discharge of capacitor C1 during the time between keying pulses. In the absence of diode D1 the charge on capacitor C1 will forward bias the collector to base junction of transistor Q1.

The capacitor C1 and resistors R3, R4 form a filter to develop a steady DC voltage for controlling overall gain of the receiver. Typical transistor circuit values are 25 V peak to peak for fly back pulses and 2 V peak to peak for video signal.

Troubles in AGC: The troubles in AGC are:

(a) too much AGC bias reduces the gain, causing a weak picture. This weak picture is without snow. Weak antenna signal can produce a weak picture but with snow; and
(b) when the AGC bias is very high it will cutoff the RF and IF amplifier, resulting in no pictures, sound is also cut off.

Advantages of Keyed AGC: These are listed below:

(a) A true representation of the peak of sync level is made by the AGC voltage. So the AGC voltage developed corresponds to the actual incoming strength.

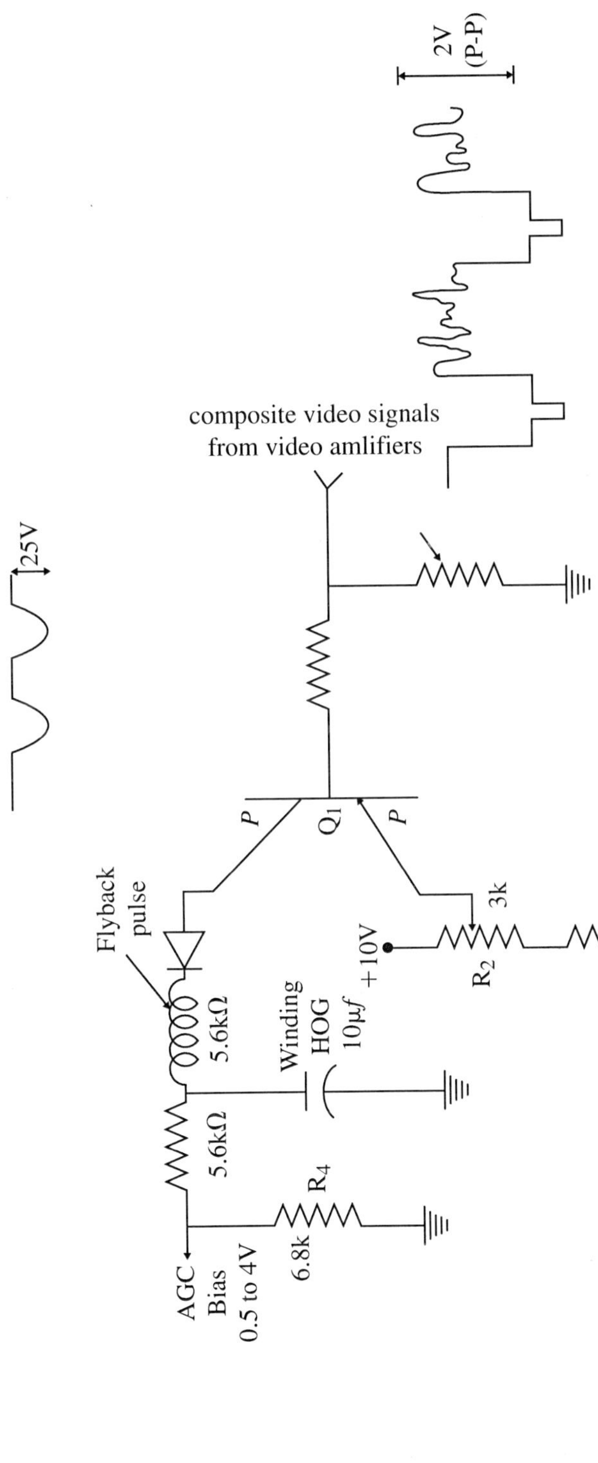

Figure 2.23 (c) Transistorised keyed AGC system.

(b) As the conduction takes place during the retrace time, long time constant are used to remove 50 Hz ripple frequency are not necessary.
(c) Since the conduction of transistor is restricted (sync pulse time), effect of noise is minimum here.
(d) The short time constant filters are used here, can enables the AGC bias to respond fast change of channels and aeroplane Flutters. So steady picture and sound output is ensured.

2.11.2.4 Types of AGC Control

The gain of any transistor amplifier can be varied by shifting the operating point either towards collector current cut-off or saturation. When a gain is changed by shifting the operating point towards cut-off region is called reverse AGC system. Similarly if the gain is varied by shifting the operating point towards collector current saturation is called forward AGC.

Reverse AGC: When a gain is changed by shifting the operating point towards cut-off region is called reverse AGC system. It requires a smaller change in VBE voltage for full gain control than the forward AGC system. The operation in this region is close to cut-off makes the receiver more susceptible to overload and cross modulation distortion on strong signals.

The circuit is shown in Figure 2.24(a) is of a single stage transistor (n-p-n) IF amplifier employing reverse AGC. The voltage divider formed

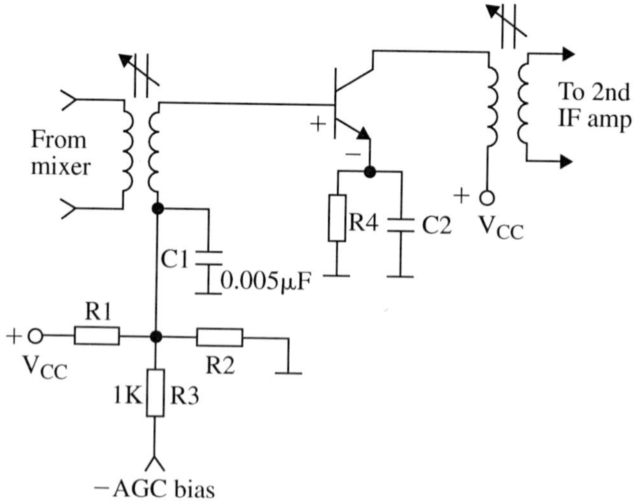

Figure 2.24 (a) Reverse AGC.

by R1 and R2 provides a suitable fixed forward bias from the VCC supply. The resistor R3 and capacitor C1 constitute the AGC decoupling network.

Forward AGC System: When the gain is varied by shifting the operating point towards collector current saturation is called Forward AGC.

Circuit Operation: A single stage tuned IF amplifier employs a forward AGC is shown in Figure 2.24(b), in this case, for any increase in signal strength the base to emitter forward bias must increase to shift the operating point of the transistor towards saturation. Similarly a decrease in forward bias. To achieve the AGC system must deliver a positive going voltage to the base of the amplifier.

If a p-n-p transistor is used. The forward AGC system would develop a negative going voltage proportional to the signal strength. As illustrated in Figure 2.24(b), amplifiers employing forward AGC often use a large resistor R9, R5 in series with the collector circuit. With this the AGC voltage varies to increase collector current, effective VCE decreases, there by allowing the transistor to approach saturation quickly for faster AGC action.

Draw Backs of Non-keyed AGC System: The AGC voltage developed across the peak rectifier load tends to increase during vertical sync periods

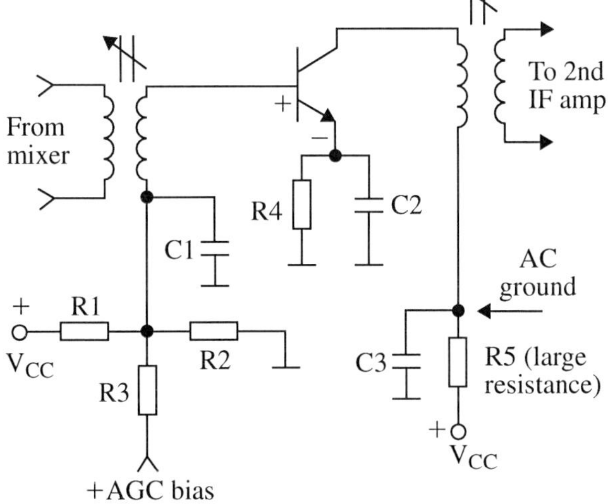

Figure 2.24 (b) Forward AGC system.

because the video signal amplitude remains almost at peak value at vertical sync pulse. The drawbacks are listed as:

(a) This result in a 50 Hz ripples over the negative AGC voltage and reduces gain of the receiver during these intervals.
(b) The reduced gain results in weak vertical sync pulses which in turn cause the vertical deflection oscillator to out of synchronization and it causes the rolling of the picture.
(c) In fringe areas noise pulses develop an additional AGC voltage which tends to reduce overall IF gain. This effect is more pronounced for dark scenes.
(d) Even when the input signal strength is low, a small AGC voltage gets developed and reduces the gain of the receiver.
(e) To over come this draw backs special AGC circuits known as keyed or gated AGC have been developed and are used in all present day television receivers.

2.12 NOISE CANCELLATION

The need for minimizing noise set up to AGC and sync separator circuits has given rise to several methods of noise cancellation. Two commonly used methods are described. In Figure 2.25, diode D2 is used as a switch which opens in the presence of noise preventing it from reaching the video amplifier. The necessary forward bias enabling it to pass noise free video signals is set by potentiometer R_3.

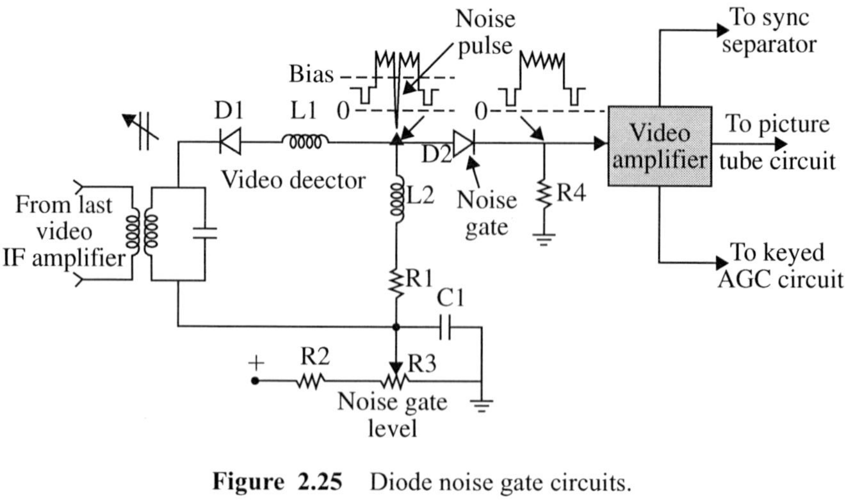

Figure 2.25 Diode noise gate circuits.

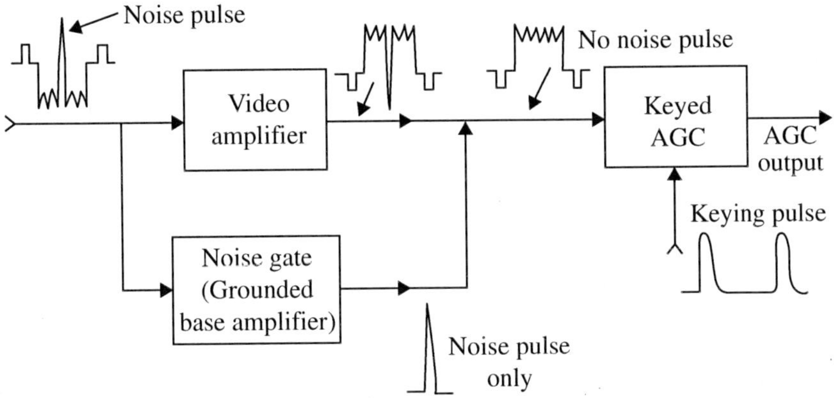

Figure 2.26 Noise cancellation by a separation of noise gate amplifier.

When a strong noise signal arrives, the diode D2 gets reverse biased there by stopping the noise pulses before reaching the video amplifier. Since video signal to both, the AGC keyed and sync separator is obtained from the output of video amplifier, noise pulses are prevented before reaching to both these circuits.

Another noise cancellation method is shown in Figure 2.26 that noise is eliminated by using a separate noise gate it separates noise from the composite video signal, amplifies it and then adds it to the inverted composite video signal.

The noise gate is a grounded base amplifier normally set to cutoff. Any incoming noise pulse of sufficient amplitude counteracts the fixed reverse bias and sets the amplified without inversion of its polarity the gain of the noise gate amplifier is set equal to that of video amplifier.

Since the two noise pulse are equal in amplitude but opposite in polarity *b*, they cancel on addition. The AGC and sync separator circuits thus remain immune to incoming noise pulses.

2.13 VIDEO AND SOUND CARRIER DETECTION

In a monochrome receiver, the same detector is used to obtain composite video signal and inter-carrier FM sound signal (Figure 2.26) in colour receivers separate detectors are used for video and sound signal detection. Such a circuit is shown in Figure 2.27.

The carrier frequency for audio and video signals is selected at a difference of 5.5 MHz. For example, channel 2 has the picture carrier frequency of 48.25 MHz and the sound carrier frequency of 53.75 MHz.

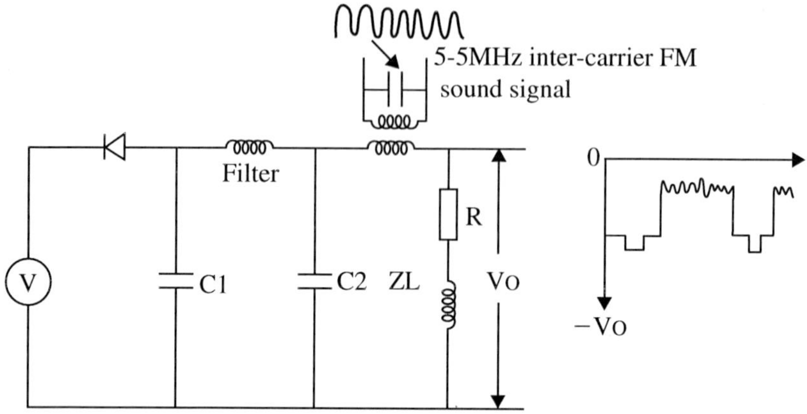

Figure 2.27 (a) Video detector and sound signal separation circuit.

These two carrier frequency are mixed with local oscillator frequency in the mixer stage of receiver to produce two IF signals 33.4 MHz for sound and 38.9 MHz for video. Thus the difference of 5.5 MHz frequency is maintained here.

At the video detector the sound IF is made to beat with picture IF frequency to transfer the audio signals to new sound IF of 5.5 MHz. This is known as 'inter-carrier sound system'.

The sounds take off circuit is used to pick up the sound IF signal (5.5 MHz). The sound trap circuit is used to prevent the sound signal entering into the picture tube.

2.13.1 Colour Receiver

It can be seen that the sound take off point is located before the video detector to minimize the 1.07 MHz interference that would be produced if the sound IF (5.5 MHz) is permitted to beat with the 4.43 MHz colour sub-carrier. This is shown in Figure 2.27(b).

Conversion 0 sound IF to inter-carrier sound (5.5 MHz) is accomplished by diode D1 which act like series detector. The input to the detector is via a series resonant circuit (C1 – L1). This circuit is tuned to about 36 MHz and is broad enough to include both picture and sound IF R1, and C2 are the load resistor and IF filter. C3 and L3 constitute a filter that separates sound IF from the composite video that also appears at the output of sound IF detector. The inductor L2 Blocks video IF and its harmonics.

Since the sound take off is before the n video detector, greater attenuation of the 5.5 MHz sound IF is possible in the video detector circuit. This is done by bridge type trap circuit located before and after the detector. The trap circuit located before the video detector (D2) is tuned to

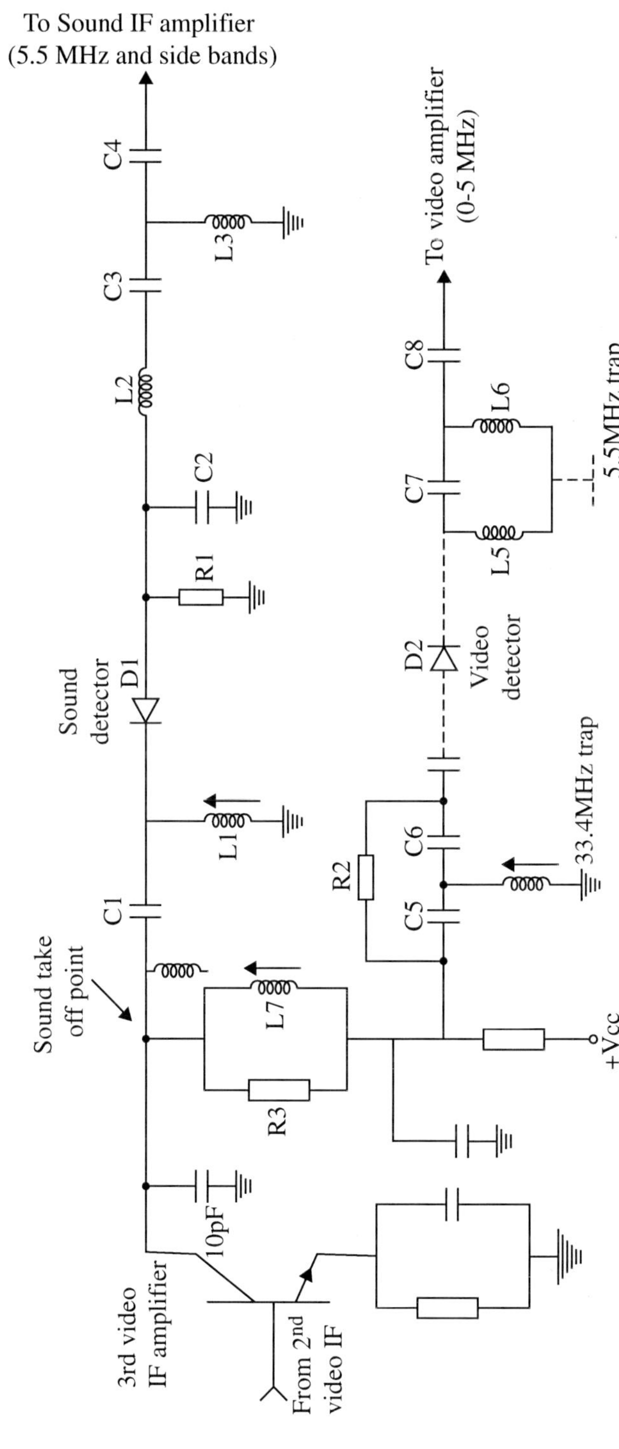

Figure 2.27 (b) Video and sound take off deflection circuit in colour receiver.

33.4 MHz and the one after the detector is tuned to 5.5 MHz. Thus signals at these frequencies are highly attenuated and do not reach the cathodes of picture tube.

In a monochrome receiver the sound IF lies at about 5% of the maximum amplitude because of inter-carrier sound detection requirements.

However in colour receiver, sound take off circuit is before the video detector and sound IF lies at almost zero level. If the sound signal present in the video detector remains for a long time, sound bars will be produced on the picture. So, additional trap circuit is provided at the input of video detector to attenuate the sound IF by 15 dB.

2.13.2 Important Design Requirements of IF Sections

(i) The combined gain of the IF amplifiers should be about 8000 to deliver a peak to peak signal of 3 to 5V at the input of video detector. For this 3 to 4 sections of amplifiers are to be cascaded.
(ii) The amplifier stages must provide necessary bandwidth and selectivity for faithful reproduction of picture, sound and colour information.
(iii) The upper skirt of the overall response curve should have the required slope for vestigial sideband correction. This is necessary to provide a uniform signal at the video detector over the entire range of video frequencies.
(iv) In monochrome receivers the lower skirt of the IF response should be designed to have a steep slope to reduce the gain at the sound IF to about 5% (−26 dB) of the maximum gain. This is essential for successful functioning of the inter-carrier sound system.
(v) In colour receivers the lower skirt of the IF response should have a slope so as to place the colour IF at about −5 dB level.
(vi) Additional trap circuit should be provided in colour receivers at the input of video detector to attenuate the sound IF by another 14 dB, because the sound signal is detected separately.
(vii) The IF section of both monochrome and colour receivers must include rejection filters to suppress adjacent channel interference.

2.13.3 Vision IF Sub System

The main subsections of vision IF sub systems are:

(i) High gain wide band IF amplifiers.
(ii) Automatic gain control circuit.
(iii) Video and inter-carrier sound signal detectors.
(iv) Automatic frequency tuning.

The various sections of the IF subsystem are divided into two parts. They are: (1) discrete circuits used in earlier receivers, and (2) Specially designed integrated circuits used by at present modern receivers.

2.13.4 Vision IF Response Curves Requirements

The overall response curve of the IF section is shown in Figures 2.28(a) and (b).

2.13.4.1 IF Amplifier Circuit

The circuit diagram is shown in Figure 2.28(c). It consist of four transistor of video IF amplifier. Each transistor has a very high cut off frequency.

At the input of the 1st IF amplifiers stage it consist of three sections of filter circuit. The first section is a series trap that blocks the lower channel interference frequency. Similarly the third section filter is used to filter the upper channel beat frequency. The second section has two filters and is designed to attenuate the sound IF frequency to about 20 dB below the picture IF level.

Coil inductor L4 is for frequency compensation and this coil is shunted by a resistor R1 to damp any oscillations. Coil inductor L5 to L6 are IF chokes inserted in the DC supply lines and AGC circuit.

The IF signals is coupled to the base of the transistor Q1 through the coupling capacitor C1. The AGC is fed only to the first amplifier and remaining two stages employs self-biasing. The collector load of transistor Q1 is a single tuned circuit is set at a suitable frequency for stagger tuning. The output from the load circuit (T1) is coupled to the base of transistor Q2 through the capacitors C2 and C3 which form a voltage divider in addition to provide impedance match between two stages.

The second stage combines a grounded emitter stage transistor (Q2) that drives a grounded base (Q3) amplifier. The output of transistor Q2 is DC coupled to the emitter Q3. The two transistors thus combine to form a cascade pair which gives a large gain compared to single stage. The transistor Q2 is not neutralized because it has low collector load impedance.

The 10Ω resistor (RE) inserted in the emitter lead of Q3 is for thermal stabilization. The primary winding of transformer T2 is a tuned circuit connected at the collector of Q3.

The third IF amplifier has a gain a common emitter configuration has a single tuned coil load (Q4). The secondary is wound for impedance matching and feeds into the cathode of the diode that is apart of the video detector.

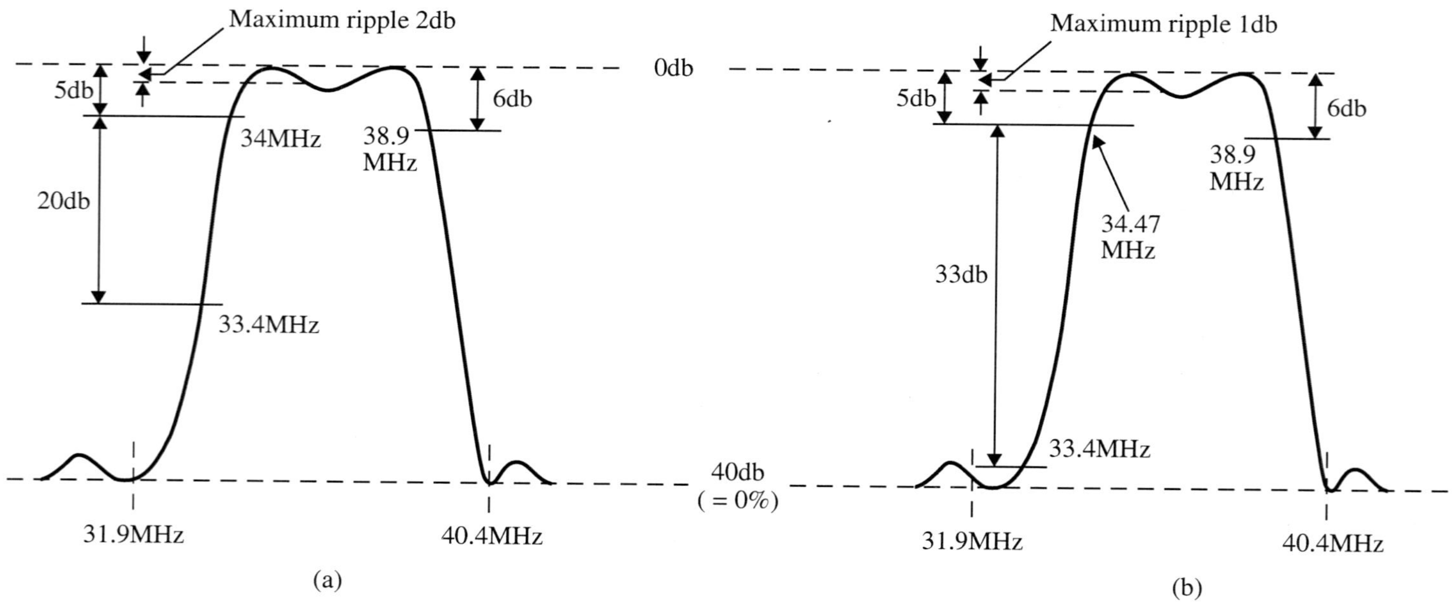

Figure 2.28 (a) IF response curve of monochrome receiver and (b) IF response curve of colour receiver.

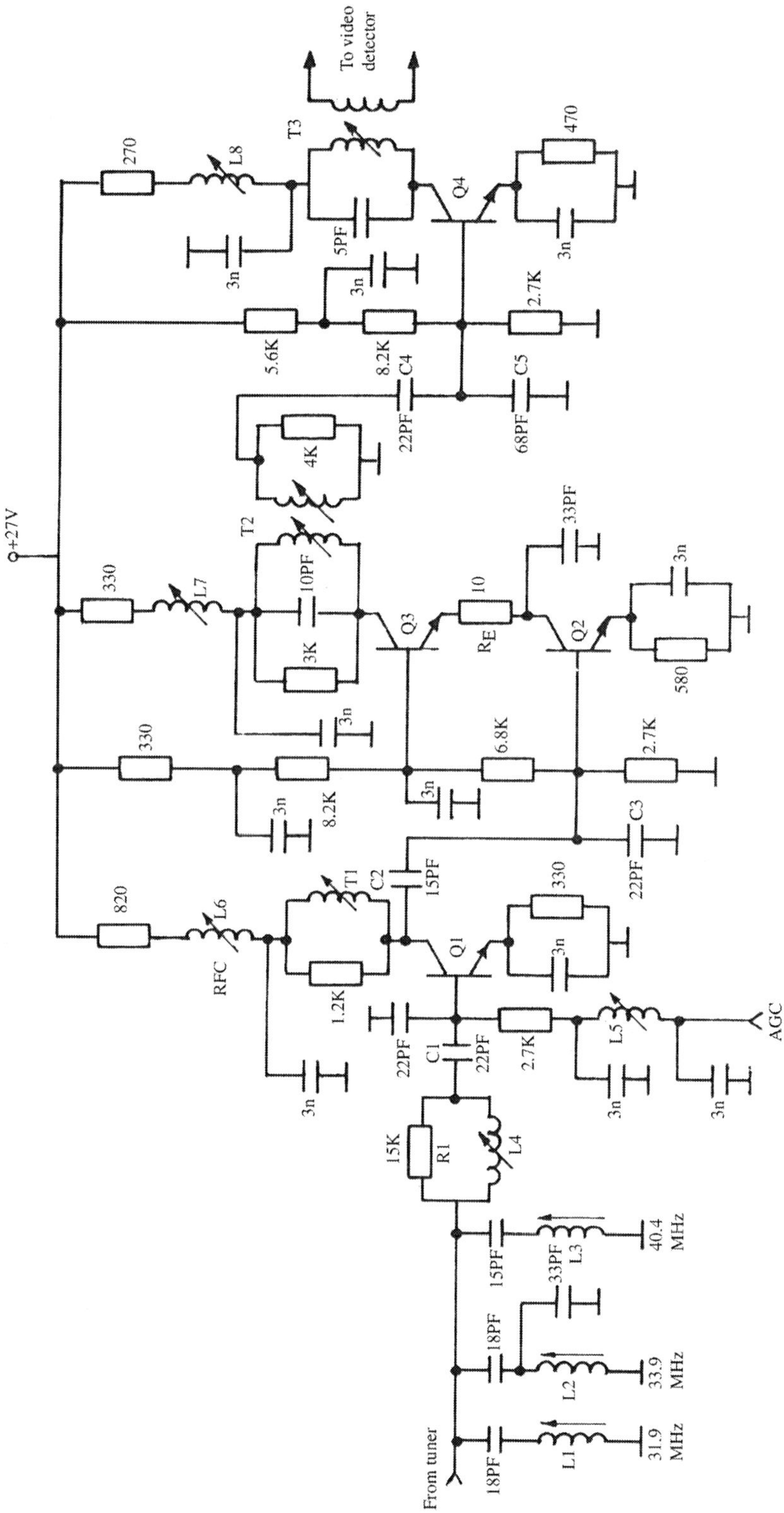

Figure 2.28 (c) Typical circuit for IF amplifier stage.

2.13.5 Vision IF Subsystem of Black and White Receivers

Many integrated circuits (IC's) are now available for use in the IF sub-system of monochrome and colour receivers. BEL CA3068 is one such popular IC designed to perform as a complete IF sub-system in black and white receivers. In practice the IC and associated components are put in a modular form and soldered on the main Printed Circuit Board (PCB).

2.13.6 Sections of IC CA 3068

The main sections of the IC CA 3068 are:

(i) High gain wide-band IF amplifiers (Overall Gain 87 dB).
(ii) Keyed AGC with noise immunity circuits.
(iii) Delayed AGC for the tuner.
(iv) Video detector.
(v) Video amplifier.
(vi) Inter-carrier sound detector.
(vii) Sound IF amplifier.
(viii) Zener regulated DC reference source.

The block diagram of vision IF subsystem is shown in Figure 2.29.

The IF signal from the tuner is fed through terminals 3 and 6 to a cathode high frequency wide band amplifier negative AGC voltage is used to control biasing of the amplifier. The output of the cascade amplifier feeds into 1st IF amplifier which is followed by another multi-stage wide bands amplifier. The combined gain of these amplifiers is enough to provide an overall gain of about 87 dB the bandwidth is controlled by external tuned circuit or through a saw filter.

The video amplifier output is fed to the AGC circuit via a noise gate. The IF AGC circuit also receives horizontal keying pulse through terminal; 3rd of the IC and develops negative AGC voltage proportional to the input signal carries strength.

The AGC output also feeds the tuner AGC delay circuit. The biasing network of delay circuit is so designed that a cross-over occurs at a predetermined RF input to top the receiver.

A feedback loop in the 2nd video IF amplifier provides low output impedance for driving the video detector thus eliminating the need for a conventional tuned circuit its input. The amplifier output is in Darlington configuration and delivers video output at a low impedance level for ease of coupling of discrete high voltage video amplifier.

Composite video IF signal from the cascade amplifier is also fed to a wide band amplifier the output of which is coupled to the inter-carrier sound detector. Frequency selective feedback is provided in this amplifier to obtain channel band pass characteristics with a peak at 5.5 MHz.

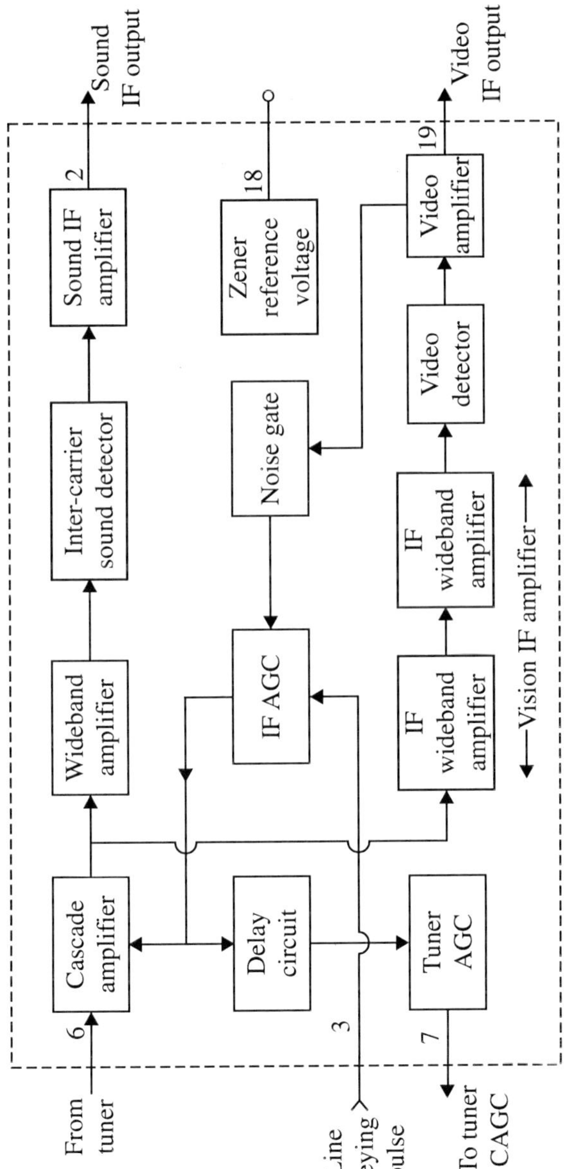

Figure 2.29 Functional block diagram of for IC CA3068.

Each stage of the IC is fed through a built in DC voltage regulator. In addition to this an isolated Zener reference DC voltage is available at terminal 18 of the IC it can be used with an external series element to provide necessary supply voltage to any terminal of the IC.

2.14 VIDEO AMPLIFIER REQUIREMENTS

To produce a required image on the screen of picture tube, the video amplifier must meet the following measurements.

(i) Gain.
(ii) Bandwidth.
(iii) Frequency distortion.
(iv) Phase distortion.
(v) Amplitude distortion.

2.14.1 Gain

The output of the video amplifier should have high amplitude to vary the intensity of electron beam in the picture tube. Then only the reproduced picture will have the full range of black and white areas and various colour shades. The monochrome picture tube needs 75 V of video signal to reproduce picture with good contrast. Colour picture tube needs 150 V to produce picture with various colour shades. But the video detector output is 2–4 V so the video amplifier should provide a gain of 25–50. So that we can get up to 150 V of video signals to drive the picture tube.

2.14.2 Bandwidth

The video amplifier should provide a flat response (constant gain) over the video frequency range of 10 Hz to (or) DC to 5 MHz.

The signal frequency corresponding to picture information scanned in the vertical direction are much lower compared with those for reproduction with a line. If the video voltage is taken from top to bottom through all the horizontal lines in a field, the variation will corresponds to half cycle of a signal with frequency of approximately 25 Hz.

2.14.3 Frequency Distortion

All the video frequency from 0 DC to 5 MHz should have uniform amplification if not, frequency distortion will be occurred if high frequency distortion occurs, and very fine details of the picture will not be reproduced. If low frequency distortion occurs the picture will be reproduced with very poor contrast.

2.14.4 Phase Distortion

The video amplifier should preserve the phase of all video frequency components. If a shift in a phase occurs then this will result in shift in position of the picture element on the screen. So the picture reproduced will be completely distorted. So to avoid a phase distortion the video amplifier should not change the phase of frequency components.

The phase angle delay should be directly proportional to thc frequency. If signal inversion is exactly 180°, it does not mean any phase distortion. There is no time delay, only the polarity of the signal is reversed.

2.14.5 Amplitude Distortion (Non Linear Distortion)

When different amplitude components get different amplifications then the amplitude distortion occurs. This is due to change in operating point of the device in the video amplifier stage. Due to this the synchronization will be lost. To prevent this we have to stabilize the operating point of the devices in the video amplifiers.

2.14.5.1 Manual Contrast Control

Any control that varies the amount of an AC video signal will operate as a contrast adjustment. So the contrast of the picture varies when the gain is varied in either picture IF section or video amplifier.

2.15 VIDEO AMPLIFIER CONFIGURATIONS

It is obvious from the preceding discussion that video amplifier must meet several exacting demands and this calls for careful and rigorous design consideration for both direct coupled and RC coupled are used and both has its own merits and demerits both type needs high frequency and low frequency compensation.

2.15.1 High Frequency and Low Frequency Compensation

The high frequency and low frequency compensations are necessary, because the gain at high and low frequency is very low compared to the mid-band frequency.

If the high frequency is poor, horizontal information will be lost. If low frequency is poor, overall contrast level will becomes poor. If there is any phase distortion, picture may be misplaced. The frequency response and phase response curve is shown in Figure 2.30(a) and (b).

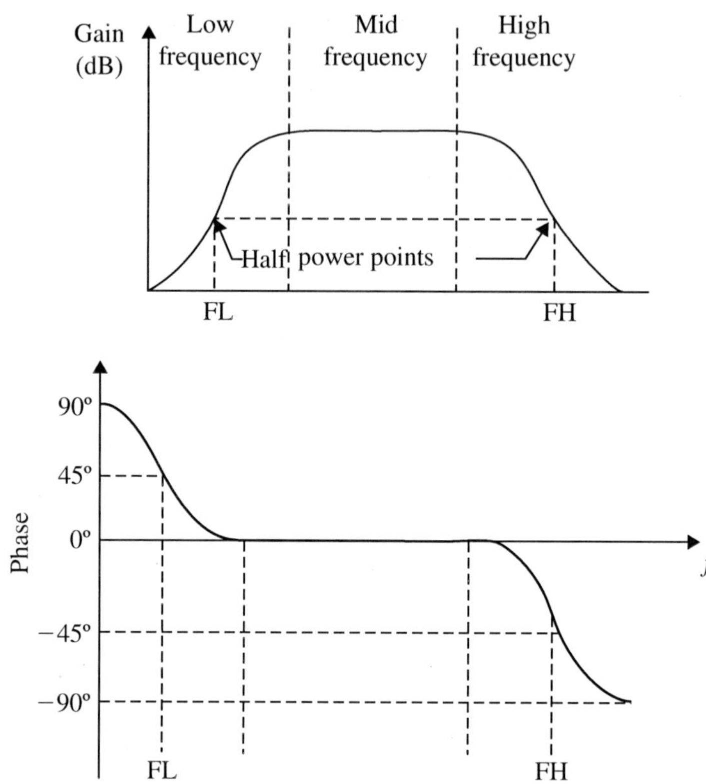

Figure 2.30 (a) Frequency response of practical amplifier, and (b) Phase response of practical amplifier.

2.15.2 High Frequency Compensation

High frequency compensation uses the following methods they are:

(i) Series peaking.
(ii) Shunt peaking.
(iii) Combinational peaking.

2.15.2.1 Series Peaking

Figure 2.31(a) shows the video amplifier with series peaking coil in this LC is in series with the two main components of CT.

Series peaking is another method used to increase the gain at high frequencies. Here, coil LC is connected in series with output. The capacitor C1 is divided into two parts, they are C_{in} and C_{out}. These two are the stray capacitances of next stage input and previous stage output respectively. This results in the shunt capacitance reduction across R_L.

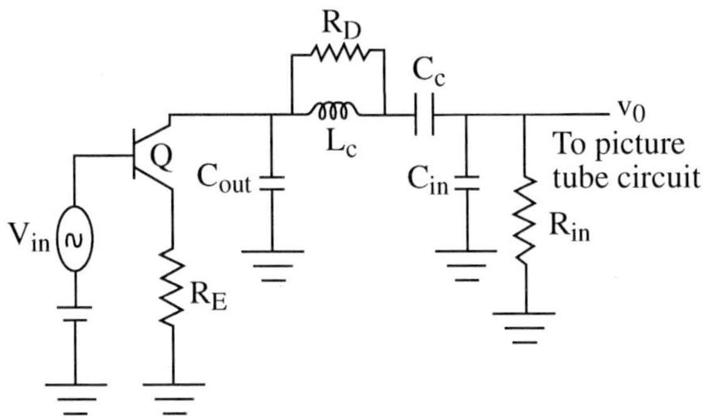

Figure 2.31 (a) Video amplifier with series peaking.

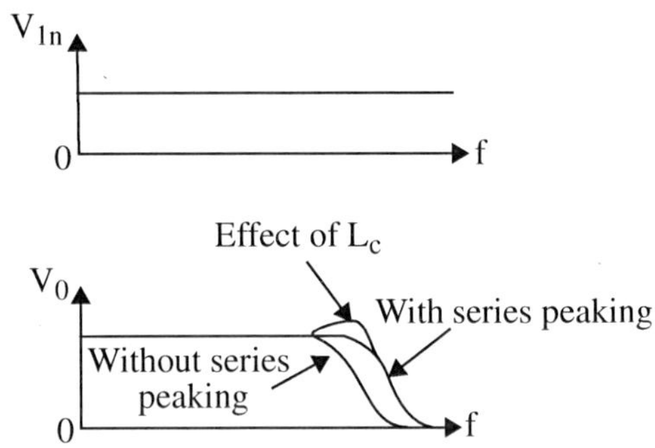

Figure 2.31 (b) Series peaking and its response.

A damping resistance R_d in shunt with peaking coil is used to prevent any oscillation in the coil.

2.15.2.2 Shunt Peaking

Figure 2.31(c) shows the video the gain of the amplifier is reduced at the high frequencies, because of shunting of input, inter electrode and stray capacitances in parallel with RL.

To boost the gain at thigh frequencies, high values of RL with peaking coils is used. This arrangement is known as 'shunt peaking'.

The smaller inductance LO is connected in series with RL. This peaking coil resonates with C1 to boost the gain at high frequencies.

2.15.2.3 Combinational Peaking

Figure 2.31(f) shows the video amplifier with combinational peaking the combinational peaking has more gain as compared with shunt and series peaking for the same frequency response RL can be increased 80% higher with 80% more gain compared with shunt peaking.

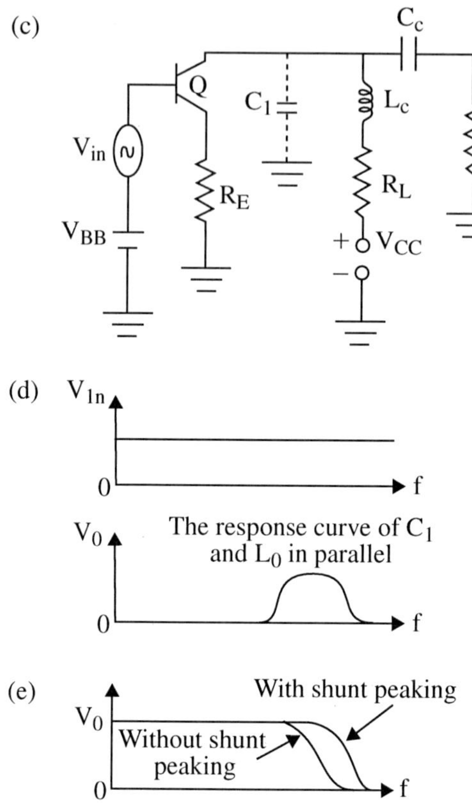

Figure 2.31 Shunt peaking. (c) Video amplifier with shunt peaking, (d) The response curve of C1 and L0 in parallel and (e) Frequency response.

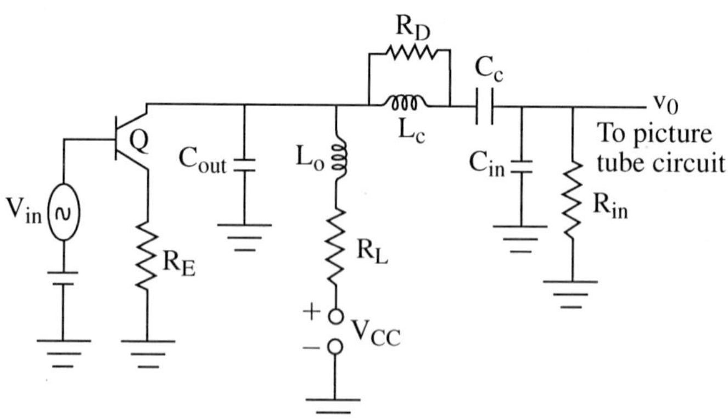

Figure 2.31 (f) Combinational peaking.

2.15.3 Low Frequency Compensation

The gain is reduced at low frequency, because of the increasing reactance of the coupling and bypass capacitances.

The de-coupling filter resistor and capacitor RF, CF in the $B+$ supply line can be used to increase the gain at low frequency and reduce the phase distortions. At low frequencies, capacitor CF will provide large reactance and resistor RF in series with R1 act as an effective load. This is shown in Figure 2.32. Due to the increased load, gain also increased at low frequencies. Phase shift due to capacitor CF and capacitor CC are opposite in nature. So phase distortion can be eliminated. CF is connected parallel with capacitor CF is used to bypass higher video frequencies.

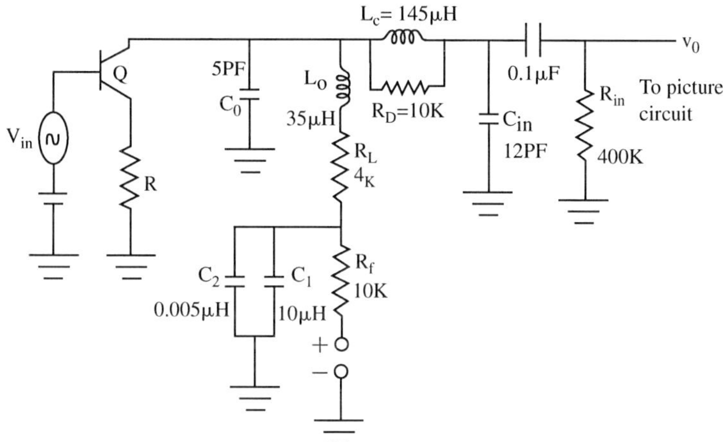

Figure 2.32 LF and HF compensation.

2.16 DC REINSERTION

The relative relationship of AC signals to the blanking and sync pulses remain same with or without the DC component. Further more, brighter, the line, greater is the separation between picture information's variations and associated pulses. As the scene becomes darker, the two components move close to each other. So they must be preserved for a true representation of the picture to be reproduced correctly.

However if we are using AC coupling between video detector and amplifier, then the DC components are completely removed, because of this, the video signal settle down with equal areas on both of zero line.

2.16.1 DC Reinsertion with a Diode

A simple transistor diode clamp circuit. For lining up sync pulses is shown in Figure 2.33. The VCC supply ICs set for a quicent value of 15 V. In

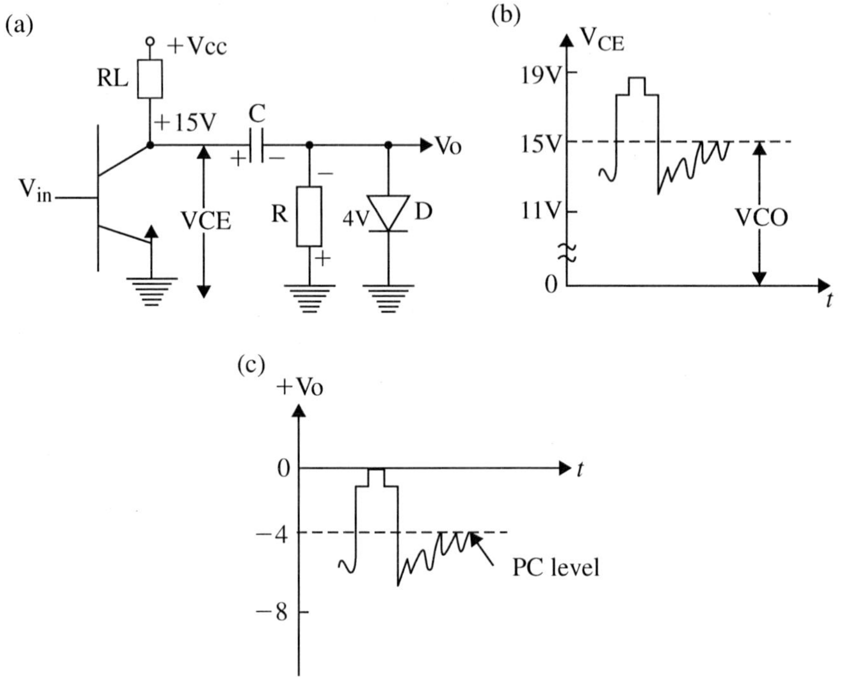

Figure 2.33 (a) Diode DC restoration of circuit (b) collector voltage and (c) output voltage.

absence of any input signal the coupling capacitor *C* changes to 5 V and so voltages across the parallel resistor *R* and diode *D* will be zero.

Assume that on application of video signal, the collector voltage swings by 8 V peak to peak the corresponding variation in collector to emitter voltage is illustrated in Figure 2.33(b). The positive increase in collector voltage is coupled through the capacitor *C* to the anode of the diode D and makes the diode to turn on.

Since a forward biased diode may be considered to be short. It effectively clamps the output circuit to ground (zero level). In effect, each positive sync pulse tip will be clamped to zero level, there by lining them up and restoring DC level of the video signal.

In case under consideration the diodes will cause the coupling capacitor to charge to a peak value of 19 V. during negative excursion of the collector voltage the capacitor fails to discharge appreciably, because now the diode is reverse biased and *R* will be large the average reverse biased across the diode is −4V.

Note that as the input video signal varies in amplitude corresponding video signal voltage appears the resistor *R* and it varies from 0 to −8V in amplitude. This is shown in Figure 2.33(c) reversing the diode in the restorer circuit will result in negative peak of input signal being clamped to zero. This mean that the DC output voltage of the circuit will be positive.

2.16.2 Need for DC Reinsertion

In the output of video detector, DC components are also present. They represent the background brightness of the picture. So they must be preserved for true representation of the picture. If we use DC coupling then the picture is reproduced correctly. Therefore DC reinsertion is needed to preserve the component and maintain the true representation of the picture.

2.17 VIDEO AMPLIFIER CIRCUITS

In earlier days black and white receivers DC coupling was used in video amplifiers despite the difficulties and high cost of providing it. However, with wide acceptance of colour receiver, demand for large screen high quality monochrome receivers declined with the result that most present day B & W receivers are low cost small screen versions. As such these employ AC or partial DC coupling in the video section. Two such representative circuits are explained.

2.17.1 AC Coupled Video Amplifier

A typical video amplifier circuit employing AC coupling and used in 37 cm B & W receiver is shown in Figure 2.34. The composite video signal from the VIF IC (pin 12 of CA7611) feeds via R21 into the emitter follower transistor (BF195c) which produces impendence match to the video amplifier. The DC voltage at corresponding pin of the IC is enough to forward bias Q1. Its output feeds at the base of video output transistor Q2 (BD115) via 5.5 MHz sound signal trap circuit. The inductor L11 in its collector circuit is for high frequency compensation. The resistor R26 in the emitter lead is un bypassed to provide negative feedback for overall wide banding. The amplified video signal is coupled to the cathode of picture tube through coupling capacitor C45 and series resistance R32.

2.17.1.1 Contrast Control

The magnitude of negative feedback is varied to control gain of the video amplifier and hence contrast in the reproduced picture. Blanking pulses are fed at the emitter of BD115 (Q2) (Figure 2.34) to switch it off during retrace intervals. This raises DC voltage at the cathode and beam current is cut off to suppress retrace line on the screen.

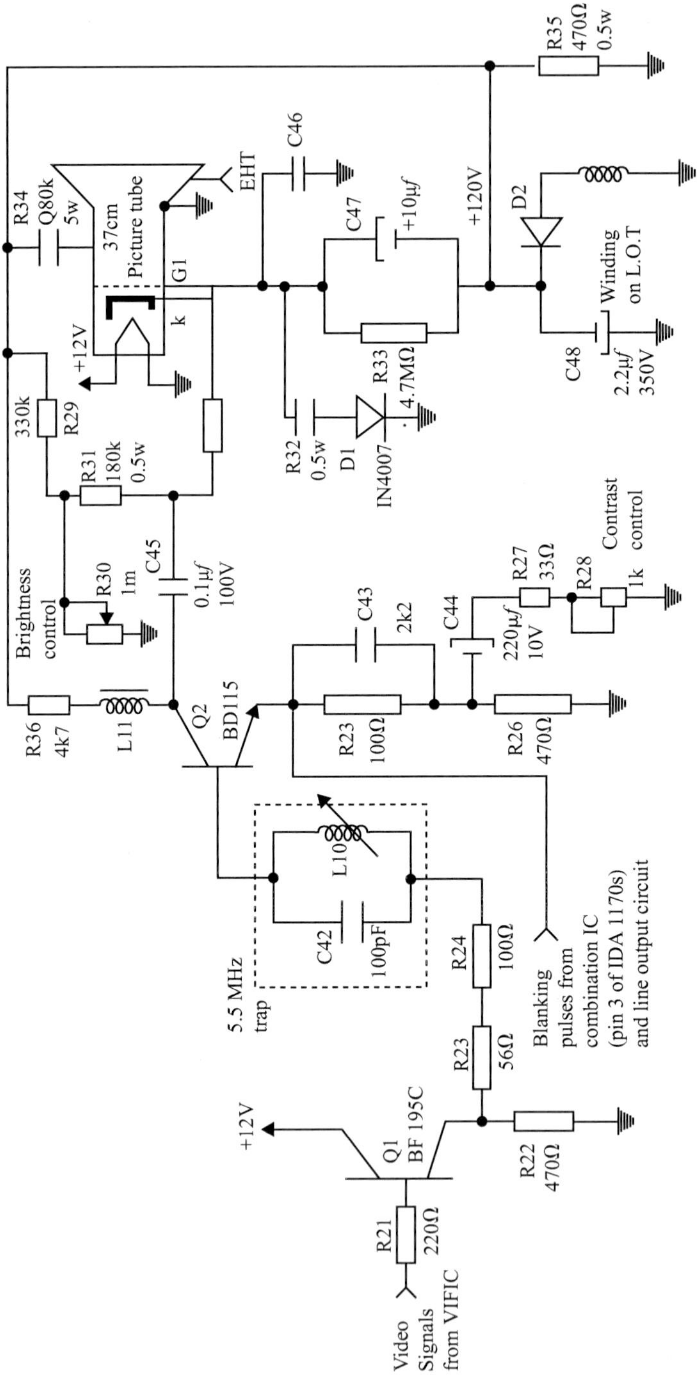

Figure 2.34 Video amplifier circuit employs AC coupling.

2.17.1.2 Brightness Control

The average brightness of the picture is controlled by varying DC potential at the cathode of picture tube. The potentiometer R30 connected from the junction of the two resistors R29 and R31 to ground is varied to change the voltage and hence it acts as a brightness control.

In order to suppress bright spot on the screen when the receiver is switched off, a large negative voltage is applied at the control grid for a short period to repel the electrons. By the time the capacitor discharges the cathode cools down and no more electrons are when available to form the beam current. During normal operation of the receiver the diode D1 remain forward biased and only about 0.7 V appear at the control grid.

2.17.2 Video Amplifier with Partial Coupling

Another video amplifier circuit is used in some 51 cm monochrome receiver is shown in Figure 2.35 the video signal is coupled to it as the base of emitter follower Q3 (BF195C) via the 5.5 MHz trap circuit and capacitors C41 and C42. The electrolytic and ceramic capacitors in parallel ensure passage of the entire wide band frequency component present in the video signal. The emitter follower output feeds at the base of video output transistor through the contrast control network that maintains a constant black level over the entire constant range.

The function of transistor Q7 is to switch off Q4 during beam blanking intervals. Normally Q7 will be off and Q4 is forward biased when blanking pulse arrive, Q7 conduct heavily with its emitter current flowing through R42. When Q4 is not conducting its collector voltage rises to 110 V with the result that these cathodes of picture tube become sufficiently positive to cutoff the beam current.

The diode D1 is for beam current limiting to prevent high brightness level if the input signal is removed. This aspect is fully shown in Figure 2.35.

The brightness control is through potentiometer R50 connected to G1 through R48. The variation of R50 enables change of potential at the picture tube control grid from +70 V to −50 V negative there by enabling full range of brightness from maximum bright to complete darkness.

2.18 SYNC SEPARATION AND PROCESSING

2.18.1 Need for Synchronizing Pulses

The synchronizing pulse is also called as "sync" and it is present in the top 25% of composite video signal. This includes both horizontal and vertical sync pulses.

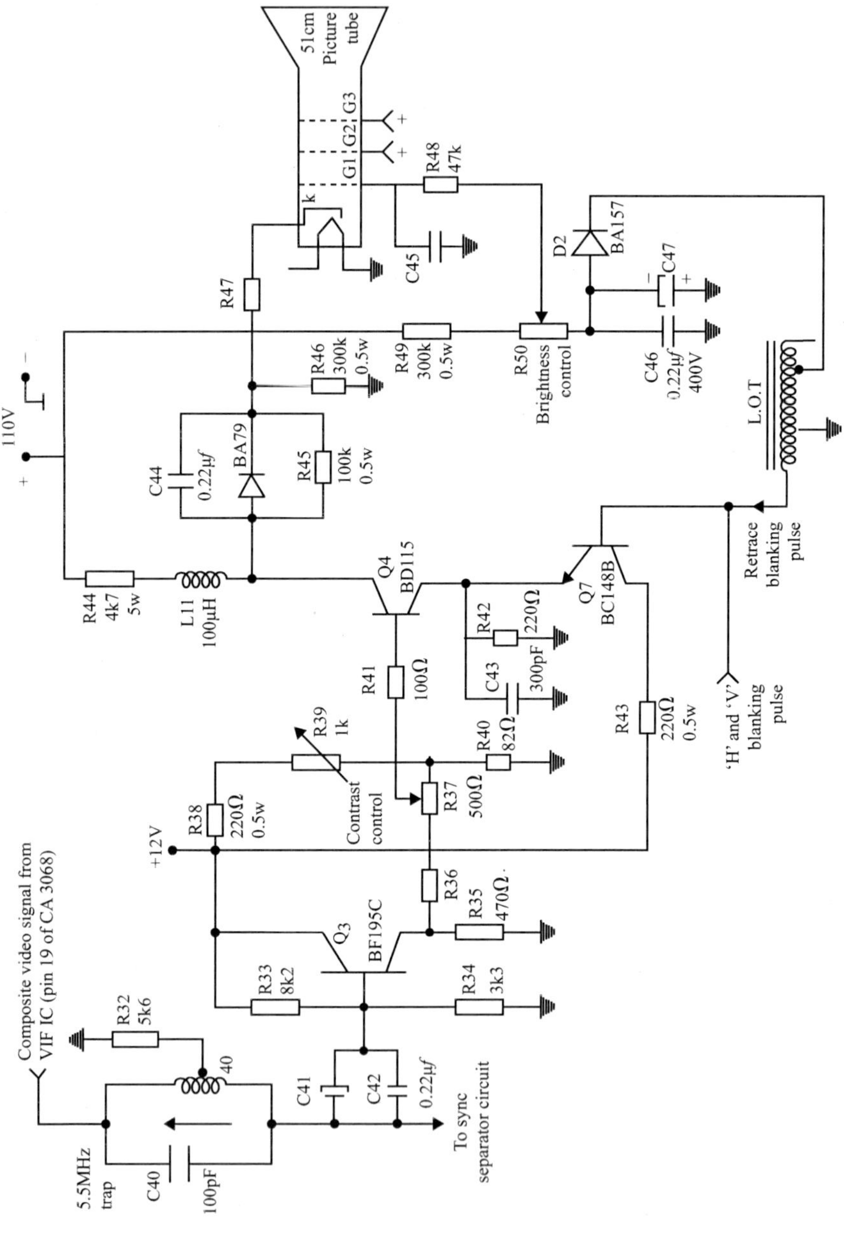

Figure 2.35 Video amplifier circuit amplifying DC coupling.

In receiver side these pulses are separated and it is used to trigger the horizontal and vertical deflection oscillators. The sync pulse separated from the composite video signal is given to integrator and differentiator to separate vertical and horizontal sync pulse respectively.

2.18.2 Synchronising Circuits

The synchronising pulses are part of the composite video signal and are:

(i) Horizontal sync pulses.
(ii) Vertical sync pulses.
(iii) Equalizing pulses.

2.18.2.1 Basic Principle

The Sync pulses are inserted at the top of the horizontal and vertical blanking pulses. The sync pulses are separated from the video signal by sync separator. Then the sync is coupled to the horizontal and vertical deflection oscillator to time with the scanning frequencies. The horizontal sync pulse holds the line structure oscillator. The vertical sync pulse holds their picture frame locked in vertically by triggering the vertical oscillator. The equalizing pulses help vertical synchronization be the same in even and odd fields for good interlacing.

2.18.2.2 Requirements of Sync Separator

(a) The sync pulses are given to low pass (integrator) and high pass filter (differentiator) to generate a triggering pulse for horizontal and vertical oscillators.
(b) The vertical field pulses produced must be identical in shape and their timing must be such that the interval between the start of the transmitter field pulse and the triggering of the vertical oscillator should be same, so that good interlaced is ensured.
(c) The sync separator output should not be affected by variations in amplitude of video signal.
(d) The output of the sync separator should not contain any interference noise pulses because it affects the horizontal and vertical synchronization.

2.18.3 Transistor Sync Separator

The transistor operates under the class C mode (*i.e.*,) biased well beyond cut-off. To provide, self bias RC combination is used between the base and emitter of the transistor.

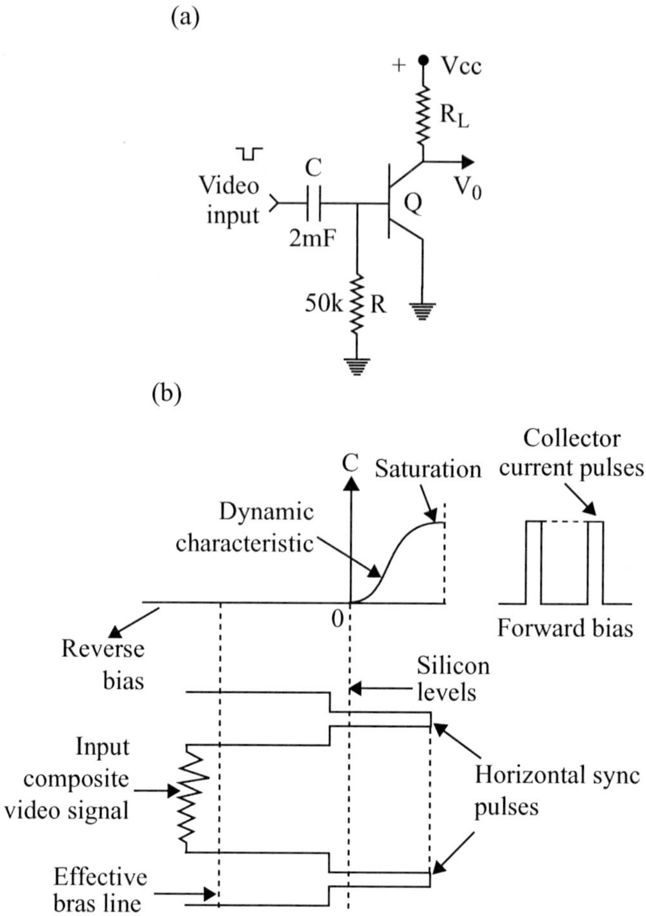

Figure 2.36 (a) Basic sync separator and (b) Output response.

The capacitor C charges, when the input signal is given to the transistor. The negative polarity of the capacitor is connected to the base of the transistor. The negative voltage reverses biase the emitter base junction. Between peaks of the input signal, capacitor C discharges through resistor R1. The R1, C1 time constant is made large to keep capacitance C1 charged to 90%.

So transistor conducts only during the positive peak of sync voltage. The separated sync pulse is at the collector terminal of the transistor.

It is shown in Figure 2.36(a) the sync pulse amplitude varies in between Vcc and collector voltage corresponding to maximum collector current.

For effective sync separation, following condition must be satisfied:

(i) Transistor should be large.
(ii) Output leakage current must be small.
(iii) Time constant RC must be long.
(iv) High frequency transistor and low power transistor can be employed.
(v) A low power transistor can be used since the transistor is off for most of the period.

2.18.4 Sync Separator with Noise Gate Transistor

The sync separator with noise gate transistor is shown in Figure 2.36(c). Here two transistors Q1 and Q2 are used.

Q1 is the sync separator. Q2 is the noise gate transistor, which is used to avoid the noise pulses in the sync separator output. The noise gate transistor Q2 is connected in series with the emitter of the sync separator Q1. The transistor Q2 is normally in saturation condition.

In absence of any noise pulse, the transistor Q1 operates normally and the emitter current completes its path through Q2. The transistor Q2 has saturated voltage of 0.6 V in between the collector and emitter.

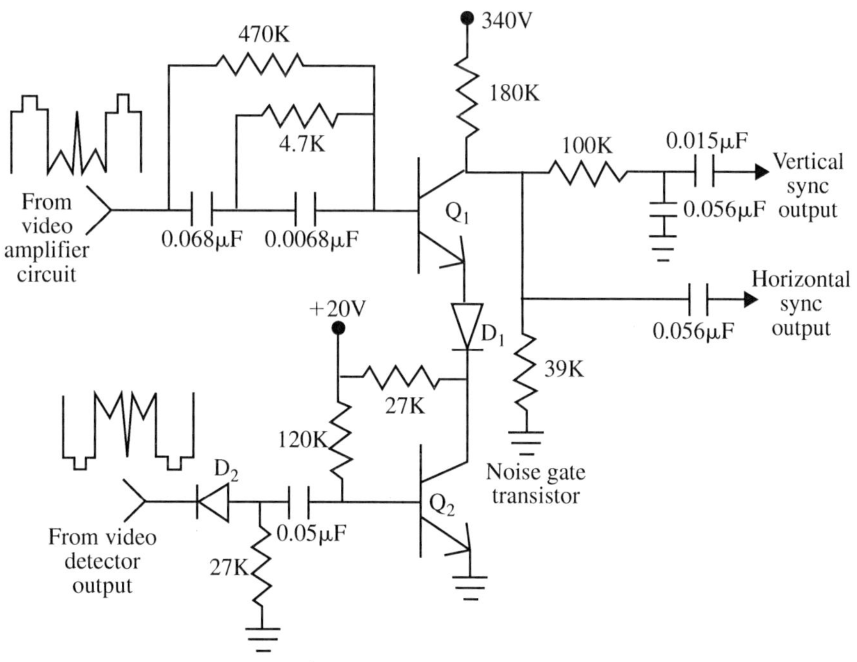

Figure 2.36 (c) Transistor sync separator with a noise gate.

If there is any noise pulse in the video signal enters the base of Q2 through diode D2. It makes the transistor Q2 to cut-off. So, the collector current of Q2 rises to +20 V, emitter of Q1 also rises to +20 V. This makes the transistor Q1 into cutoff. Hence no separation of sync pulse occurs.

The normal amplitude of negative video signal at the input of Q2 is not sufficient to turn it off. So the noise pulse of amplitude is less than the sync pulse amplitude will not turn off the transistor Q2.

2.18.5 Block Diagram of Sync Separator

The block diagram of sync separator and waveform at different points are shown in Figure 2.37.

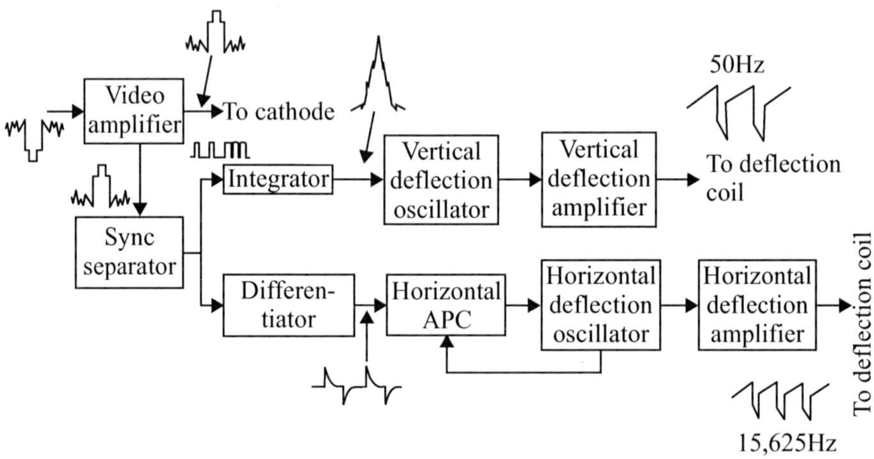

Figure 2.37 Block diagram of sync separator.

As shown in Figure 2.37 the video signal from the video amplifier is fed to the sync separator circuit. The sync separator separates the vertical and horizontal sync pulses. So at the output of sync separator we have a train of horizontal and vertical sync pulses. These are then given to low pass and high pass filters to generate triggering pulse for horizontal and vertical deflection oscillators.

The vertical deflection oscillator is triggered by the integration out. After proper amplification they are fed to the vertical deflection coil.

2.18.6 Separation of Horizontal and Vertical Sync

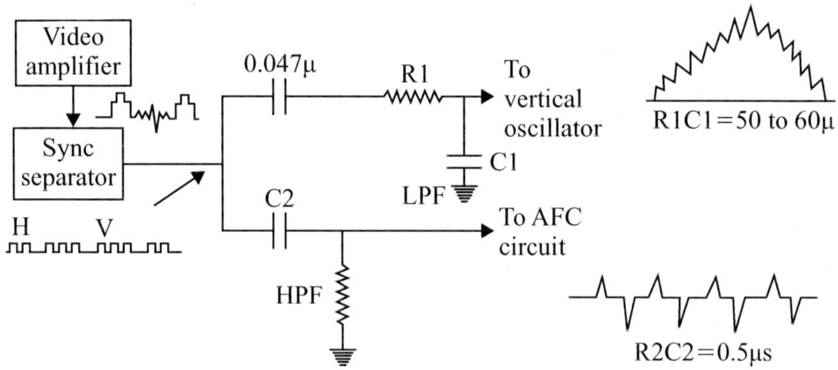

Figure 2.38 Horizontal and vertical sync separation.

2.18.6.1 Separation of Vertical Sync

An integrator R1, C1 is used as low pass filter to separate the vertical sync pulse, by choosing a value of 0.01μf for capacitor C1 in the low pass filter it bypasses the horizontal sync pulses. The time constant R1, C1 of this circuit is very high ($60 - 100\mu$s).

The serrated vertical sync pulse consist of five individual pulses each of 27.3μsec. When V vertical sync pulse appear at the input of integrator the capacitor C1 charges $27.3\mu s$ of the applied voltage through resistor R1. The Vertical sync pulse duration is 27.3μs. After this the capacitor C1 discharge for 4.7μs. Before completely discharging the capacitor C1 begins to charge again for 27.3μs. Thus the voltage across the capacitor increases with spikes as shown Figure 2.38. The horizontal H sync and equalising pulse cannot produce any output at the integrator, because the very small ON time duration of these pulse cannot charge the capacitor to produce output. So at the integrated output we get a voltage corresponding to V vertical sync pulses only.

This operation repeated for five all five serrated pulses. This makes the gradual increase in voltage across the capacitor with small spikes super imposed on it. If the five pulses are passed to integrator, the capacitor C1 discharges almost zero. The time constant required to produce integrator output is 50 to 60μs.

2.18.6.2 Separation of Horizontal Sync

In a differentiator circuit a very low time constant is used. The time constant of R2 and C2 of high pass filter network is 0.82μs. The H horizontal sync pulses has a 'ON' time duration of 4.7μs arrive at the input. Now the capacitor C2 charges exponentially at the rate determined by time constant R2 and C2. Then, before the trailing edge the output voltage across capacitor C2 discharge to zero volts.

When the trailing edge comes, the capacitor charges with opposite polarities and quickly discharges. The output is thus in form of spikes which are given automatic frequency control (AFC) system for correcting the horizontal oscillator.

When *V* vertical sync and equalising pulses come to the input of the differentiator they do not produce any output. Because the duration of the time constant is $0.82\mu s$ is too short for vertical sync to produce any appreciable voltage across R2 resistor. Thus differentiator produces output only corresponding to the *H* horizontal sync pulse.

2.19 DEFLECTION CURRENT WAVEFORM

Figure 2.39 illustrates the required nature of current in deflection coils. As show there, it has a linear rise in amplitude which will deflect the beam at uniform speed without squeezing or spreading the picture information. At the end of the ramp the current drops sharply for a fast retrace or fly back. Zero amplitude on the saw tooth waveform corresponds to the beam at the centre.

The peak to peak amplitude of saw tooth wave determines the amount of deflection from the centre. The saw tooth waveform can be positive

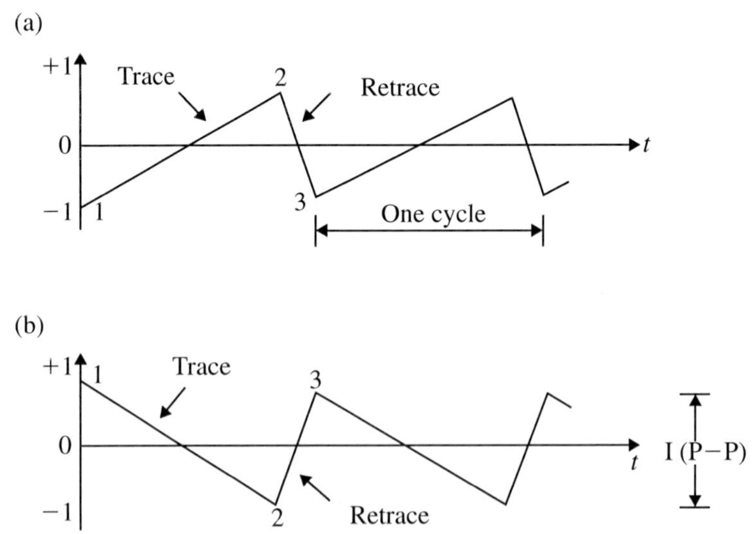

Figure 2.39 Deflection current wave form. (a) Positive going trace, and (b) Negative going trace.

or negative going, depending on the direction of winding on the yoke for deflecting the beam from left to right and top to bottom. In both cases, the trace include linear rise from start at point 1 to the end at point 2, which is the start of retrace finishing at point 3 for a complete saw tooth cycle.

2.19.1 Need for Trapezoidal Voltage

To deflect the electron beam, it requires to provide currents to deflection coils. The current which flows into the horizontal and vertical deflection coil must have a sawtooth waveform to obtain linear deflections of beam during trace periods. Because of the inductive nature of the

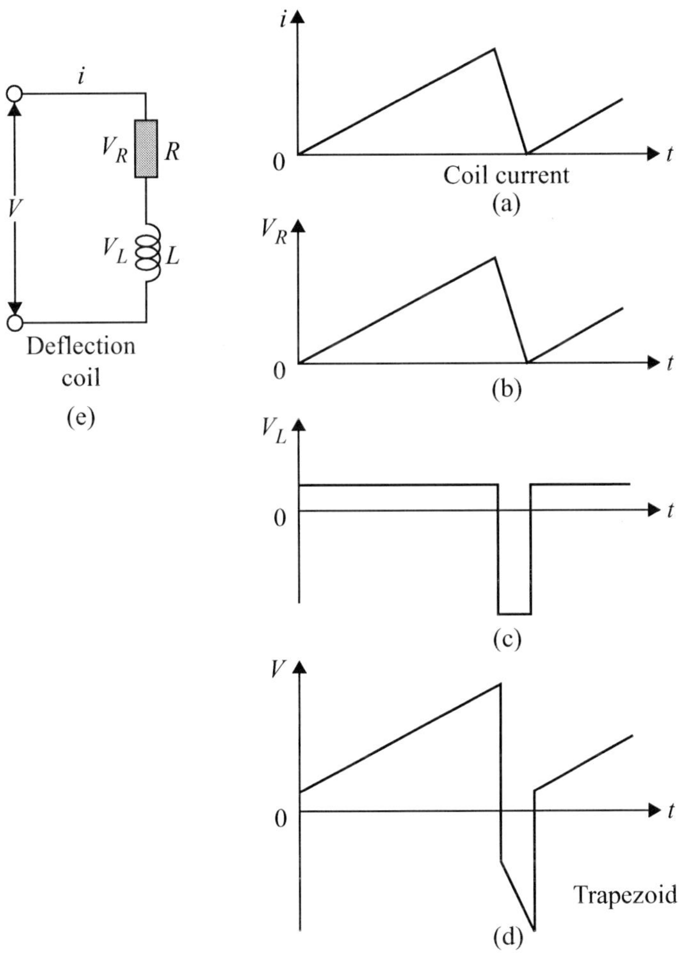

Figure 2.39 (a) Current and voltage wave shapes in a deflection coil. (b) Voltage drop across resistor, (c) Voltage drop across inductor, (d) Resultant voltage V (V_R+ V_L) across input terminals of coils and (e) Equivalent circuit of deflecting coil.

deflecting coil, a modified saw tooth voltage must be applied across the coils to achieve a saw tooth current through them.

Consider an equivalent circuit of deflecting coil consists of resistance R in series with a pure inductance is shown in Figure 2.39(f).

When the saw tooth input current is applied to the resistor (R) and capacitor (C) the voltage drop across the resistor (R) is in the form of saw tooth waveform. But the voltage across L depends on rate of change of current.

$$Vc = L\frac{di}{dt}$$

A rapid change in inductor IL produces more self induced voltage. As a result voltage in inductance VL in at relatively low level during trace period, as shown in Figure 2.39(c), because of fast current drop in IL during trace period, a spike appears across the coil.

Therefore a saw tooth current in inductance (L) produces a rectangular pulse. So to produce saw tooth current in an inductor, a rectangular voltage should be applied.

When the voltage drop across resistor (R) and inductor (L) is added together, the result is a trapezoidal waveform. So, to get the saw tooth current in deflection coil, we need to apply the trapezoidal voltage as the input this is shown in Figure 2.39(d).

2.20 DEFLECTION OSCILLATOR

The oscillator commonly used in direct deflection circuit earlier receivers are:

(i) Blocking oscillator.
(ii) Multi-vibrator.
(iii) Complementary pair relaxation oscillator.

However, in modern receivers where deflection drive is a part of an integrated circuit, the techniques employed to generate 50 Hz and 15,625 Hz are quite complex through based on above listed type of oscillator. In some IC's the oscillator is around an operational amplifier where a miller integrator is used to develop sweep output. In another form it is a current controlled multi-vibrator circuit. In yet another type it is voltage controlled oscillator where an external L-C resonant circuit is provided to set the reference frequency. The basic circuit of such an oscillator as found in the IC CA3223 is shown in Figure 2.40.

It is designed to operate at a multiple of line frequency $(n \times f_L)(n \times 15,625\,\text{Hz})$. The output is divided to obtain horizontal line frequency (F_R) and vertical line frequency (F_V) for line and frame deflection circuits. The

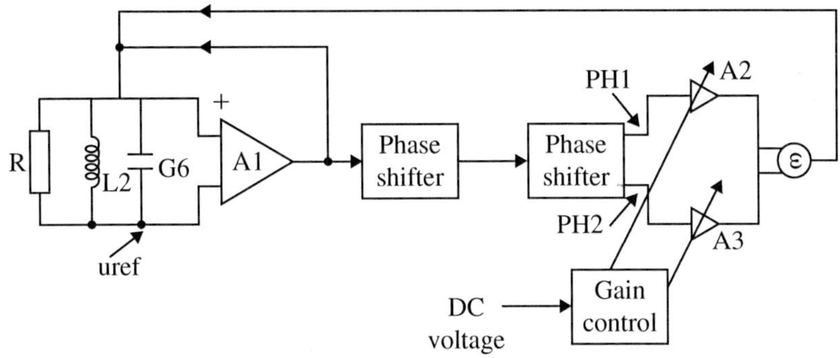

Figure 2.40 Block diagram of voltage controlled oscillator.

centre frequency of the oscillator is determined by the resonant frequency of parallel tuned circuit formed by inductor (L2) and capacitor (C16). The amplifier A1 (Figure 2.40) provides regenerative gain in the feedback loop. The output of A1 feeds a 90° phase shifter and then a splitter that produces two components PH1 and PH2 which are 180° out of phase with each other. Depending on which phase of the summed signal is feedback to the resonant circuit, the tank ckt is parallel with inductive or capacitance impedance.

If the parallel impedance is inductive, the resonant frequency increases and if it is capacitive, resonant frequency (F_r) decreases. Thus feedback loop controls the oscillator frequency to stay fixed at $n \times f_L$.

The wave shaping circuits forms part of each type of oscillator circuit and is followed by a pre-amplifier the output thus obtained are directly applied to the vertical and line output stages.

2.21 HORIZONTAL AFC

A deflection oscillator triggered by sync pulses for each cycle is cable of exact synchronization, if there is no noise interference.

2.21.1 Need for AFC

Horizontal and vertical sync pulses are separated using high pass and low pass filter. Instead of sync pulse the noise pulses from the external interference will trigger the oscillator at a wrong time. In order to make the synchronization and more immune to noise, an automatic frequency control (AFC) is used for horizontal deflection oscillator in all television receivers. AFC is generally used for vertical deflection oscillation. The main reason is that the vertical RC integrator is a low pass filter that removes the high frequency noise pulses.

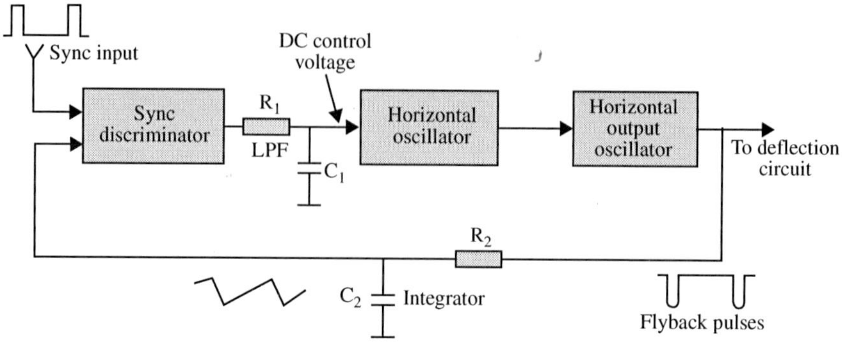

Figure 2.41 (a) Black diagram of AFC.

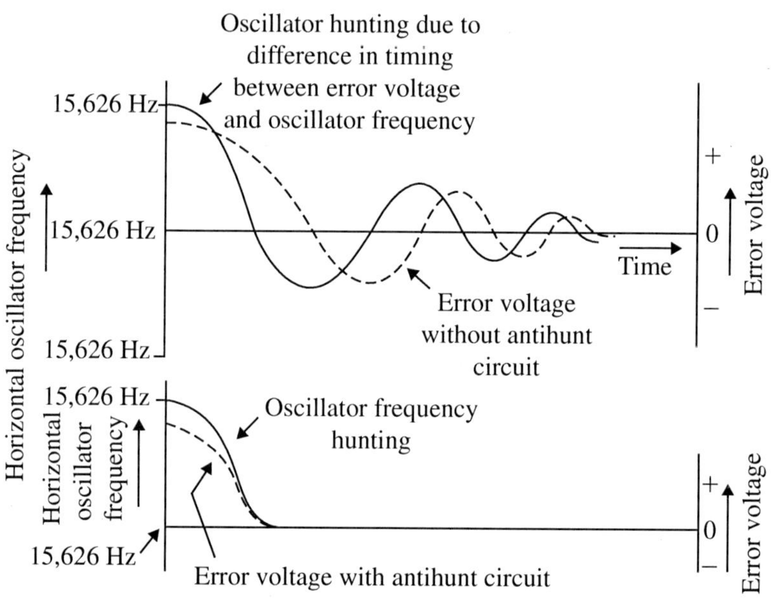

Figure 2.41 (b) Hunting in AFC.

The block diagram of automatic frequency control (AFC) is shown in Figure 2.41(a) the sync pulse signal separated in the sync separator is compared with the horizontal oscillator waveform in the phase discriminator, this control voltage is filtered by a low pass filter and fed to the voltage controlled oscillator (VCO) like blocking oscillator. This DC control voltage corrects the frequency and phase of the horizontal oscillator and makes coincidence with that of sync pulses.

2.21.2 Hunting in AFC Circuits

The filtering circuit that follows the diode section of the discriminator that controls the performance of the AFC circuit. Too large time constant makes the control sluggish, while insufficient damping, on account of two small a time constant, causes the oscillator to hunt while returning to the correct frequency. Excessive hunting in the AFC circuit appears as a weaving or gear tooth effect on the picture. This is shown in Figure 2.41(b).

2.21.3 Anti-Hunting Networks

In order to prevent this double section filter is often used for the DC control voltage. In this network shown in Figure 2.42(a) the R1C1 time constant 0.005μ s is large enough to filter the noise, horizontal sync and fly back pulse effects. The second section, R2 and C2 in series, is known as the 'anti-hunt network'. The relatively low resistance of R2 serves as a damping resistance across capacitance C1 making the output voltage more resistive and less capacitive, there by reducing the time delay.

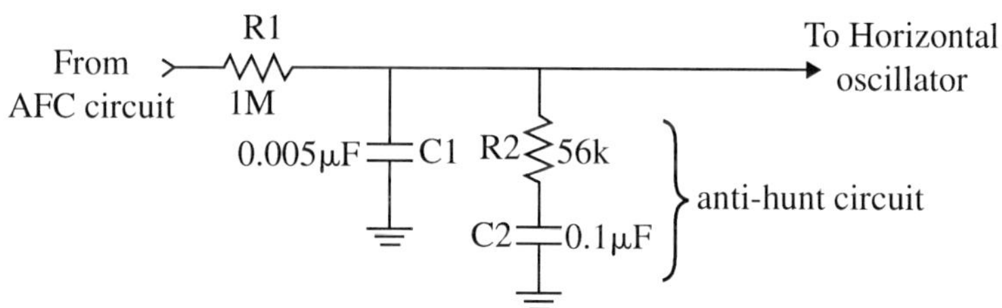

Figure 2.42 (a) Anti-hunt network.

2.22 FRAME DEFLECTION CIRCUIT-REQUIREMENT

The function of frame deflection circuits in television receiver is used to produce a linear vertical deflection of picture tube electron gun. Vertical deflection or scanning circuits starts with an oscillator frequency of 50 Hz to produce rectangular pulses and synchronized with incoming frame (vertical) sync pulses.

Wave shaping circuit follows the oscillator and shapes the rectangular pulses top trapezoidal form for causing linear rise of current in vertical deflection waveforms.

Amplifier circuit follows the wave shaping network to produce enough power to drive frame coils in deflection yoke. To prevent rolling of picture a vertical hold controls are provided. Similarly to obtain full height of

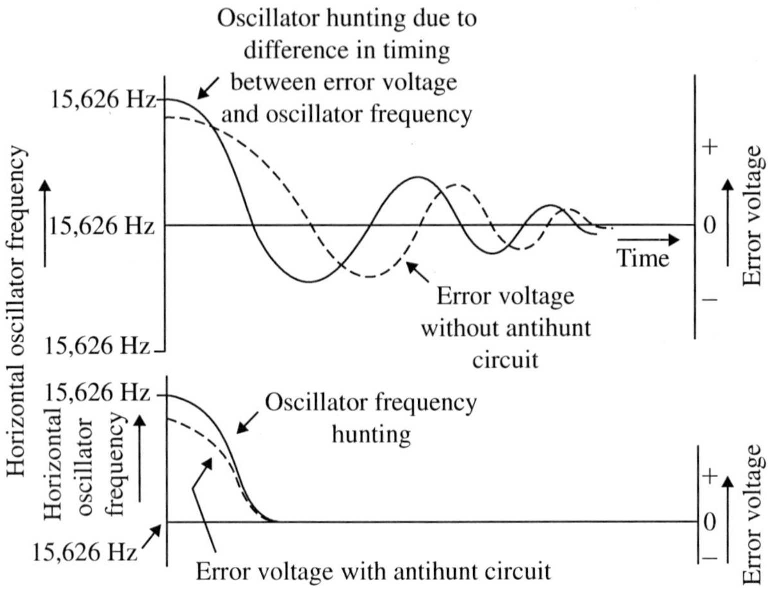

Figure 2.42 (b) Correction in control voltage.

raster and to ensure linear deflection a vertical height and vertical linearity adjustments provided in frame deflection circuits.

2.22.1 Requirement of Vertical Deflection Circuits

In order to have a discrete or integrated vertical deflection circuit it is necessary to know different requirements that such a circuit must meet. The block diagram of vertical system shown in Figure 2.43 will enable location of various circuit block.

Requirements of vertical deflection circuits:

(i) Vertical rolling of picture.
(ii) Height of the raster.
(iii) Suppression of undesired oscillations.
(iv) Correct aspect ratio.
(v) Amplifier configuration.
(vi) Vertical deflection coils.
(vii) Vertical linearity.
(viii) Internal vertical blanking.

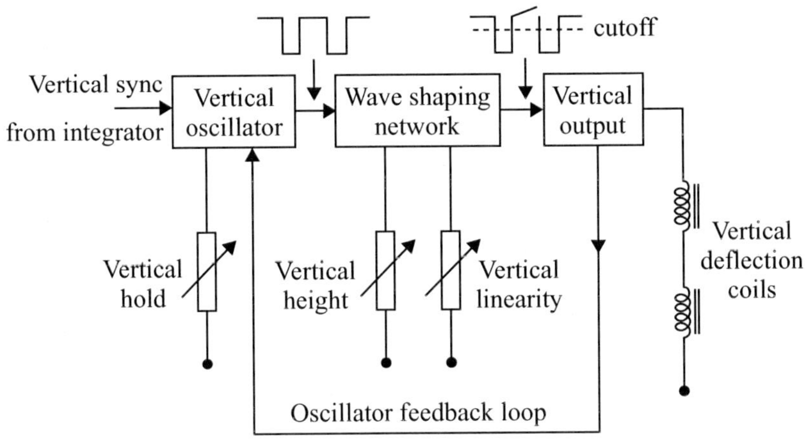

Figure 2.43 Block diagram of the frame deflection section of a TV receiver.

2.22.1.1 *Vertical Rolling of the Picture*

Rolling of picture upwards or downwards on screen occurs due to incorrect vertical scanning frequency. In the block diagram shown in Figure 2.44 the vertical oscillator frequency is higher than the 50 Hz.

Relative timings of vertical blanking in the composite video signal and vertical retrace period in saw tooth deflection current are not same. This result in vertical blanking occurs during trace time instead of retrace time.

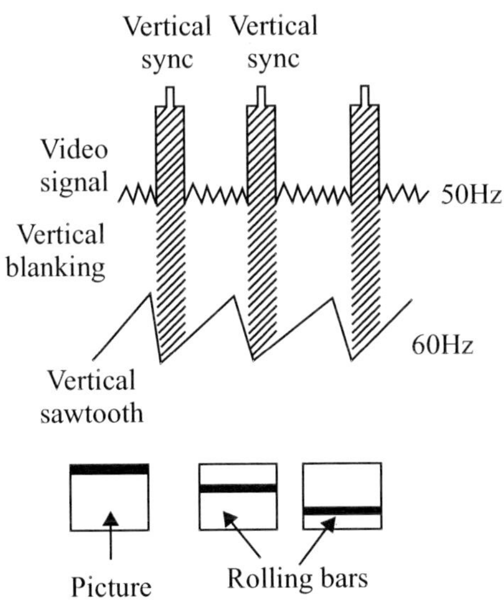

Figure 2.44 Vertical rolling when the deflection frequency is more than 50 Hz.

For vertical frequency higher than 50 Hz for each saw tooth current advances into the trace time for succeeding the blanking pulses. This results formation of black bar across the screen during vertical blanking pulse. And appears lower down the screen for successive cycles.

Picture rolls much faster rate if the oscillator frequency deviates from 50 Hz. When the sync pulse frequency and a saw tooth frequencies are same, then every vertical retrace occurs within the blanking time.

Due to this blanking bars are produced at top and bottom edges of raster and are not visible.

When the vertical oscillator is locked by sync pulses each frame is reproduced over the previous and picture holds. Oscillator frequency is controlled by hold control.

2.22.1.2 Height of the Raster

To fill the raster from the top to bottom a definite amplitude of saw tooth current must flow into the deflection coils. So the output amplifier is designed to meet these requirements. The height of ramp is adjusted and this control is called 'height control'. This is shown in Figure 2.45.

In some circuits designs a part of emitter resistance changes of output amplitude is variable and left by passed. due to this variations of resistances changes the negative feedback and hence gain of amplifier and through this height of the raster changes.

2.22.1.3 Suppression of Undesired Oscillator

During retrace time the current through the deflecting coils suddenly drops to zero. And the energy stored in the collapsing magnetic field starts frequency oscillations.

Frequency of ringing is produced due to the disturbed capacitance and inductance of deflection coils.

This effect of oscillations are not suppressed, it tends to pull the beam upwards and downwards alternatively and results in narrow strip at the top of the screen.

To prevent this ringing a shunt damp resistance is connected parallel across the two halves of vertical scanning. This is shown in Figure 2.46.

2.22.1.4 Correct Aspect Ratio

In receiver a low DC voltage from horizontal line output stage by rectifying and filtering line pulses.

This makes the frame drive s to saw tooth amplitude dependent on line scan amplitude.

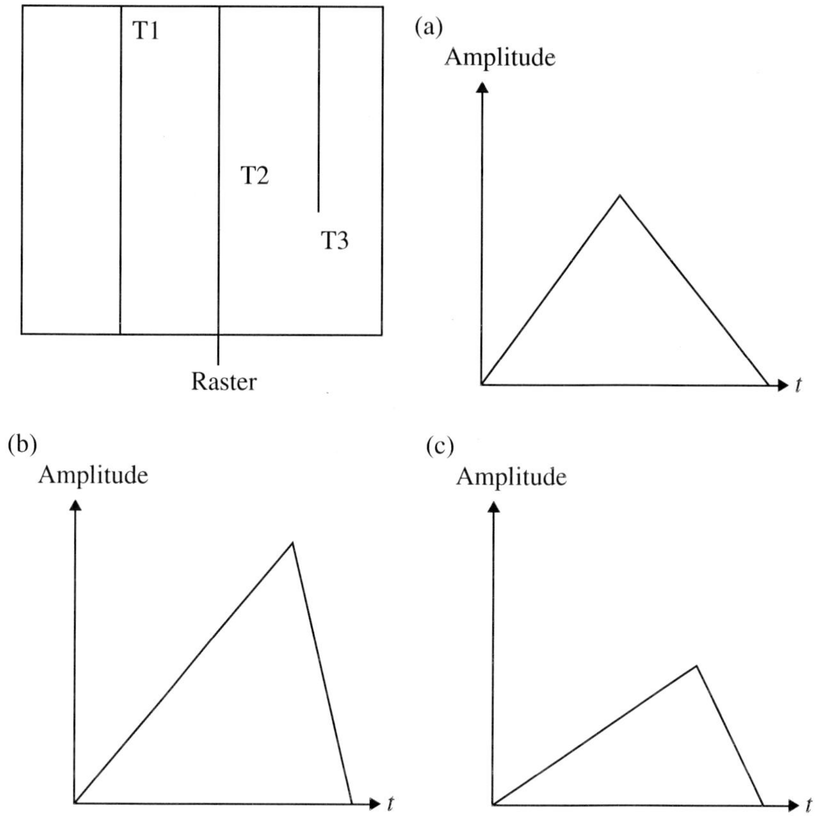

Figure 2.45 (a) Correct amplitude, (b) Excessive amplitude, and (c) Insufficient amplitude.

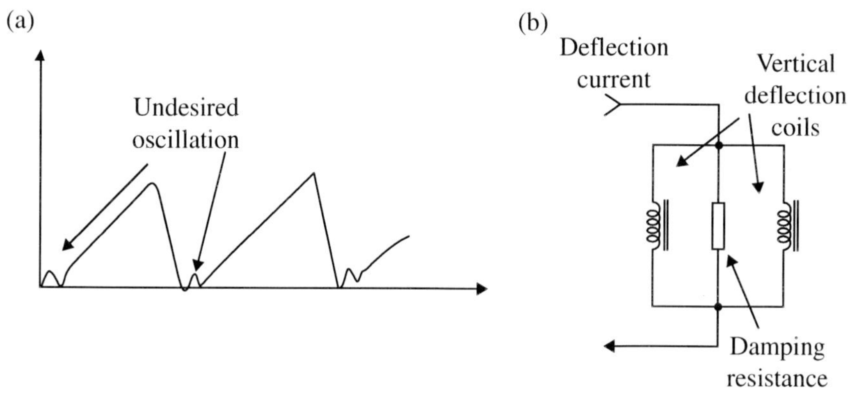

Figure 2.46 (a) Effect of oscillation on the vertical, and (b) Shunt resistor across the delation coil deflection current.

If excessive load current reduces horizontal scanning width, vertical height also decreases, thus to maintain a correct aspect ratio of the raster.

2.22.1.5 Amplifier Configuration

(a) The purpose of field scan amplitude is to convert input trapezoidal voltage waveform into saw tooth current in field deflection coils of sufficient magnitude to scan the picture.
(b) Many amplifier configuration is in use, but all operate under class *A* or *AB* mode condition to ensure linear operation
(c) Impedance of coils is mainly resistive because of low frequency spectrum that constitutes 50 Hz trapezoidal voltage.
(d) The design of output circuit basically the same as that an audio output stage to match the loud speaker.
(e) To amplify 50 Hz saw tooth with no distortions a high voltage transistor is needed for amplification.

2.22.1.6 Vertical Deflection Coils

(a) Magnetic field for vertical deflection is developed by two coils mounted 180° apart on the neck of the picture tube.
(b) In vertical deflection period the magnetic flux generated by saw tooth current in deflection windings to move the electron beam down and return to the top for a quick retrace.
(c) Scanning coils are designed with *L* varies between 3 mH and 30 mH with resistance value between 3 to 15Ω. Impedance and no of turns depends on the required deflection angle.
(d) Wider deflection angle tubes require a strong magnetic field to cover entire raster. A short neck 37 cm picture tube requires a 30 mH deflection coil and a peak to peak driving current of over 1 amperes.
(e) Two halves of deflecting coils are connected parallel to provide impedance match for maximum efficiency

2.22.1.7 Vertical Linearity

The effect of three different current waveform reproduced as shown in Figure 2.47.

When the saw tooth wave is linear, vertical scan allows an equal spacing between horizontal lines and there is no distortion. This is shown in Figure 2.47(a).

The saw tooth tends to become flat towards its close, hence the magnetic field increases at a reduced rate and results in horizontal lines are crowd together at bottom of the raster. This is shown in Figure 2.47(b).

Horizontal lines are crowded at the top of raster because of slow rate of change of current at the beginning of saw tooth wave. This is shown in Figure 2.47(c).

In picture tube, the distance from the electron gun to four corner is much greater than the distance to the centre of the screen. Due to this it remains more deflection at the edges and results in a streching effect. This is shown in Figures 2.47(e) and (f).

2.22.1.8 Internal Vertical Blanking

(a) The voltage pulses are produced in retrace intervals, in vertical output circuit are coupled to the picture tube to provide blanking during vertical retrace time.

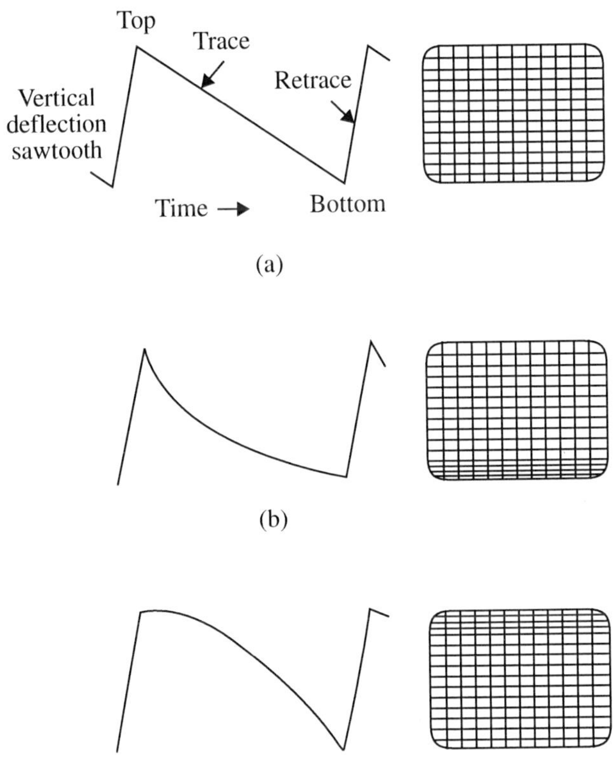

Figure 2.47 (a) Linear sawtooth causes equal spacing between the lines, (b) Non-linear sawtooth causes the bottom crowded lines, and (c) Non-linear sawtooth causes top crowded lines.

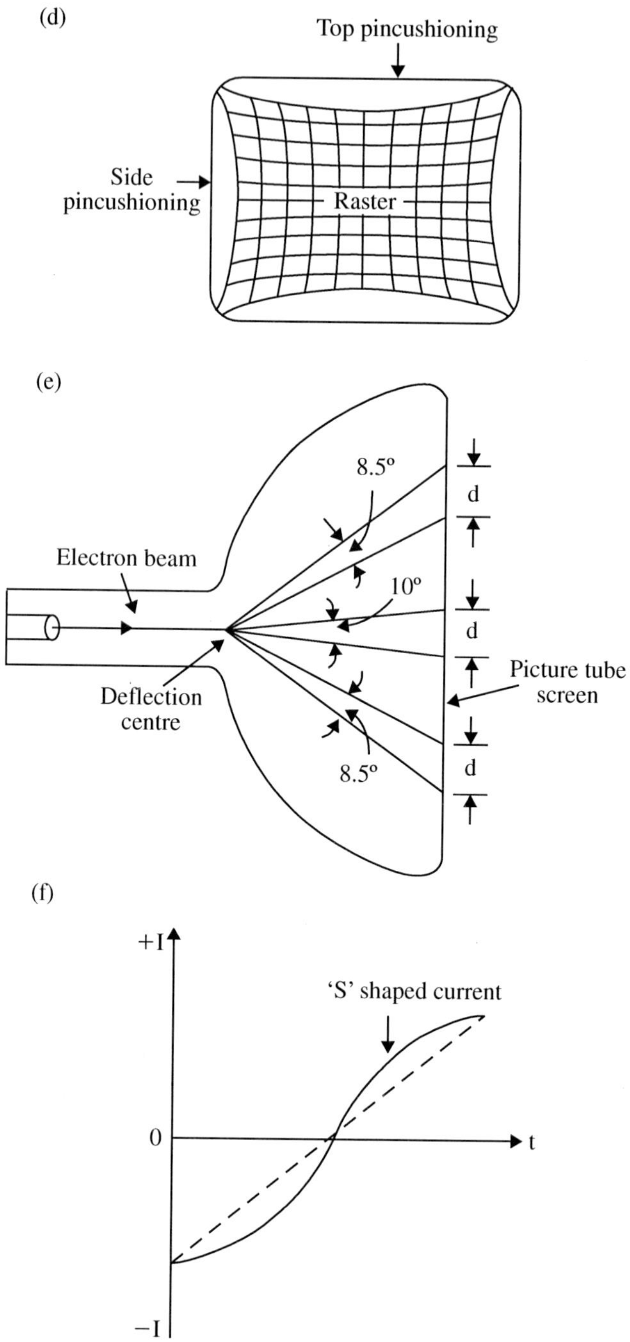

Figure 2.47 (d) Pincushion distortion, (e) Effect of flat and screen, and (f) Modified current.

(b) During the vertical retrace period's starts. The horizontal line gets traced, and it appears on the screen.
(c) All receivers use a vertical retrace suppression circuit to turn off the electron beam.

2.23 LINE DEFLECTION CIRCUIT

2.23.1 Basic Line Deflection Circuit

The Figure 2.48(a) shows the circuit of horizontal (line) output stage. The corresponding input and output wave form are shown in Figure 2.48(b).

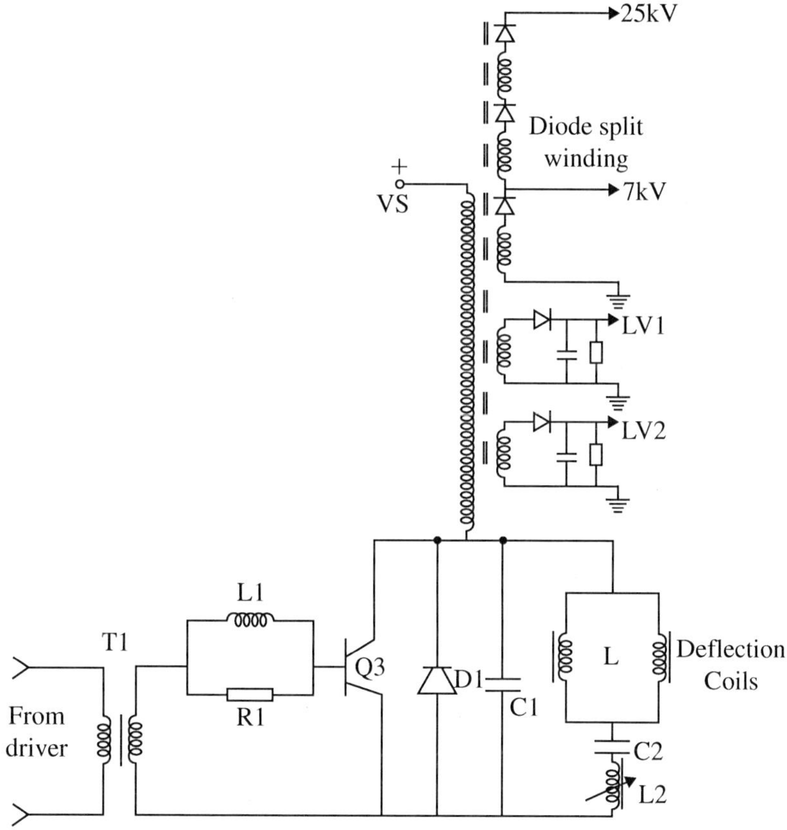

Figure 2.48 (a) Basic circuit of line output stage.

From the above figure

Q1 – horizontal amplifier.

D1 – damper diode

C1 – fly back capacitance

C2 – *S* correction capacitance.

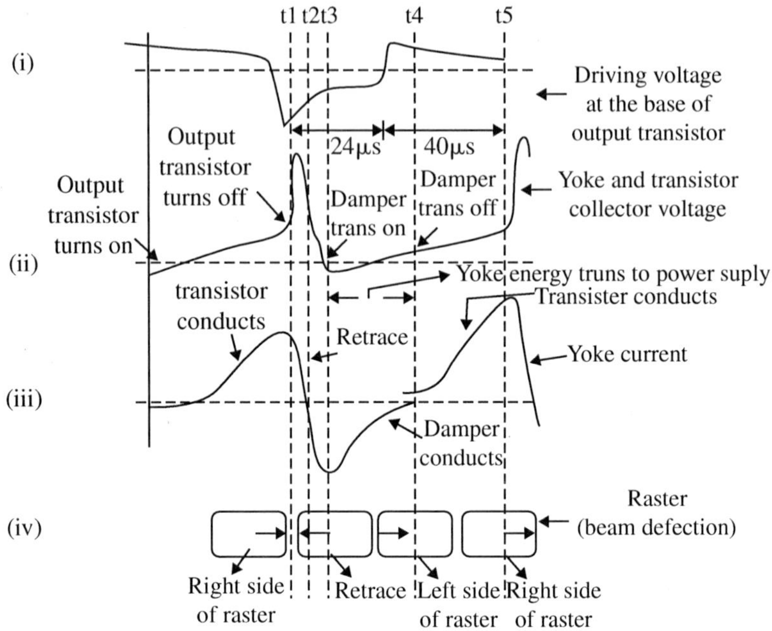

Figure 2.48 (b) Waveforms corresponding *t*.

2.23.1.1 Operation

(a) During the time period t_1, the transistor Q1 conducts and the resulting current through the yoke deflects the beam to the right of the raster. During this interval, diode D1 is reverse biased by +Vcc supply.

(b) At an instant time t_1, the sharp negative pulse appears at the base of transistor Q1 makes it to turn off and retrace occurs. And now the beam returns to left of the raster during first half of the self-oscillations. The self induced voltage across the coil keeps the damper diode to reverse biase during the time period t_2.

(c) At an instant time t_3, diode D1 conducts to deflect the beam to the centre of the screen.

(d) At an instant time t_4, transistor Q1 conducts and deflects electron beam from centre to right of the raster. Then the diode D1 immediately cuts off. And the cycle repeats to scan the raster.

(e) The '*S*' correction capacitor is used to block the DC to yoke coil which causes the heating of coil.

2.24 GENERATION OF EHT VOLTAGE

The circuit diagram of EHT voltage generator is shown in Figure 2.49. From the driver stage, the driving voltage is fed to the base of transistor Q1.

During the trace period, transistor Q1 conducts heavily and collector current increases at a steady rate. Due to this increase in current through the coil a strong magnetic field is built around the secondary winding.

The horizontal output stage is used to produce the EHT Voltage. When the transistor Q1 conducts, collector current increases. Due to this increase, current through the coil also increases. This produces the strong magnetic field around the primary coil. This is shown in Figure 2.49.

When the transistor Q2 comes to off state, the collector current falls to zero. The magnetic field around the coil will be collapsed. Due to this collapsing of magnetic field EMF will be induced. In secondary winding of fly back transformer it is stepped up and given to rectifier diodes.

The fly back voltage is used to provide high voltage to picture tube and other circuits.

About 18,000 V AC is available at the anode of stack diode. EHT rectifier diode converts this high AC voltage in to DC voltage. Diodes D3 and D4 are used to produce DC voltage of 400 and 80 V, C4 and C5 are used to remove the 15,625 Hz ripples from the resistor and used to discharge the capacitor C4 and C3 when ever the receiver is switched OFF.

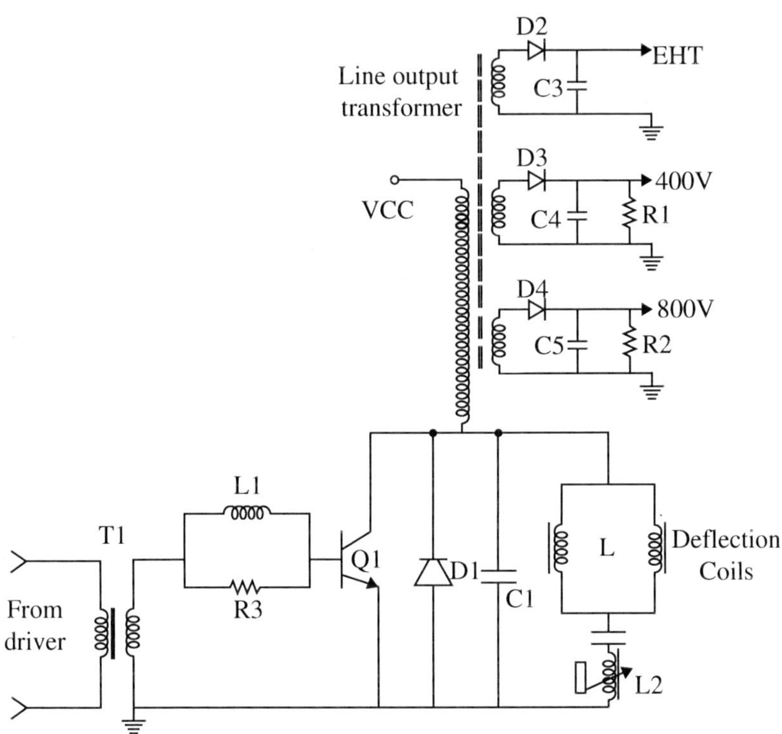

Figure 2.49 Generation of EHT.

2.25 RECEIVER ANTENNAS

The TV receiving antennas is the front end converter in TV receiver. The video signal (RF modulated electromagnetic signals travels in the line of sight model. These electromagnetic signals are picked up by the receiving antenna. The FM waves generates some current and voltage in the dipole of TV antennas.

For both UHF and VHF television channels, one half wavelengths is a practical size. Hence an under grounded resonant dipole is the basic antenna often used for reception of television signals.

A matched transmission lines joins the antenna to the input terminals of the receivers. It is to be noted that the signal picked up by the antenna consists of both sound and picture components. A signal antenna can be designed to pick up signals from several channels that be close to each other.

2.25.1 Functions of Receiving Antennas

(i) The antenna should collect the RF modulated video signal as maximum as possible and attenuate the interference due to switching, adjacent channels, reflections *etc.*

(ii) The antenna should be sufficiently wide band, so that it can be used for sufficient number of channels.

(iii) The effect of environment like temperature, wind, raining dust should be minimum on antenna.

2.25.2 Antenna For VHF Channel

Most of the receiver can develop a picture with sufficient contrast even with a weak signal. But for a picture with no snow and ghosts, the needed antenna signal strength ranges between 100 and 2000μV. A half wave dipole will deliver satisfactory signal for receivers located close to the transmitter. Elaborate arrays are needed for locations for away from the transmitter.

2.25.3 Yagi-Uda Antenna

This is a widely used antenna for television receivers. Generally for location with in 40 to 60 km from the transmitter is the folded dipole with one reflector and one director. This is commonly called Yagi-Uda antenna or Yagi antenna. The array is shown in Figure 2.50. The elements are arranged collinearly and close together.

The resultant improves in the front to back ratio of the signal pick-up makes the antenna highly directional. Hence it can be oriented for efficient

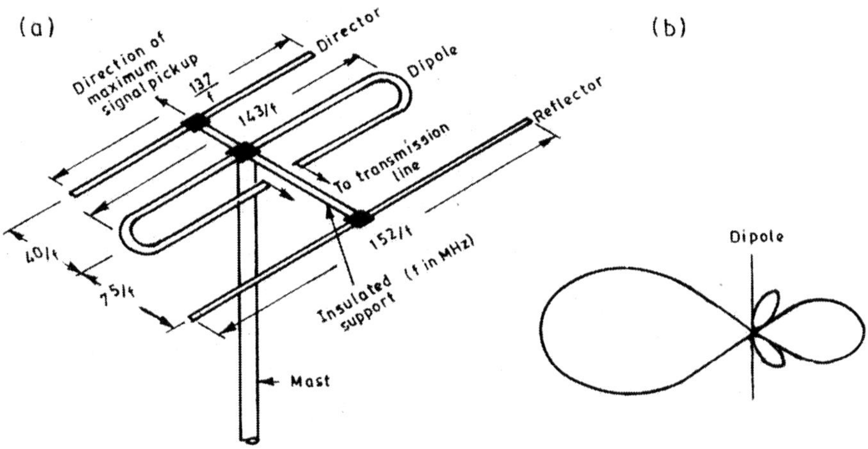

Figure 2.50 (a) Yagi-Uda antenna and (b) Radiation pattern.

pick up for a particular station. But bring the radiators close has the negative effect of lowering the input impedance of the array.

2.25.3.1 Antenna Length

The antenna length can be calculated from the following formula

$$\text{Wavelength } (\lambda) = 3 \times \frac{10^8}{f_{ch}(Z)} \text{ meters}$$

practically the length is kept 6% less than the calculated value.

2.25.3.2 Antenna Mounting

The receiver antenna is mounted horizontally for maximum reception for the transmitting antenna, horizontal polarized results in more single strength, less reflection and reduced ghost images.

The thickness of the pipe should be selected in a way that the antenna structure does not get bend or twisted during strong wind or occasional sitting and flying off of birds. The antenna is mounted on a suitable structure at a height above 10 metres from ground level.

2.25.4 UHF Antenna

The basic principal of antenna for collecting signals from stations that work in UHF band is more or less the same as that in VHF band. Due to higher attenuation suffered by UHF signals, it becomes important to have very high gain and directive antennas. As we have developed a large number of microwave antennas for specific applications, the two types that find extensive application for television reception are Bow-Tie antenna and parabolic reflector antenna.

The antenna gives a gain close to 7 dB and it is relatively unidirectional as seen from its radiation pattern. These characters are more suitable for reception from television transmitters of medium capacity. To avoid receiving unwanted signals, the back lobe of the radiation pattern can be reduced by bringing the radiators close to each other.

2.25.5 Indoor Antenna

In strong signal areas it is sometimes feasible to use indoor antennas provided the receiver is enough sensible. These antennas are available in a variety of shapes.

2.25.6 In Line Antenna

Figure 2.51 shows another combination antenna which is called in-line antenna. It consist of half wave folded dipole with reflector VHF band, that is in line with a shorter half wave folded dipole used for the upper VHF

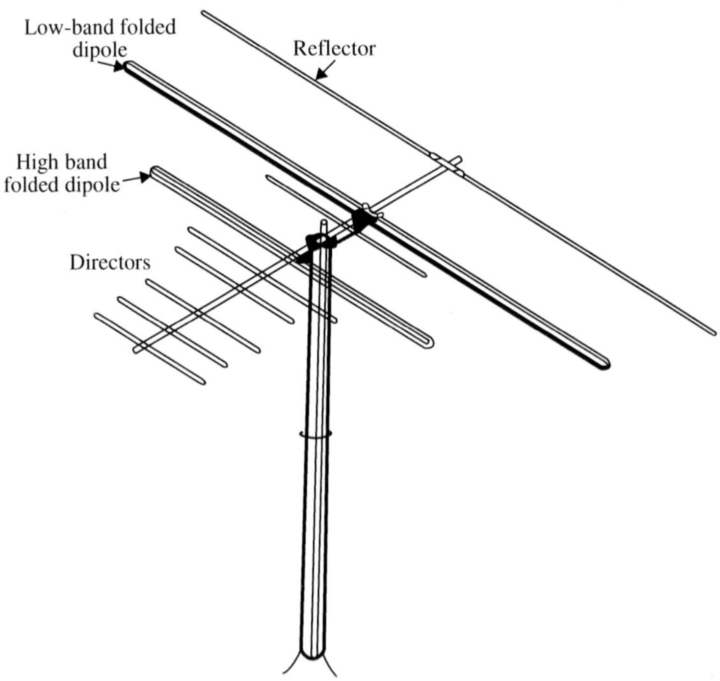

Figure 2.51 In line YAGI array for lower and upper VHF (band I and III).

band. The distance between the two folded dipole is equal to approximately one quarter wavelength at a high band dipole frequency this is the length of the line joining the short dipole to the long dipole, here the transmission line to the receiver is connected.

The directivity pattern of the inline antenna is relatively uniform on VHF channels with a unidirectional broadside response pattern its input resistance is around 150Ω.

2.25.7 Parabolic Reflector Antenna

Parabolic reflector antenna is shown in Figure 2.52. In this antenna the dipole is placed at the focal point of a parabolic reflector. The principle is same as that of parabolic reflection of the headlight of the vehicle in reverse way. The incoming electromagnetic waves are focused by the reflector towards the dipole. This offers both high gain and directives.

Various combinations of different VHF and UHF antenna are available in use. A signal lead in line gives signal to the receiver through the use of special coupling device which is mounted directly on the antenna itself.

2.25.8 Bow-Tie or Die Fan Antenna

This Bow-tie or Di-fan half-wave dipole is the simplest type of UHF antenna. As shown in Figure 2.53 the dipoles are triangular in shape made out of metal sheet, instead of rods. This unit has a broadband response with radiation pattern looking as the figure of eight. When a screen reflector is

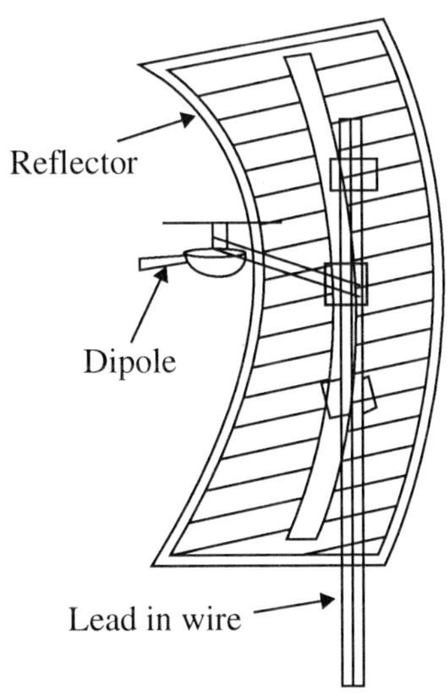

Figure 2.52 Parabolic reflector antenna.

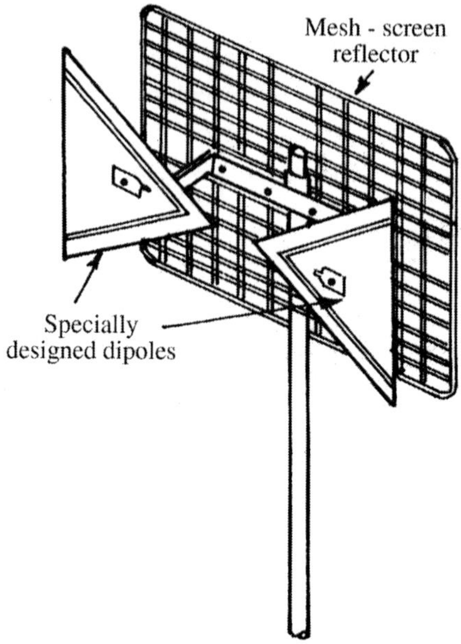

Figure 2.53 Dipole UHF antenna.

available at its back the response becomes unidirectional. For higher gain two or four sets of dipoles can be put together to form a array it is to be noted that the use of a mesh screen reflected instead of a rod as used in VHF antennas. Screen reflects are more efficient than rods. But their big size and bulks makes it impossible.

2.25.9 Fringe Area Antennas

Infringe areas where the signal level is very low, high gain antenna arrays are required, the gain of the antenna increases with the number of elements used. A Yagi antenna with a large number of directors is most commonly utilized for VHF band.

2.25.10 Multi-band Antenna

It is impossible to receive all the channels of lower and higher VHF band with a single antenna. The major problem in using a single dipole for both VHF band is difficult maintain the broadband response. The reason is the directional pattern of a low band dipole splits into side lobes at the third and fourth harmonic frequencies in the range 174 to 233 MHz band.

Apart from a high band dipole, cut off for a half wavelength in the frequency 174 to 233 MHz band is not suitable. As a result, antennas for the both VHF bands generally utilize separate dipoles for each band. Or a dipole for the lower VHF band can be modified to have broadband unidirectional response in the upper VHF band also.

2.25.11 Diplexing of VHF Antennas

When it is needed to combine the outputs from the lower and upper VHF band antennas to a common lead in wire called feeder set-up, it is desirable to employ a filter circuit. The filter network should match with the signal sources to the common feeder and isolates the signals in the antenna from each other. This filter arrangement is called a diplexer. The circuit will be designed for band I (47 to 68 MHz) and band III (174 to 236 MHz) as shown in Figure 2.54. In the same way, a tri-plexer filter can be used when three different antennas are to be given their outputs to common socket in the receiver. The combining arrangement can be further improved to connect UHF antenna to the same fedder line.

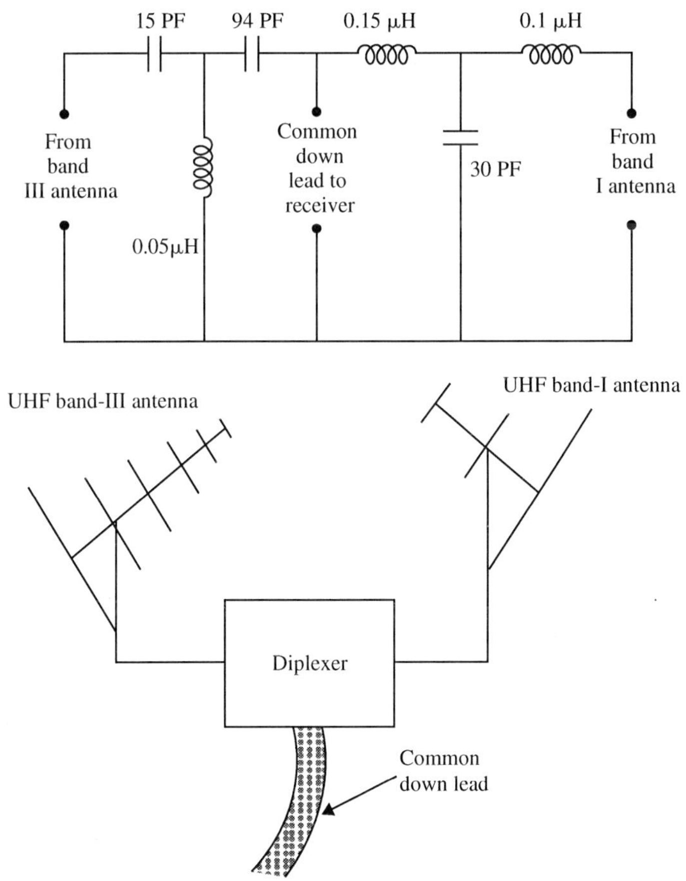

Figure 2.54 Diplexer network.

KEY TERMS AND CONCEPTS

IF	Intermediate Frequency. The local oscillator is adjusted above the channel frequency such that the subtraction of the two frequency produce two intermediate frequency.
PIF	Picture intermediate frequency is 33.9 MHz.
SIF	Sound intermediate frequency 33.4 MHz.
AM	The amplitude modulation is used for picture signal transmission. The bandwidth of the video signal is from 0 (DC) to 5 MHz.
FM	The frequency modulation is used for sound signal transmission. The bandwidth required for sound signal transmission is 150 kHz.
UHF	Ultra High Frequency.
VHF	Very High Frequency.
CIN diplexer	Constant Impedance Notch.
Ghost interference	Ghost interference arises as a result of discrete reflection of the signals from the surface of buildings, bridges, hills, and towers.
Leading ghost	The ghost picture, caused by direct signal appears displaced to the left and is known as 'leading ghost picture'.
Trailing ghost	The picture due to the reflected signal that arrives late, appears displaced to the right. Such displaced pictures are known as 'trailing ghost pictures'.
Guard band	A small frequency band introduced between two consecutive channels in order to reduce the interference.
AFC	The AFC generates trisecting pulses for the horizontal oscillator.
AGC	Hence an automatic gain control (AGC) is used to automatically control the gain of amplifiers.
AFT	Automatic frequency tuning is necessary for high quality colour TV reception.

EHT	The horizontal output generates B ++ supply and very high voltage for final accelerating anode of the picture tube *i.e.*, called EHT.
Trap circuit	Trap circuit is used to separate the sound from the composite video signal.
PLL	The phase lock loop control compares two signals and develops a control voltage, proportional to thc frequency or phase different between them.
DC reinsertion	DC reinsertion is needed to preserve the component and maintain the true representation of the
Sync separation	The composite video signal at the output of video amplifier is given to the sync separator the sync separator separates the horizontal and vertical sync pulses.
Vertical & horizontal sync pulse	These are then given to low pass (integrator) and high pass filter (differentiator) to generate a triggering pulse for horizontal and vertical oscillators.
Frame deflection circuit	The function of frame deflection circuit of a TV receiver is to cause linear vertical deflection of the picture tube.
Anti-hunt network	To avoid hunting problem in AFC circuit anti-hunt network is used.
Balun transformer	Balun transformer is used for matching the impedance of the antenna (or) ribbon wire (300r's) with the input independence of amplifier (75).
Co-channel interference	If two channels are operating at the same carrier frequency and located nearby then they will interface with each other. This is called co-channel interference and it is common in fringe areas.
Adjacent channel interference	It is the interference due to stations located nearby and allocated as adjacent channels.

REVIEW QUESTIONS

Part A

1. Why is a simple tuner circuit is used in electronic tuning?
2. Define image rejection ratios.
3. What are the requirements of an AGC system?
4. What are the different types of coupling methods available?
5. What is a trap circuit?
6. What is the need for the DC restore?
7. State the disadvantages of high level modulation.
8. Define low level modulation.
9. What is the need for sync pulse separator?
10. State how the vertical sync pulse is separator?
11. What is the need for synchrozation?
12. Define diplexer.
13. Write short notes on hunting.
14. What is mean by anti-hunt circuit?
15. Draw the block diagram of sync separator.
16. Define the Balun transformer.
17. Why the deflection current is high in colour receiver?
18. Define the directivity of antenna.
19. What is the function of AFT control in a TV receiver?
20. Which type of circuit is used to separate horizontal and vertical sync pulse?
21. Mention the major difference between the monochrome and colour television.
22. What is ghost image?
23. Mention the factors that influence the choice of IF.
24. Define frequency distortion.
25. List out the major purpose of using VHF tuner.
26. What are essential components of RF section?
27. What do you understand by polarization?
28. What are sky waves?
29. Why AM is preferred over FM broadcasting the picture signal?
30. What is dipole array antenna?
31. What do you refer by Yagi Uda antenna?
32. Why is the sound signal amplitude attenuated to about 5% of the maximum output voltage?
33. Which causes 'smear' in the picture and it is eliminated?
34. Why do we prefer horizontal polarization for television receiving antenna?
35. Where can you employ indoor receiver antennas?

36. Write short notes on AGC circuit.
37. What do you mean by Peak AGC system?
38. What is forward AGC control?
39. What is reverse AGC?
40. List the drawbacks of non keyed AGC.
41. List the merits of keyed AGC system.
42. State troubles in AGC?
43. Why it is necessary to cancel the noise pulses present at the AGC voltage?

Part B

1. Describe briefly the factors that influence the choice of picture IF = 38.9 MHz and sound IF = 33.4 MHz in the 625B monochrome television system.
2. What is electronic tuning? Explain with circuit diagram how different channel can be selected by this method.
3. With necessary diagram explain the operation of forward and reverse AGC.
4. Illustrate the operational details of vision IF sub system.
5. Write a short notes on propagation of TV signal.
6. Give the constructional details of a turnstile antenna and explain by drawing the radiation pattern its suitability for television transmission.
7. Explain with a circuit diagram how the high voltage pulses induced in the output transformer windings, are used to generate EHT supply.
8. Draw the amplified block diagram of UHF tuner and explain the functions of each section why its outputs connected to VHF tuner, amplifier mixer.
9. Explain the principle of operation of a synchronous video detector with the help of a circuit diagram present its merits.
10. With a neat block diagram explain the operation of TV receiver.
11. Explain the video amplifier circuit with high frequency and low frequency compensation.
12. What are the essential functions which are assigned to the IF section of the receiver? Show by sketching output voltage verses frequency response of the IF section.
13. Draw the block diagram of monochrome TV receiver and briefly review the nature of inputs and output signals of various sections.
14. With a circuit diagram describe the IF section of a TV receiver. Explain how the use of a saw filter simplifies the design of IF amplifier.
15. Describe the horizontal deflection stage of a TV receiver. How EHT voltage is generated from this section?

16. With a circuit diagram describe the transistorized video driver and explain how DC restoration and wide bandwidth are achieved in this stage.
17. What do you understand by inter-carrier sound system? Explain why any shift in the LO frequency does not affect the frequency of inter-carrier beat signal. Where and how the inter-carrier sound signal separated from the video signal?
18. Explain the operation of AFT control in a TV receiver.
19. Draw the simplified block diagram of UHF tuner and explain the functions of each section. Why its output is connected to UHF tuner mixer?
20. Explain the principle of operation of a synchronous video detector with the help of a circuit diagram. Present its merits.

QUESTIONS AND ANSWERS

1. **State any two factors that affect the TV signal transmission?**
   The two factors that affect the TV signal transmission are:
   - TV uses line of sight transmission. Therefore the transmission range is limited to very small distance. Also the transmitted power goes on reducing with increasing distance. The signal strength should be sufficiently large to reduce the effect of noise and produce good quality picture in TV transmission.
   - The height of the antenna is also an important factor. By increasing the height of the antenna the transmitted power will reduces. So the transmitting antenna should be, mounted on the centre point. The point should be high as possible.
2. **Why FM is not used for picture signal transmission?**
   FM is not used for picture signal transmission because it contains more number of sidebands. So it will increase the channel bandwidth. The increased channel bandwidth is not affordable since no of channels transmitted in particular frequency band may reduce, therefore FM is not suitable for picture signal transmission.
3. **Why AM is preferred for picture signal transmission?**
   Negative modulation is preferred because the effect of noise on the picture is less. In order to reduce the channel bandwidth vestigial side band transmission is a standard practice. So the overall channel bandwidth can be accommodated with of 7 MHz.
4. **State advantages of FM sound signal transmission.**
   The advantages of FM sound signal transmission are:
   - The increase in channel bandwidth of 7 MHz is insignificant even though more bandwidth is needed for FM.

- The effect of noise interference is minimum in FM. Since the amplitude of modulated carrier remains constant. The noise can be easily removed by placing a amplitude limiter at the receiver end.
- Transmission efficiency is high in FM. Because the amplitude of FM is independent of modulation. So the low level modulation also can be used in FM.
- The effect of adjacent channel interference is reduced by providing a guard band between two channels.

5. **Why it is necessary to employ a de-emphasis circuit in the receiver?**
   Since the pre-emphasis is used at the transmitter side so it is necessary to high frequency noise in the receiver side. So we have to employ de-emphasis at the receiver.
6. **Mention the requirements of high level modulation.**
   In high level modulation, the video signal has to be modulated by the picture carrier in the final power amplifier which has a high power level. So normally grid bias modulation is employed.
7. **List the drawbacks of high level modulation.**
   The drawbacks of high level modulations are:
   - The video modulation is needed to supply high video voltages in the order of several hundred volts to the load that are highly capacitive. These loads need large current.
   - The load available to the driver stage is variable since grid modulation is employed.
   - To provide vestigial side band properties, VSB filters are employed. It requires high power at the operating channel. So the linearity and high video driving powers introduce design problems.
8. **Why low level modulation is most commonly used?**
   In low level modulation, the signal is modulated before amplification. The power level of this transmitter is less than 10 W so, it is easy to demodulate at the receiver. So low level modulation is most commonly used.
9. **State the advantages of low level IF modulation system.**
   The advantages of low level IF modulation system are as follows:
   - As the modulation is done at low power level and at IF, solid state device can be utilized for the construction of a transmitter which becomes more reliable and efficient.
   - The VSB filters can be used immediately after the modulation stage lumped components can be employed for its design to shape both the lower and higher side band response.
   - In the UHF and VHF bands, the video transmitters are identical this is very cost effective.
   - The stages next to the VSB filters can be permanently tuned for a wide band linear operation. This result in simplified tuning procedure.

10. State the factors that influence the choice of intermediate frequencies.

The factors that influence the choice of intermediate frequencies are:

- Image rejection ratio.
- Pick up due to local oscillator radiation from TV receivers.
- Image frequencies should not lie in the FM band.
- Interference or direct pick up from bands assigned for other services.
- Gain.

11. Name the essential parts of TV transmitter.

The essential parts of TV transmitter are:

- Video processing unit
- Visual modulator which is a diode bridge modulator
- VSB filter
- Phase compensator or delay equalizer
- Frequency up converter
- Linear and power amplifier

12. Name the essential components of RF section

RF tuner section consists of:

- RF amplifier.
- Mixer.
- Local oscillator and is normally mounted on a separate subchasis is, called the front end.

13. What is the purpose of using RF tuner?

RF tuner section is in the front end of the receiver. This section receives the signal, from the antenna. The signal received from the antenna may have less amplitude, so RF amplifier is used to raise the amplitude of signal to the required level and then fed to mixer. By mixing two frequencies, intermediate frequencies will be generated.

$$\text{IF} = \text{LO frequency} - \text{RF signal frequency}$$

14. State the major functions of RF tuner.

The major functions of RF tuner are as follows:

- It selects the channel to be received by switching pre tuned circuits in the RF stage and the local oscillator.
- It converts the RF into IF by mixing it with the local oscillator frequency and feed the signal to the video IF amplifier.
- The RF tuner isolates the local oscillator signal from the antenna. the RF amplifier acts as a buffer, preventing radiation and interference to other receivers.
- RF tuners blocks the interference antenna pick up signals in the IF ranges and prevents them from entering the receiver and mixes with the local oscillator and then to the IF amplifier.
- It rejects the image frequencies by means of RF selective circuits.

15. Define guard band.

Guard band can be defined as a small frequency band introduced between two consecutive channels in order to reduce interference.

16. Define co-channel interference.

Two stations operating at the same frequency if located closely will interfere with each other. This phenomenon which is common in fringe areas is called 'co-channel interference'.

17. Define ghost interference.

Ghost interference arises as a result of discrete reflections of the signal from the surface of hills, bridges, buildings, towers *etc.*, electromagnetic waves from the transmitter to receiver. Since reflected path is longer than the direct path, the reflected signal takes a longer time to arrive at the receiver.

18. What do you understand by ground waves?

Vertically polarized electromagnetic waves are radiated at zero or small angles with ground. They are guided by the conducting surface of the ground along which they are propagated, such waves are known as ground or surface waves. As the ground waves travel along the surface of the earth, their attenuation is proportional to frequency. The attenuation is reasonably low below 1500 kHz.Therefore, all medium wave broadcast and long wave telegraph and telephone communication is carried out by ground wave propagation.

19. What are sky waves?

In ground wave propagation, a frequency above 1600 kHz does not serve any useful purpose as the signal gets very much attenuated within a short distance of its transmission. Therefore, most radio communication in short wave bands up to 30 MHz is carried out by sky waves. When these waves are transmitted high up in the sky, they travel in the straight line until the ionosphere is reached. This region begins about 120 km above the surface of the earth. The region consists of large concentrations of charge gaseous ions, free electrons and neutral molecules. The ions and free electrons cause to band all passing electromagnetic waves.

20. Describe briefly about space wave propagation.

Propagation of radio waves above about 40 MHz is not possible through either sky wave or surface wave propagation. Therefore, the only alternative for transmission in the VHF and UHF bands, despite large attenuation is by radio waves which travel in a straight line from transmitter to receiver. This is called 'space wave propagation'.

21. Which causes 'smear' in the picture and it is eliminated?

The effect of phase distortion is severe at low frequencies, because for a small phase delay at low frequencies will cause a large time delay.

This low frequency phase distortions can be observed by the eye as smear on the picture.

22. Why do we prefer horizontal polarization for television receiving antenna?

We prefer horizontal polarization for television receiver antenna because it results in more signal strength, less reflection and reduced ghost images.

23. Where can you employ indoor receiver antennas?

In strong signal areas it is sometimes feasible to use indoor antennas provided the receiver is sufficiently sensitive.

24. Define diplexer.

The outputs of both the video and the audio transmitter are combined by the diplexer circuit and given to a common broadcast transmitting antenna.

25. What do you mean by Yagi Uda antenna?

This is a widely used antenna for television receivers. Generally, for locations within 40 to 60 km from the transmitter is the folded dipole with one reflector and one director. This is commonly called Yagi antenna or Yagi Uda antenna. A Yagi antenna with a large number of directors is commonly used with success in fringe areas for stations in the VHF band.

26. What is dipole array?

Dipole antenna is used for band I and III transmitters. It consists of dipole mounted on the four sides at the top of the antenna tower. Each panel has an array of full wave dipoles mounted in front of reflectors. To get an unidirectional pattern, the four panels mounted on the four sides of the tower are so fed that the current in each lags behind the previous by 90°. This is done by changing the field cable length by $1/4$ to the two alternate panels and by reversal of polarity of the current.

27. Define Turnistile antenna.

For equal TV signal radiation in all directions, an arrangement known as 'turnstile array' is often used. In this type of antenna two crossed dipole of length $\lambda/2$ are used in turnstile array arrangement. The resultant field in all direction is equal to the square root of sum of the squares of fields radiated by each conductor in that direction. To increase the directivity in vertical plane turnstile antenna are stacked one over the other. The distance between two antenna in vertical plane is also $\lambda/2$.

28. State the functions of receiving antennas.

The functions of receiving antennas are as follows:

- The antenna should collect the RF modulated video signal as maximum as possible and attenuate the interference due to switching, adjacent channels, reflections *etc.*

- The antenna should be sufficiently wide band, so that it can be used for sufficient number of channels.
- The effect of environment like temperature, wind, raining, dust should be minimum on antenna.

29. Define adjacent channel interference.

Adjacent channel interference occurs as a result of beats between any two of the frequencies or between a carrier and any side bands. A course dot structure is generated on the screen if picture carrier of the channel beat with sound carrier of lower adjacent channel. The next most prominent source of interference is the one produced by picture sideband components of upper adjacent channel.

30. Define trailing and leading ghost picture.

The direct signal is usually stronger and assumes control of the synchronizing circuitry and so the picture due to the reflected signal that arrives late, appears displaced to the right. Such displaced pictures are known as 'trailing ghost' pictures. Then the ghost picture, now caused by direct signal appears displaced to the left and is known as 'leading ghost picture'.

31. Define image rejection ratio.

Image rejection ratio is defined as the output due to desired station divided by output due to image signal.

32. What is the main purpose of using VHF tuner?

The purpose tuner unit is to amplify both picture and sound signals picked up by the antenna and to convert the carrier frequencies and their associated side bands into intermediate frequencies.

33. What is the purpose of Balun transformer?

To match the impedance of the antenna and receiver a balancing unit called "balun transformer" is used in between these two. Because, the transmission line has 300Ω impedance, but the receiver has only 75Ω impedance. So the impedance matching is done by balun transformer.

34. What are the different digital tuning techniques?

There are two digital tuning techniques:

- Phase-locked loop control technique.
- Frequency synthesis technique.

35. What are the major tasks to be done by detector?

The video detector is designed to recover composite video signal and to transform the sound signal to another lower carrier frequency.

36. Why is video amplifier required?

The video amplifier is required as:

- The amplitude of the composite video signal at the output of video detector is not enough to drive the picture tube directly.
- It produces a picture with contrast, about 80 to 100 volts at the picturc tubc.

37. What is the principle of inter-carrier sound system?

At the video detector the sound IF is made to beat with picture IF frequency to transfer the audio signal to new sound IF of 5.5 MHz. This is known as 'inter-carrier sound system'. The sounds take off circuit is used to pick up the sound IF signal (5.5 MHz). The sound trap circuit is used to prevent the sound signal entering into the picture tube.

38. What are Sync pulses?

The Synchronizing pulses called 'Sync' are part of the composite video signal as the top 25% of the signal amplitude. The sync pulses include horizontal, vertical and equalizing pulses. Sync separator separates these signals from the composite video signal.

39. Define frequency distortion.

The inequality in gain at different frequency components of the received signal is called frequency distortion.

40. Why high frequency and low frequency compensations and low frequency compensation is necessary?

The high frequency and low frequency compensations are necessary, because the gain at high and low frequency is very low compared to the mid band frequency. If the high frequency is poor, horizontal information will be lost. If low frequency is poor, overall contrast level will becomes poor. If there is any phase distortion, picture may be misplaced.

41. Write short notes on AGC circuit.

AGC circuit is used to control the gain of RF and IF amplifiers. The change in gain is achieved by shifting the operating point of transistors used in the amplifiers. The operating point is changed by a bias voltage that is developed in the AGC circuit.

42. What do you mean by peak AGC system?

The system based on sampling the sync tip levels is known as "peak" AGC system. The Peak AGC system is also called as non-keyed AGC system.

43. What are the two types of AGC control?

Forward AGC control and Reverse AGC control are two types of AGC control.

44. What is reverse AGC?

In any transistor amplifier, gain is varied by shifting the operating point either towards collector current cutoff or saturation. This actually varies beta of the transistor and hence the stage gain changes. When gain is changed by shifting the operating point towards current cutoff, then it is called "Reverse AGC".

45. What is forward AGC?

In any transistor amplifier, gain is varied by shifting the operating point either towards collector current cutoff or saturation. This actually varies beta of the transistor and hence the stage gain changes. When gain is changed by shifting the operating point towards collector current saturation, then it is called "Forward AGC".

46. List the drawbacks of non keyed AGC.

The AGC voltage developed across the peak rectifier load tends to increase during vertical sync pulse periods because the video signal amplitude remains almost at the peak value every time vertical sync pulses occur. This results in a 50 Hz ripple over the negative AGC voltage and reduces gain of the receiver during these intervals. The reduced gain results in weak vertical sync pulse which in turn can put the vertical deflection oscillator out of synchronism causing rolling of the picture.

47. List the merits of Keyed AGC system.

The merits of Keyed AGC system are:

- A true representation of the peak of sync level is made by the AGC voltage. So the AGC voltage developed corresponds to the actual incoming strength.
- Hence the conduction takes place during the retrace time, long time constant are used to remove 50 Hz ripple frequency are not necessary.
- Since the conduction of transistor is restricted (sync pulse time), effect of noise are minimum here.
- The short time constant filters are used here, can enables the AGC bias to respond fast change of channels and aeroplane Flutters. So steady picture and sound output is ensured.

48. State troubles in AGC.

The troubles in AGC are:

- Too much AGC bias reduces the gain, causing a weak picture.
- This weak picture is without snow. Weak antenna signal can produce a weak picture but with snow.
- When the AGC bias is very high it will cutoff the RF and IF amplifier, resulting in no pictures the sound is also cut off.

49. State any three requirements of AGC.

The three requirements of AGC are as follows:

- The AGC automatically adjust the gain of RF stage and IF stages of the receiver in order to maintain a constant level of signal at the video receiver.
- The AGC circuit should produce a DC bias voltage from the incoming video signal *i.e.*, provision should be made in the circuit to convert video signal AC to a DC bias.
- The AGC circuit should act very fast to reduce the gain of the amplifier for strong signals.
- When incoming signals are been, the AGC should apply a delayed DC bias to the RF amplifier.

- According to the amplitude of the incoming signals, the AGC circuit should produce a varying DC bias voltage.

50. Why it is necessary to cancel the noise pulses present at the AGC voltage?

Depending on the AGC bias voltage, the gain of RF and IF stages at the receiver are controlled automatically. Therefore if there is any noise pulse, then it will increase the gin of RF and IF stages.

51. Why do we need AFC?

The frequency of horizontal sync pulse is 15 625 Hz. so the noise pulse can easily pass through the high pass filter. This causes out of synchronization. To maintain the constant frequency we need AFC.

52. Define hunting in AFC.

The low pass filter followed by the discriminator section controls the performance of the AFC circuit. If the constant is low, the oscillator is hunt, while returning to the correct frequency. Excessive hunting in AFC circuit appears as 'wearing' or 'gear tooth' effect on the picture.

53. State the requirements of vertical deflection circuits.

The requirements of vertical deflection circuits are:

- Vertical rolling of picture.
- Height of the raster.
- Suppression of undesired oscillations.
- Correct aspect ratio.
- amplifier configuration
- Vertical deflection coils.
- Vertical linearity
- Internal vertical blanking.

54. Why does AC coupling increase the cost of the receiver?

If AC coupling is used, than DC level of the video signal must be reinserted using reinsertion technique. Since additional circuit is needed at the receiver, and the cost get increased.

55. What is the need for trapezoidal voltage waveform to derive the deflection coil?

To produce a saw tooth current in inductor L produces a rectangular pulse. When the voltage drop across R & L is added together, the result is a trapezoidal waveform. So, to get the saw tooth current in deflection coil, we need to apply the trapezoidal voltage as the input.

56. Mention the reason for DC reinsertion in a TV receiver.

In the output of video detector, DC components are also present. They represent the background brightness of the picture. So they must be preserved for true representation of the picture. If we use DC coupling then the picture is reproduced correctly. Therefore DC reinsertion is needed to preserve the component and maintain the true representation of the picture.

CHAPTER 3

ESSENTIAL OF COLOUR TELEVISION

Key Topics

- ✓ Colour compatibility
- ✓ Colour perception
- ✓ Three colour theory
- ✓ Grass man's law
- ✓ Luminance, Hue and saturation
- ✓ Colour TV – Camera Tubes
- ✓ Different types of colour picture tube
- ✓ Automatic degaussing circuit
- ✓ Gray scale tracking
- ✓ Colour signal transmission
- ✓ Colour burst
- ✓ Weighting factor
- ✓ Formation of chrominance signal

Learning Objectives

After reading this chapter, the learner should be able to:

- ♦ The compatibility and colour perception implies that the colour television signal must produce a normal black and white picture on monochrome receiver without any modification of the receiver circuit.
- ♦ Three colour theory implies that TV system is based on the principle of our eye. Optic nerve system of human eye can be able to separate the colour and recombine them. To get different colours we have to mix the primary colour these are two types: (1) Additive mixing; and (2) Subtractive mixing.
- ♦ Grass mans law is used to find the proportion of each colour that combines with other colour to create the additive effect of new colour.
- ♦ The colour picture has three main characteristics to specify the visual information they are luminance, Hue, and saturation.
- ♦ Chromacity diagram implies a space co-ordination representation of all the spectral colours.
- ♦ Colour TV camera is used to split the received optical images into the three primary colours.
- ♦ Based on the electron gun configuration and phosphor coating, we have three types of picture tubes they are: (1) Delta gun picture tube. (2) Precision-In-Line (P.I.L) and (3) Trintron picture tube.
- ♦ Colour purity, static convergence and dynamic convergence are explained briefly.
- ♦ Automatic degaussing is used to remove magnetic flux from metal parts and it demagnetize iron and steel parts of the picture tube.

3.1 COMPATIBILITY

Colour compatibility implies that the colour TV signal must produce a normal black and white picture on a monochrome TV receiver and the main requirement of colour television was to make it compatible with existing monochrome TV system.

A colour TV receiver must be able to produce a black and white picture using the normal monochrome signal to achieve compatibility. This is referred as "reverse compatibility". The following condition should be satisfied by the colour system:

(i) Bandwidth of colour signal should be equal to monochrome signal.
(ii) The location and spacing of picture and sound carrier frequency should be maintained as 5.5 MHz as in monochrome system.
(iii) The brightness (luminance) information should be same for both monochrome and colour system.
(iv) The composite colour signal must contain colour information and necessary signals to decode it at the receiver.
(v) The colour TV system must use the same deflection frequency and sync signals as used in monochrome TV system.
(vi) The colour information should be carried in a way that it does not affect the picture reproduced on the screen of the monochrome receiver.
(vii) The picture IF and sound IF should remain same for monochrome as well as colour signals. The picture and sound IF carrier frequencies should not be changed.
(viii) The colour signals should be accommodated in the same channel bandwidth 7 MHz this bandwidth should not be increased. The synchronization and blanking signals should be carried as well without any effect of colour signals.

3.2 COLOUR PERCEPTION

All the objects we have seen are focused sharply by the lens system of the eye on its retina. The retina which is located on the back side of our eye has light (optical) sensitive organs which are connected to optic nerve to the optic centre of brain. This can be able to sense the light variations.

As per the Helmhotz theory, the light sensitive organs which measure the visual sensations are of two types called as rods and cones. The rods are sensitive to brightness sensation, thus perceive objects in black and white. So all the colours perceived are identified by the rod as various shades of gray from black to white.

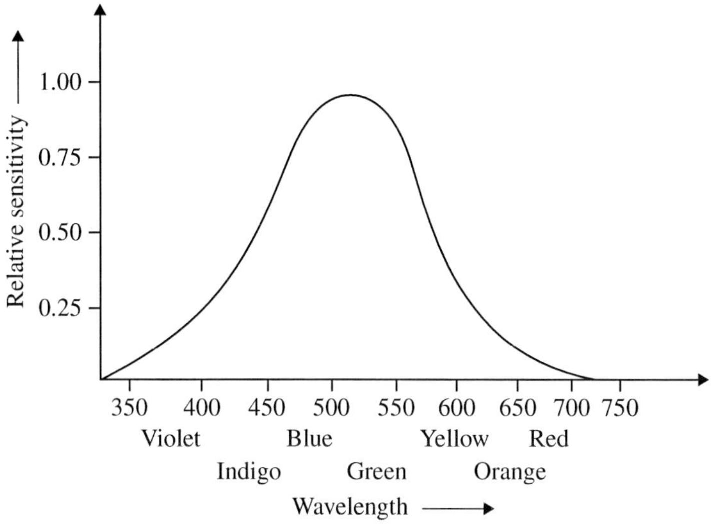

Figure 3.1 Sensitivity of colour.

All colours we see are additive mix of red, green and blue in suitable colour proportions. The cones are used to identify the colour information. At the retina there are three sets of cones, used to identify the 3 primary colours. Accordingly they are called red cone, blue cone and green cone. So all the colour are separated into three primary colours by these cones one set of cones detects the blue colour in the object focused on the retina, the second set perceives red colour and third is sensitive for green colour. The separated colours are transferred to the optical centre of the brain using optical nerves then they are mixed together, to identify the original colour.

The approximate relative response of the eye to different colours is shown in Figure 3.1. It will noted from the different colour wavelength from that we find that green colour has more sensitivity than any other colour. Thus rods and cones enable the optical centre of brain to perceive brightness and colour of any object seen by the eye.

3.3 THREE COLOUR THEORY

In monochrome (black and white) system we are transmitting only the luminous (brightness) information. But in colour system, we have to transmit information about the colours also.

In general by mixing of two or more colour wavelengths produce some other colour regarding to our human eye, all light sources are separated into three colours. Our nerves system of the human eye are able to separate the colour and once again recombine them. All the colour TV system is based on this principle of our eye.

For example, a sky blue colour is formed by mixing the proportion of blue and white colour. The three basic colours are called primary colour, and are: (i) red, (ii) green and (iii) blue.

To get different colour shading, we have to mix these primary colours. There are two types of colour mixing called (a) additive mixing and (b) subtractive mixing.

3.3.1 Additive Mixing

In this method the principle of reflection does not exist. Here the two or three primary colour lights sources is combined together to generate a new different colour. This is by mixing of pure colours, we can generate new colour. Thus with the help of red, green and blue as primary colours, it is possible to generate wide range of colours with additive mixing. If we are mixing all the three primary colours with a particular intensity then we get white colour.

The colours obtained by mixing any two primary colours are called complementary colours.

Primary	Complementary
Red + Green	Yellow
Red + Blue	Magenta
Blue + Green	Cyan

The Figure 3.2(a) shows the method of additive colour mixing. From this method we find that a white colour is obtained by mixing a complementary colour and a primary colour which is not contained in complementary colour. According to Grassman's law, by mixing 30% red, 59% green and 11% blue, we can get white colour.

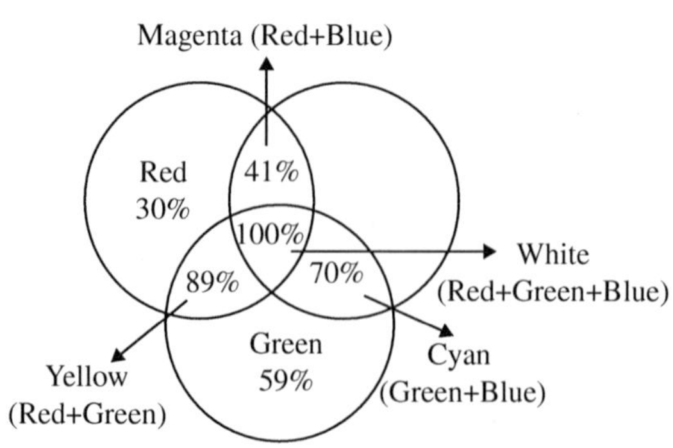

Figure 3.2 (a) Additive mixing.

3.3.1.1 Grassman's Law

The human eye does not have the capability to differentiate the each colour that they mix to form a new colour. But, it perceives only the resultant colour. So it is necessary for colour television system to know the quantum of each colour that combines with other colour to create the additive effect of new colour.

So based on the sensitivity of human eye to various colours, the reference white for colour television system has been chosen to be a mixture of colour light fluxes in the following ratio.

$$\begin{aligned}
100\%\ \text{White} &= 30\%\ \text{red} + 59\%\ \text{green} + 11\%\ \text{blue}\\
100\%\ \text{Yellow} &= 30\%\ \text{red} + 59\%\ \text{green}\\
100\%\ \text{Magenta} &= 30\%\ \text{red} + 11\%\ \text{blue}\\
100\%\ \text{Cyan} &= 59\%\ \text{green} + 11\%\ \text{blue}
\end{aligned}$$

Thus the eye perceives the new colours depending on the algebraic sum of percentage of red, green and blue light fluxes. This forms the basic of colour signal generation and it is known as "Grassman's Law".

3.3.2 Subtractive Mixing

In subtractive mixing two colours are mixed the reflecting properties of colour pigments are used. These two colours absorb all the wavelength but reflect their characteristic colour wavelength. That is when two or more colours are mixed they reflect the wavelength which is common to both. This method of mixing is generally used in colour printing and colour painting by mixing primary colours, we can get black colour as shown in Figure 3.2(b).

Different colours are obtained by subtracting primary and secondary colours from white, so the mixing is called subtractive mixing.

$$\begin{aligned}
\text{Red} &= \text{white}-\text{blue}-\text{green}\\
\text{Blue} &= \text{white}-\text{red}-\text{green}\\
\text{Green} &= \text{white}-\text{blue}-\text{red}\\
\text{Yellow} &= \text{white}-\text{blue}\\
\text{Magenta} &= \text{white}-\text{green}\\
\text{Cyan} &= \text{white}-\text{red}\\
\text{Black} &= \text{red}+\text{green}+\text{blue}
\end{aligned}$$

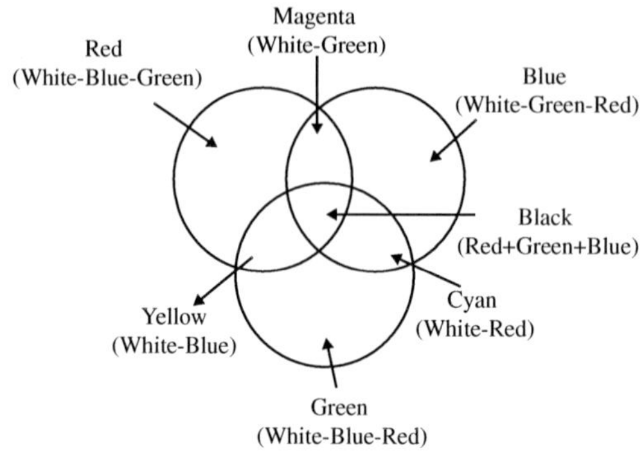

Figure 3.2 (b) Subtractive mixing.

It is noted in Figure 3.2(b) the additive mixing of three primary colours generates white, their subtractive mixing gives the black colour.

For example, to get sky blue, white and blue are mixed. Sky blue reflects the wavelength of white as well as blue. The colours shown are colours perceived by the eye on reflection from that white background.

3.3.3 Chromaticity Diagram

Chromaticity diagram is a convenient graphical representation of all spectral colours and their mixtures.

Figure 3.2(c) shows the two dimensional representation of hue and saturation in the X–Y plane. If the three dimension representation is taken, the 'Z' axis will show relative brightness of the colour.

The chromaticity diagram is formed by all the rainbow colour along a horse shoe shaped triangular curve. The various saturated pure colours are represented along the perimetre of the curve. The corner represent the three primary colours red, green and blue. The central area of the triangular curve represents the white colour.

The white colour lies on the co-ordinates $X = 0.31$ and $Y = 0.32$, the wavelength of the primary colour are as follows:

$$\text{Red} = 700\,\text{nm}$$

$$\text{Green} = 546.1\,\text{nm}$$

$$\text{Blue} = 438.8\,\text{nm}$$

The advantage of this diagram is to determine the result of additive mixing of two or more colours. The 'Z' axis shows the brightness of the picture. If the brightness of the picture increases, the chromaticity diagram will be enlarged.

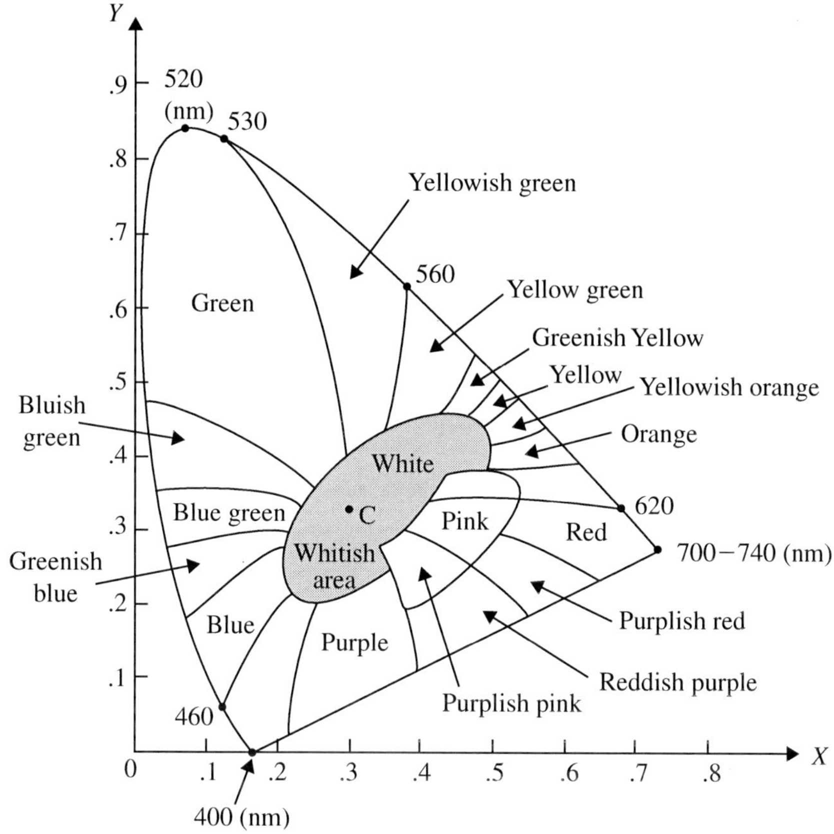

Figure 3.2 (c) Chromaticity diagram. Note that red, green and blue have been standardized at wavelengths of 700, 646.1 and 438.8 nanometers respectively. X and Y denote colour coordinates.

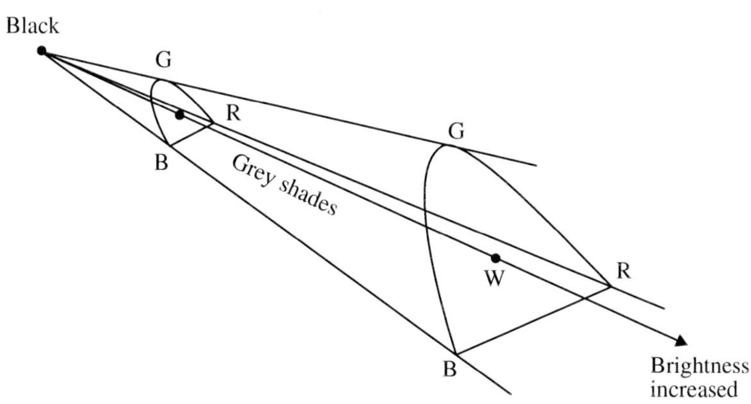

Figure 3.2 (d) Representation of luminance (brightness).

3.4 LUMINANCE, HUE AND SATURATION

The colour picture has three characteristics to specify the visual information. These are:

(i) Luminance or brightness
(ii) Hue or Tint
(iii) Saturation

They are discussed below.

3.4.1 Luminance or Brightness

It denotes the intensity level of light. It is the measure of brightness (light intensity) as received by the eye. Every colour has some luminance. In the black and white pictures only brightness information is observed by the eye and not the colour. If the colour picture is observed, both luminance and colour information is perceived by the eye.

For example, in black and white pictures the brighter area has more luminous than the dark areas. Different colours have different shade of luminous for example, in monochrome TV systems receive colour information, the areas corresponding to red colour appear black and yellow colour areas appear as white it means red has less luminance than yellow.

As shown in Figure 3.1(a) green and yellow has more luminance level compared to violet and red. Thus on monochrome TV system, green and yellow will appear as light grey or white, where as dark red will appear as black. Blue will appear as grey.

3.4.2 Hue or Tint

This is the predominant spectral colour of the received light. Thus the colour of any object is distinguished by its hue or tint. Different hues have different wavelength of spectral radiation. In most of the cases colour is decided from hue. If there are two dominant colours, then hue represents both the colours. For example, green leaves have green hues and red tomatoes have red hue bananas have greenish or yellow hue.

3.4.3 Saturation

This is the spectral purity of the colour light. It is the percentage of white diluted in a colour. Since hue colours rarely alone it indicates the amount of other colours present. The saturation may be indicated as how little the colour is diluted by white. A fully saturated colour has no white. As an example, vivid green is fully saturated and when diluted by white becomes light green.

The hue and saturation information of colour is combined, then it is called 'chrominance' (C) and it does not conation brightness information. Brightness is represented as luminance (Y) signal.

3.5 COLOUR TELEVISION ANTENNAS

The colour television antenna requirements are some what different than close for monochrome receivers. In monochrome receivers antennas, the emphasis on higher gain while it may vary from channel to channel because of the wide frequency range.

Normally manufactures design antennas to deliver more gain on higher frequency channels than on lower channel in order to compensate for higher transmission losses lower receiver sensitivity at very high frequency in such antennas the gain not only vary from channel to channel but also from one end of the channel on the other. In Figure 3.3 the typical response curve of wide range antenna for monochrome channel 4 is illustrated channel 4 is equal to 61–68 MHz.

As shown the Figure 3.3, the gain changed about 4 dB from beginning top end of channel is this antenna is used for colour TV reception from the same channel the colour signal spectrum, which lies around 66.8 MHz will received almost 2 dB less gain that the video carrier and most of its side bands.

In colour television, the relative phase angle of the combined colour difference signal phaser determines the colour in the picture. Any change is

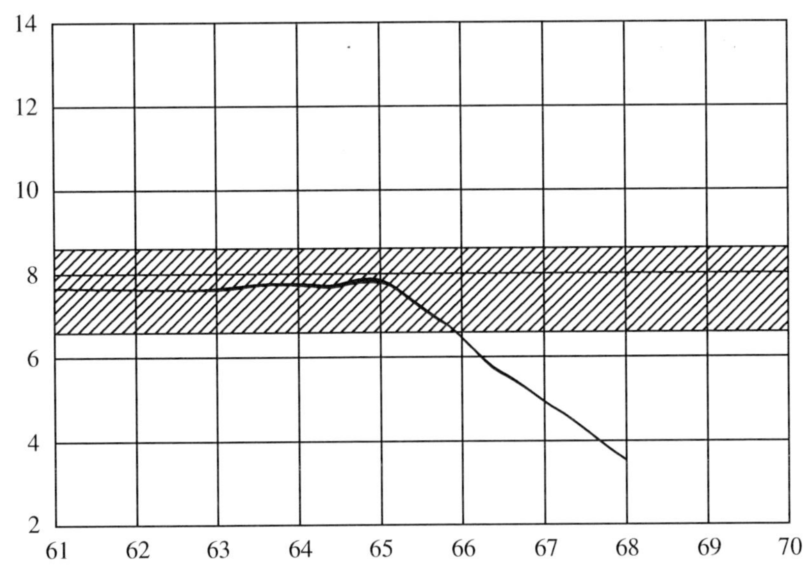

Figure 3.3 Typical response curve of a wide range antenna for channel 4.

gain is accompanied by the phase shift of the incoming signal. Thus change in gain in the region of colour signal spectrum tends to change the hues in the picture. If large phase shift occurs no colours get reproduced.

The most important requirement colour TV antenna is a flat response over the entire channel, as shown in Figure 3.3 the antenna output should not vary by more than 1 dB over the frequency range of any one channel for satisfactory reproduction of colour details.

3.6 VALUE OF LUMINANCE (Y)

In a colour camera, the luminance are Y signal is formed by adding VR, VG, VB in the following proportion

$$Yy = 0.30VR + 0.59VG + 0.11VB$$

Dropping V as usually done,

$$Y = 0.30R + 0.59G + 0.11B \text{ volts.}$$

Note that, these percentages correspond to relative brightness sensitivity of the eye to these primary colours. The matrix for obtaining Y is shown in Figure 3.4. The values of R1, R2, R3 and RC are so chosen that the three colour voltages add up in the ratios of 0.3R : 0.59G : 0.11B to yield $Y = 1$.

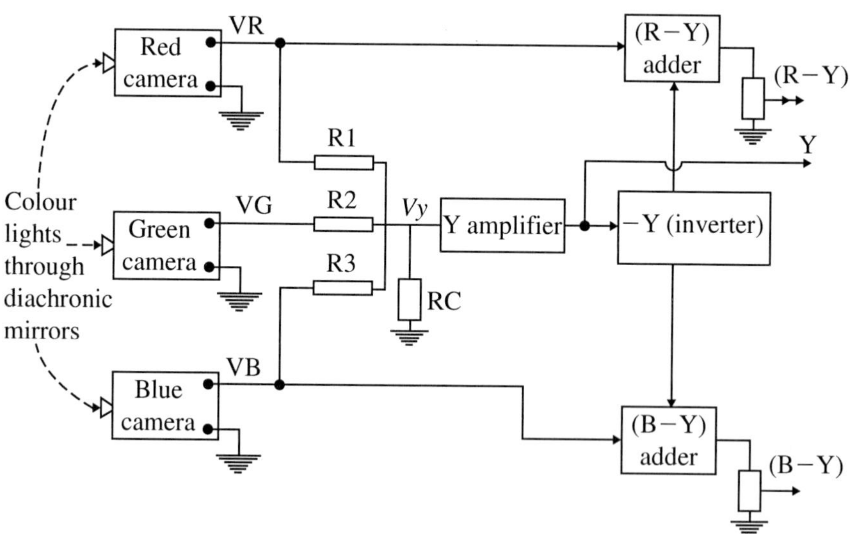

Figure 3.4 Resistance matrix for obtaining Y and circuit blocks colour difference signals.

When $R = G = B = 1\ V$; for bright white,

$$Y = 0.30(1) + 0.59(1) + 0.11(1) = 1\ V.$$

When $R = G = B = 0\ V$; for darkness,

$$Y = 0.30(0) + 0.59(0) + 0.11(0) = 0V.$$

Thus magnitude varies from 100% ($1V$) to zero as light intensity changes from maximum bright to darkness. At 50% brightness *i.e.*, $R = G = B = 0.5V$,

$$Y = 0.30(0.5) + 0.59(0.5) + 0.11(0.5) = 0.5V.$$

Now for a colour object say a painted red and fully illuminated, $Y = 0.30(1) + 0.59(0) + 0.11(0) = 0.3V$. This is so because for red, there is no output from green and blue pick up.

Similarly for magenta $= 0.3(1) + 0.59(0) + 0.11(1)$

$$Y \text{ for magenta} = 0.41V$$
$$Y \text{ for cyan} = 0.3(0) + 0.59(1) + 0.11(1) = 0.7V$$
$$Y \text{ for yellow} = 0.89V$$
$$Y \text{ for green} = 0.59V$$
$$Y \text{ for blue} = 0.11V$$
$$Y \text{ for black} = 0.00V$$

3.6.1 Formation of VR, VG, VB

Colour	Additive mixing of colours	Amplitude for VR in (volts)	Amplitude for VG in (volts)	Amplitude for VB in (volts)
White	Red + Green + Blue	1	1	1
Yellow	Red + Green	1	1	0
Cyan	Blue + Green	0	1	1
Green	Green	0	1	0
Magenta	Red + blue	1	0	1
Red	Red	1	0	0
Blue	Blue	0	0	1
Black	Red-Green-Blue	0	0	0

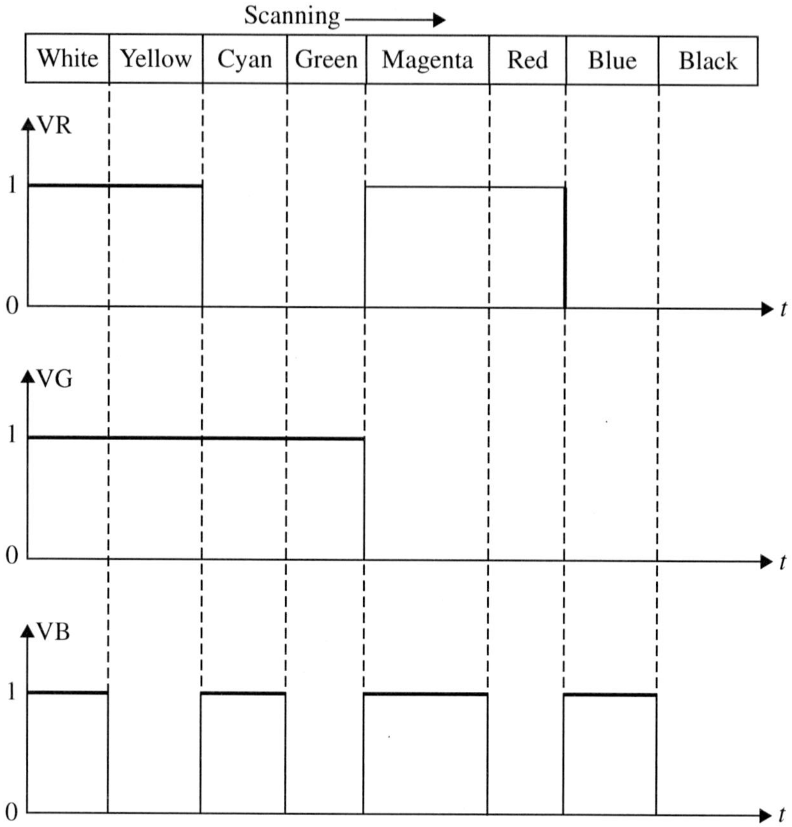

Figure 3.4(a) The magnitude varies from maximum 1 to 0 V. For VR, VB and VG.

3.6.2 Formation of Y signal

Colour	Formation of Y signal $Y = 0.30R + 0.59G + 0.11B$	Y signal amplitude in (volts)
White	$Y = 0.30(1) + 0.59(1) + 0.11(1)$	1
Yellow	$Y = 0.30(1) + 0.59(1) + 0.11(0)$	0.89
Cyan	$Y = 0.30(0) + 0.59(1) + 0.11(1)$	0.70
Green	$Y = 0.30(0) + 0.59(1) + 0.11(0)$	0.59
Magenta	$Y = 0.30(1) + 0.59(0) + 0.11(1)$	0.41
Red	$Y = 0.30(1) + 0.59(0) + 0.11(0)$	0.30
Blue	$Y = 0.30(0) + 0.59(1) + 0.11(0)$	0.11

The staircase patterns for the Y signal for different colours are shown in Figure 3.4(b). The lumination on the bar is varied from maximum

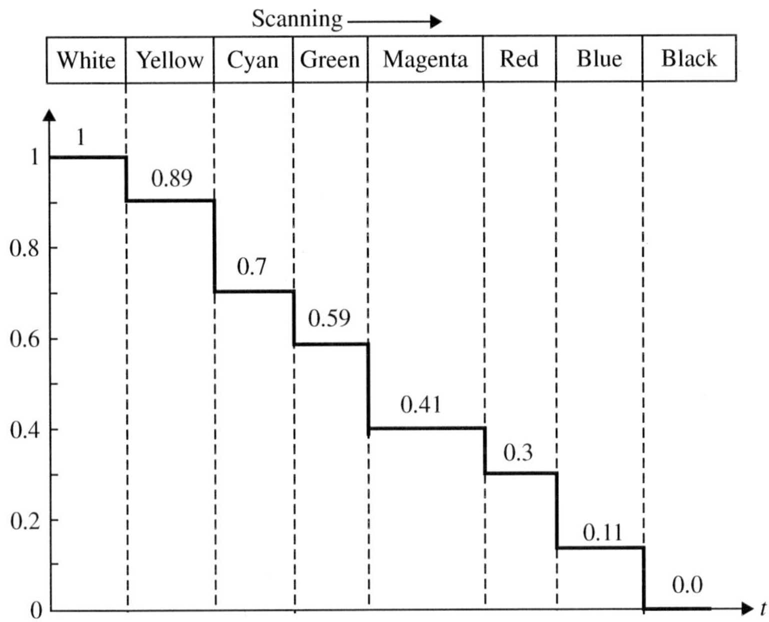

Figure 3.4 (b) The magnitude varies from 1 to 0.0 for Y signal.

brightness to darkness, the magnitude varies from $1V$ (maximum) to $0V$ (minimum), similarly for Red the magnitude of R varies from 0.30 to $0V$. Such a behavior is true for all the colours. Hence the brightness of various colours perceived by the human eye. And the (Y) amplitude is same for monochrome and colour camera under the conditions of illumination.

The simple optical separator causes excessive light loss this problem has been solved by using di-chromic mirror.

3.7 COLOUR DIFFERENCE SIGNALS

The three colour voltages developed by the camera tube cannot be transmitted separately because due to the channel bandwidth limitations. This problem can be solved by generating two colour—difference signals with a carrier frequency called 'colour subcarrier frequency' and are accommodated in the range of 3.5 to 5 MHz of the channel bandwidth where the amplitude of the Y signal is very small. This process is called 'encoding'.

3.7.1 Formation of Colour Difference Signals

The colour difference signal (R–Y) and (B–Y) are shown in Figures 3.4(c) and (d). The Y signal obtained from the Y signal matrix is low in amplitude and so therefore it is amplified and then it is passed to inverter to obtain ($-Y$) signal.

And then $(-Y)$ signal is passed to the adder circuits (as shown in Figure 3.4). In one adder circuit combines the red camera output with $-Y$ to produce (R–Y) signal. And in the second adder combines blue camera output with $-Y$ to produce (B–Y).

The two colour difference signal can be expressed as

$$(R - Y) = R - Y$$

where

$$Y = 0.30R + 0.59G + 0.11B$$

$$(R - Y) = R - (0.30R + 0.59G + 0.11B)$$

$$= R - 0.30R - 0.59G - 0.11B$$

$$(R - Y) = 0.7R - 0.59G - 0.11B \tag{3.1}$$

3.7.1.1 Formation of (R–Y) signal

Colour	**Formation of (R–Y) signal** $(R - Y) = 0.7R - 0.59G - 0.11B$	**(R–Y) signal amplitude in (volts)**
White	$(R - Y) = 0.7(1) - 0.59(1) - 0.11(1)$	0
Yellow	$(R - Y) = 0.7(1) - 0.59(1) - 0.11(0)$	+0.11
Cyan	$(R - Y) = 0.7(0) - 0.59(1) - 0.11(1)$	−0.7
Green	$(R - Y) = 0.7(0) - 0.59(1) - 0.11(0)$	−0.59
Magenta	$(R - Y) = 0.7(1) - 0.59(0) - 0.11(1)$	+0.59
Red	$(R - Y) = 0.7(1) - 0.59(0) - 0.11(0)$	+0.7
Blue	$(R - Y) = 0.7(0) - 0.59(0) - 0.11(1)$	−0.11

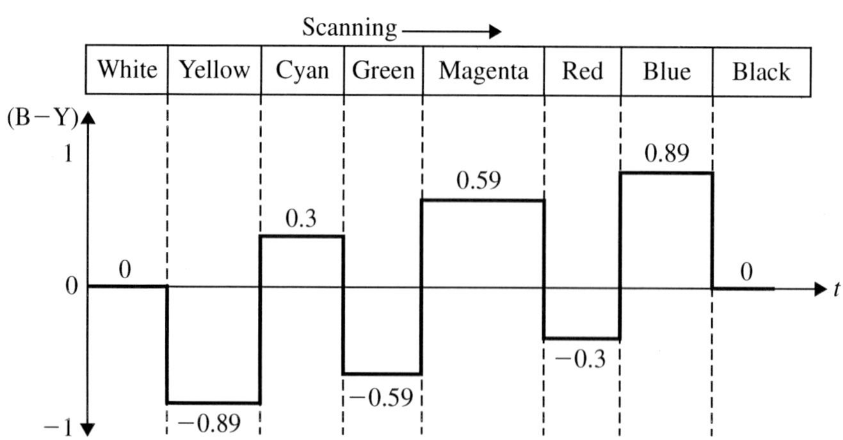

Figure 3.4(c) Formation of (R–Y) signal.

$$(B - Y) = B - Y$$

Substituting Y value in Equation (3.1)

$$(B - Y) = B - (0.30R + 0.59G + 0.11B)$$
$$= B - 0.30R - 0.59G - 0.11B$$
$$(B - Y) = -0.30R - 0.59G + 0.89B \quad (3.2)$$

3.7.1.2 Formation of (B–Y)

Colour	Formation of (B–Y) signal $(B - Y) = -0.30R - 0.59G + 0.89B$	(R–Y) signal amplitude in (volts)
White	$(B - Y) = -0.30(1) - 0.59(1) + 0.89(1)$	0
Yellow	$(B - Y) = -0.30(1) - 0.59(1) + 0.89(0)$	−0.89
Cyan	$(B - Y) = -0.30(0) - 0.59(1) + 0.89(1)$	+0.3
Green	$(B - Y) = -0.30(0) - 0.59(1) + 0.89(0)$	−0.59
Magenta	$(B - Y) = -0.30(1) - 0.59(0) + 0.89(1)$	+0.59
Red	$(B - Y) = -0.30(1) - 0.59(0) + 0.89(0)$	−0.3
Blue	$(B - Y) = -0.30(0) - 0.59(0) + 0.89(1)$	+0.89
Black	$(B - Y) = -0.30(0) - 0.59(0) + 0.89(0)$	0.00

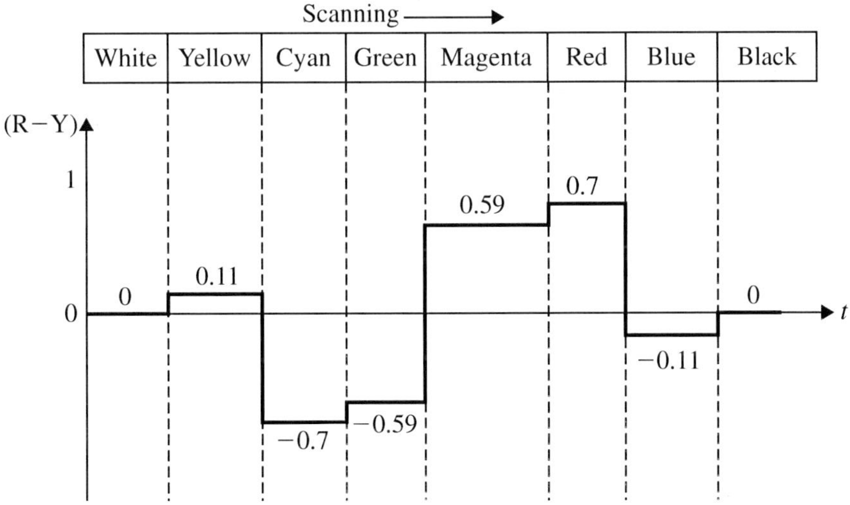

Figure 3.4(d) Formation of (B–Y) signal.

3.7.2 Suitability of (R–Y) and (B–Y) for Transmission

The luminance Y signal

$$0.30R + 0.59G + 0.11B \tag{3.3}$$

By substituting $R = (R - Y)$, $B = (B - Y)$ and $G = (G - Y)$ and rearranging we can write it as:

$$0.30(R - Y) + 0.59(G - Y) + 0.11(B - Y) = 0 \tag{3.4}$$

Therefore

$$\begin{aligned}(G - Y) &= -\frac{0.30}{0.59}(R - Y) - \frac{0.11}{0.59}(B - Y) \\ &= -0.51(R - Y) - 0.186(B - Y)\end{aligned} \tag{3.5}$$

from Equation (3.5) we get (G–Y) from (R–Y) and (B–Y) by placing a suitable resistor matrix at the receiver.

3.7.3 Reason for Not Choosing (G–Y) for Transmission

(i) The proportion of G (59%) in Y (100%) is relatively large and hence the amplitude of (G–Y) is small in comparison with (R–Y) and (B–Y). This causes signal to noise ratio problems at the transmitting end.
(ii) The amplitude of (R–Y) and (B–Y) is small so it becomes necessary to use amplifiers in the matrix network.

3.7.4 Bandwidth of Y Signal

The bandwidth of Y signal is same for monochrome and colour TV systems (*i.e.*, 0 to 5 MHz). The frequency components beyond 3.5 MHz are small in amplitude (colour difference signals) and they are necessary to produce the fine details of the picture. The colour information is interleaved in the region (3.5 to 5 MHz) is to avoid the interference between the two signals [*i.e.*, Y (luminance) and C (chrominance)].

3.7.5 Compatibility with Colour Difference Signals

In order to maintain the compatibility between the monochrome and colour TV system is done by the use of colour difference signals. Also,

(i) For any picture elements having no colour value, the magnitude of R, G, B will be equal.
(ii) During monochrome signal transmission the colour difference signal value becomes zero (*i.e.*, (R–Y) and (B–Y) = 0). And only the Y signal

transmitted. Hence there is no interference from colour difference signal in the reproduction of black and white parts of the picture.

(iii) The colour difference signal (R–Y) and (B–Y) are small in amplitude when compared to primary colours and Y signal. It makes their transmission and processing at the receiver much easier

(iv) The colour difference signal simplifies the design of chroma filter section in the receiver (because due to small in amplitude it reduces the bandwidth).

3.7.6 Encoding of Colour Difference Signal

The formation of chrominance (C) can be obtained by combining the two colour difference signal (R–Y) and (B–Y) as shown in Figure 3.19.

The colour difference signals (R–Y) and (B–Y) are modulated by using a single subcarrier frequency for the modulation of chrominance signal the technique employed here is quadrature modulation.

3.8 COLOUR TELEVISION DISPLAY TUBES

The basic arrangement of a colour camera using diachronic mirrors is shown in Figure 3.5.

3.8.1 Principle and Construction

It is essentially composed of three monochrome camera tube systems, in which each camera tubes receives selectively filtered colours to produce an electrical signal proportional to the respective colour. Light from the scene processed by the objective lens is directed to mirrors. That splits the light into three basic primary colours R, G and B.

The mirrors are located with dichroic material so that they reflect a specific colour and other spectral frequencies to pass through.

3.8.2 Working Principle

The first diachronic mirror reflects blue light but passes the remainder. The front silvered mirror M1 reflects it on the target plate of the blue pick up tube through a relay lens. The light that passes the first diachronic mirror impinges on a second similar mirror.

Here the red colour component is reflected and remaining passes through. This being the green light component passes on to the green picks up tube through corresponding relay lens.

The red light falls on the target plate of red picture tube on reflection from the silvered mirror M2 thus reflected light from the scene gets split its R, G, B components to develop proportionate voltages at the vidicon tubes.

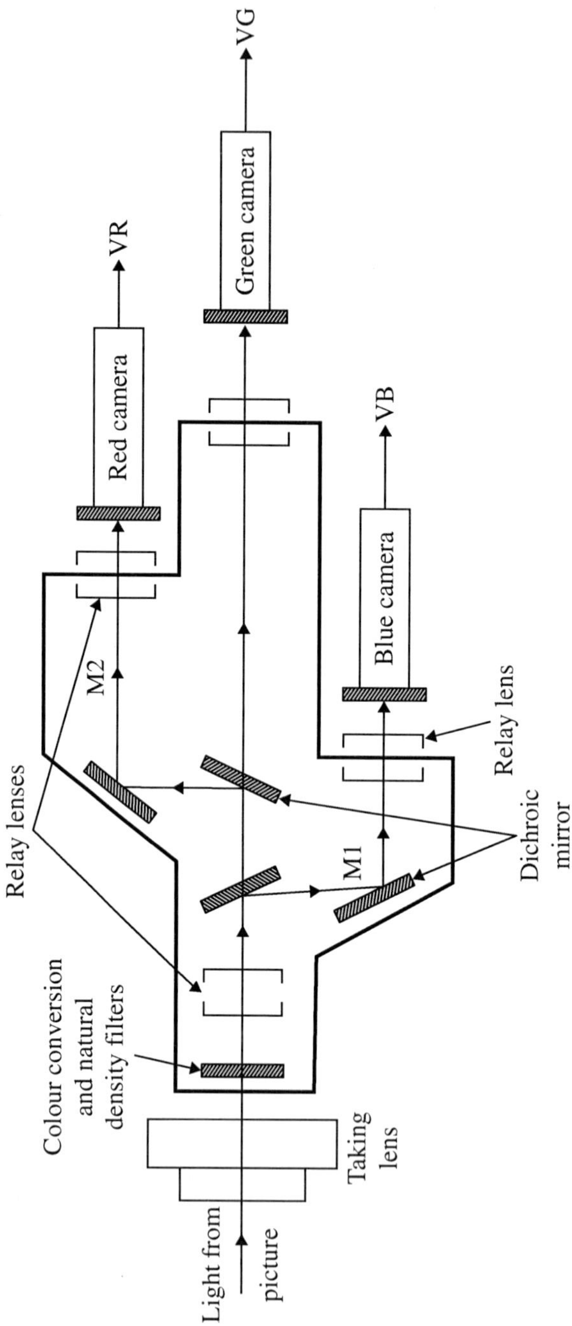

Figure 3.5 Basic colour camera arrangement using diachronic mirror.

The scanning occurs in the usual way and the video signals VR, VG and VB of carrying intensities become available. These are processed suitably before modulation and transmission.

3.9 SINGLE TUBE COLOUR CAMERA

It is possible to use one pick up tube instead of three and yet develop VR, VG, VB separately. In one such camera Hitachi of Japan, a striped target is used instead of conventional uniform target. The face of the vidicon is bonded with a filter having three stripes of red, green and blue those are separated from each other. The target of the pick up tube also has three sections each sections is located in such a way that it is just behind and in parallel with the corresponding colour filter selection. Thus when the camera tube beam scans the three section target, red, green and blue video outputs are developed depending on the intensity of light falling on them.

3.10 COLOUR PICTURE TUBE

The colour picture tube screen area is coated with small dots of three different colour phosphorus. There are three independent electron beams for red, green and blue. The electron beam hit their respective phosphorus generate red, blue, green lights. These red, blue and green light emitting dots are very close to each other. Their light mix additively to produce the required colour for that pixel. The construction of colour picture tube is little different compared to monochrome picture tubes. The three colour phosphor dots can be arranged in various ways. Depending on the electron gun configuration and phosphor coating, we have following three types of colour picture tubes:

(i) Delta gun picture tube.
(ii) Precision-in-Line (P.I.L) or gun-in-line picture tube.
(iii) Trintron (three-in-line cathode) picture tube.

3.10.1 Delta Gun Colour Picture Tube

Radio Corporation of America was first developed this picture tube as the name implies, this tube employs, three separate guns, one for each colour, in this type the colour phosphor dots are arranged in triangular (Δ) form and hence its called 'delta gun picture tube'. There are 1×10^6 dots on the screen divided in three colours. A perforated metal is placed behind the screen the construction details of delta gun picture tube is shown in Figure 3.6.

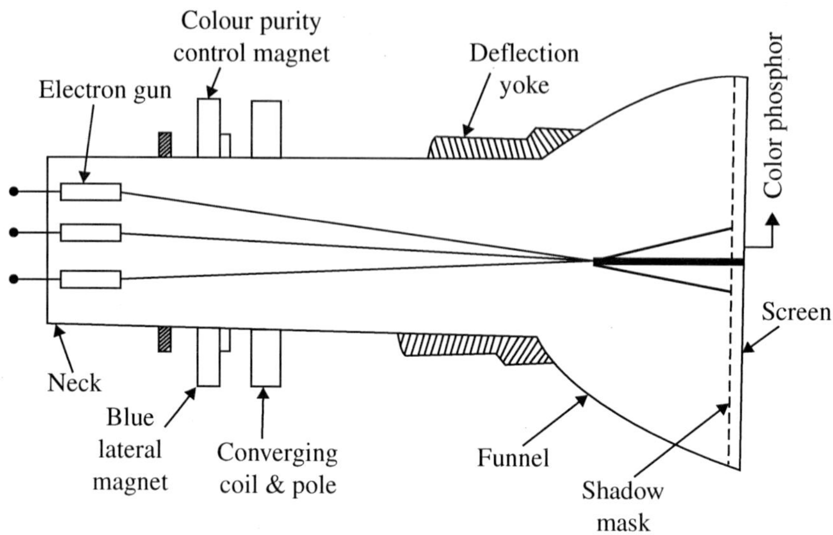

Figure 3.6 Construction of delta gun tube.

3.10.1.1 *Electron Gun Section*

Shadow mask picture tube employs three separate guns one for each phosphor Figure 3.7 shows this arrangement the guns are equally spaced at 120° interval with respected to each other. And tilted towards to the centre line of the picture tube. They form an equilateral triangle (Δ).

Three independent electron beams for each primary colour is coming out of three guns. Each gun has a heater filament, cathode, control grid and accelerating grid, focusing and convergence electrodes are also present separately for each gun. So that individual colour signal can be coupled to the grid cathode circuit to modulate the beam intensity. Each gun had separate screen grid G2, the voltage at which is adjusted for the desired cut off characteristic of the gun.

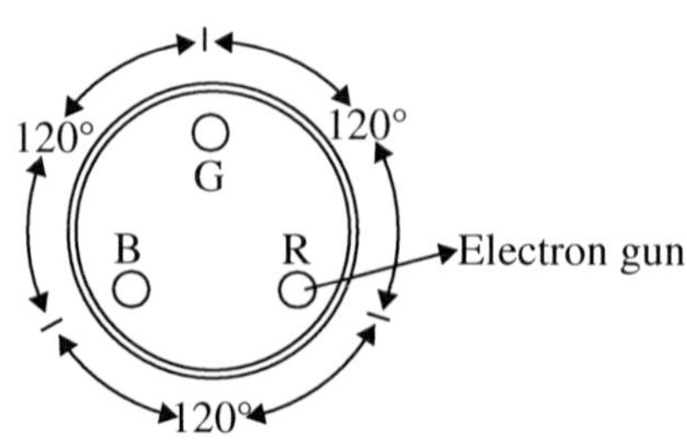

Figure 3.7 Gun structure.

3.10.1.2 Shadow Mask Picture Tube

As shown in Figure 3.8 the tube employs a screen where three phosphor dots are arranged in a groups are known as triads (Δ). Each phosphor dots corresponds to one of the primary colours. Triads are repeated and depending upon the size of the picture tube approximately 1,000,000 such dot forming nearly 333,000 triads are deposited on the glass face plate. About one cm behind the tube screens a thin perforated metal sheet known as 'shadow mask'. The mask has hole for every phosphor dot triad on the screen.

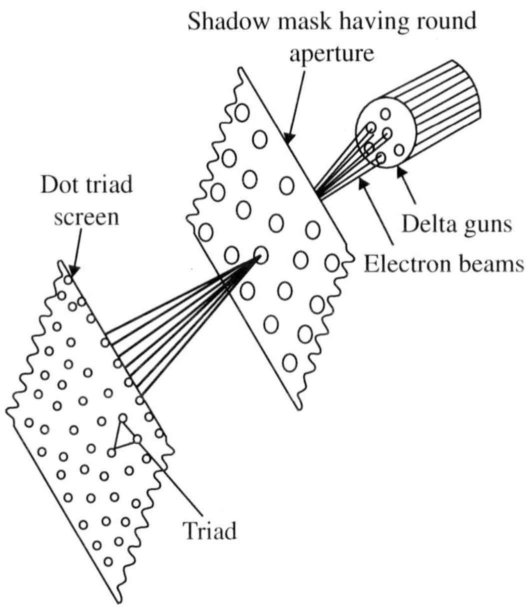

Figure 3.8 Shadow mask and triad structure.

3.10.1.3 Operating Principle

The various holes are so oriented that electrons of the three beams passing through any one hole will hit only the corresponding colour phosphor dots on the screen. The beams are blanked or masked out while moving between the aperture.

Most of the part of electron beams is blocked by the shadow mask since the mask is conductive; it samples this beam and returns to the ground. The holes only allow to pass the electron beam to screen. Only 20% of electron beam passes to screen through the mask. Therefore to maintain the required brightness, the beam current is increased. To achieve this, a higher final anode voltage is used. Thus all colour picture tubes use high final anode voltages compared to monochrome picture tubes.

If the electron beams are not deflected properly then the beam carrying red colour signal may hit blue phosphor dot on the screen. Because of this, reproduced colours are incorrect. Such problems occurs due to incorrect convergence.

3.10.1.4 Disadvantages of Delta Gun Picture Tube

(a) The convergence problems are very difficult to recover. Very complex arrangements are required to make the entire electron beam incident on their corresponding phosphor dots at every part of the screen.
(b) The focus is not sharp over the complete screen (about 20 fit only).
(c) Since the electron transparency of the shadow mask is low, lot of energy is wasted in the TV receiver.

3.10.2 Precision-in-Line Colour Picture Tube

The precision-in-line (P.I.L) colour picture tube, the name itself denotes that it has three guns which are aligned precisely in a horizontal line. The gun and mask structure of the P.I.L tube are illustrated in Figure 3.9. The inline gun configuration helps in simplifying convergence arrangements. The phosphor is arranged in vertical strips on the screen in the form of vertical strips in triads (R, G, B) and are repeated along the breadth of the tube. There is a continuous strip from the bottom of the screen for each colour. They are placed one after another on the screen. A transparent shadow mask is placed in front of the screen.

One vertical line of slots for one group of fine strips of red, green and blue phosphor since all the three electrons beam on the same plane, being in the centre (green) moves along the axis of the tube however, because of inward tilt of the right and left guns the blue and red beams travel at an angle and meet the centre beam at the aperture grille mask. The slots in the mask are so designed that each beam strikes its own phosphor and is prevented from landing on other colour phosphors.

The PIL tube carries with it deflection yoke magnet assembly for purity and convergence these are placed on the neck of the picture tube the PIL tube is a more efficient. And it has higher electron transparency, but it needs fewer convergence adjustments.

3.10.2.1 Advantages of Precision in Line Colour Picture Tube

(a) Since the shadow mask is slotted, the electron transparency is better than delta gun therefore efficiency of PIL tube is better.

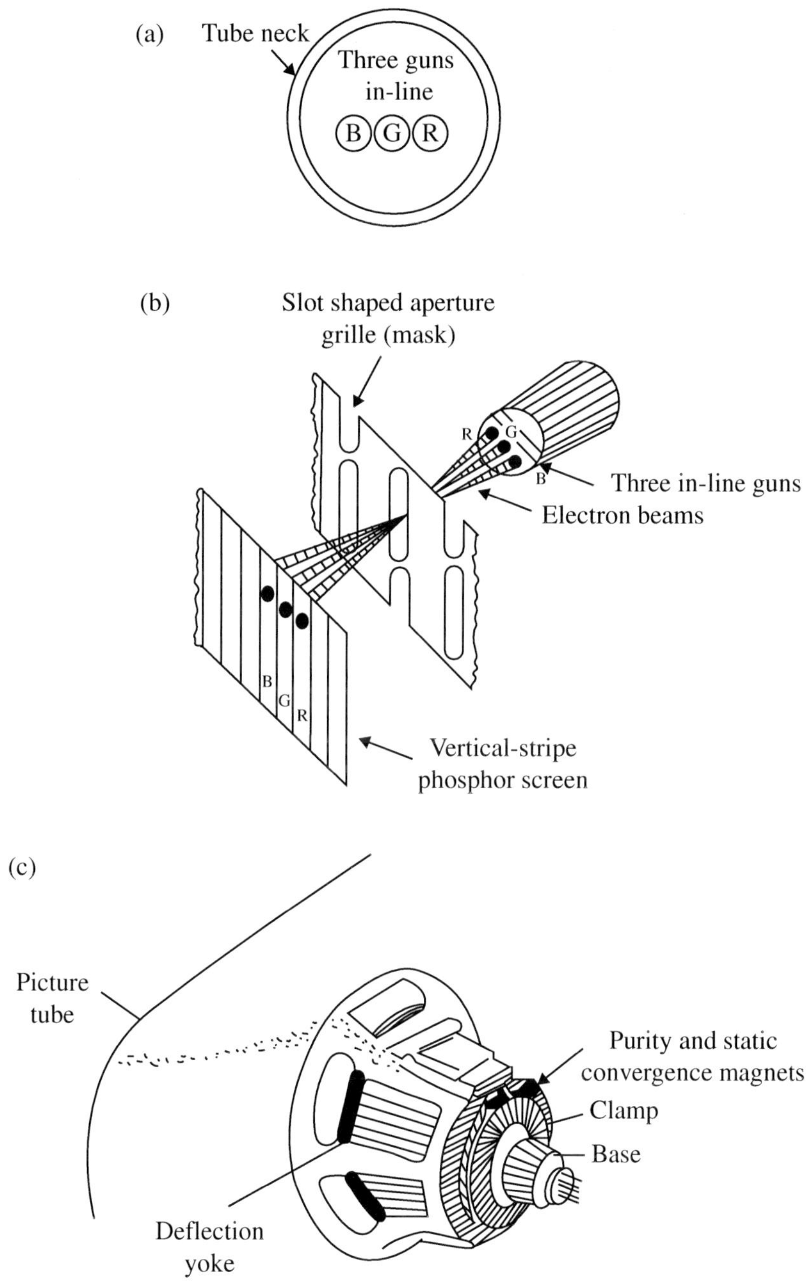

Figure 3.9 Gun and mask structure of P.I.L tube. (a) In-line gun, (b) Electron beam, aperture grille and striped three colour phosphor screen, and (c) Yoke and magnet assembly.

(b) Because of the in-line guns the convergence adjustments becomes simple.
(c) The manufacturing of the screen becomes simple.

3.10.3 Trintron Colour Picture Tube

The Trintron are or three in line cathodes colour picture tube was developed by ‘Sony’ Corporation of Japan. The main feature of in-line gun with vertical phosphor stripes is that the convergence and purity are simplified first all the beams are in the same plane. Then it is necessary to adjust the outside beams left and right with respect to the centre beam. Also the vertical stripes makes this type less sensitive to beam landing errors caused by the earths magnetic field.

3.10.3.1 Electron Gun Section

The electron gun assembly of the Trintron is illustrated in Figure 3.10. The tube has only single gun structure with three cathodes for red, green and blue beams. The control grid G1 is a single cup with three apertures for three beams. The screen grid G2 is common one for initial focus and acceleration focus grid is used for all the three beams. The electrostatic convergence is achieved by four convergence plate as shown in Figure 3.10.

3.10.3.2 Convergence Arrangement

The green beam passes through the centre plate while the other two beams between the outer plates both the centre plates connected internally to anode. The red and green beams passing between the outer plates can be moved horizontally for convergence adjustment by varying the voltage to the outermost plates. The voltage for outer convergence plate is about 450V less than the anode voltage.

3.10.3.3 Phosphor Screen

The phosphor screen is in the form of vertical stripes as shown in Figure 3.10(b). Instead of 3 dot delta structure shadow mask tubes. The aperture grille similar to the shadow mask instead of holes, the grille one vertical slots for each set of red, green and blue stripes. The green is at centre directly behind each slot. Because of slotted aperture grille the beam current reaching the phosphors is increased which increases the brightness capability of three tubes. The line in being and vertical stripe arrangement simplifies and eliminates many of the in-convergence problems. Instead of passing through separate smaller lenses it passes through smaller portion

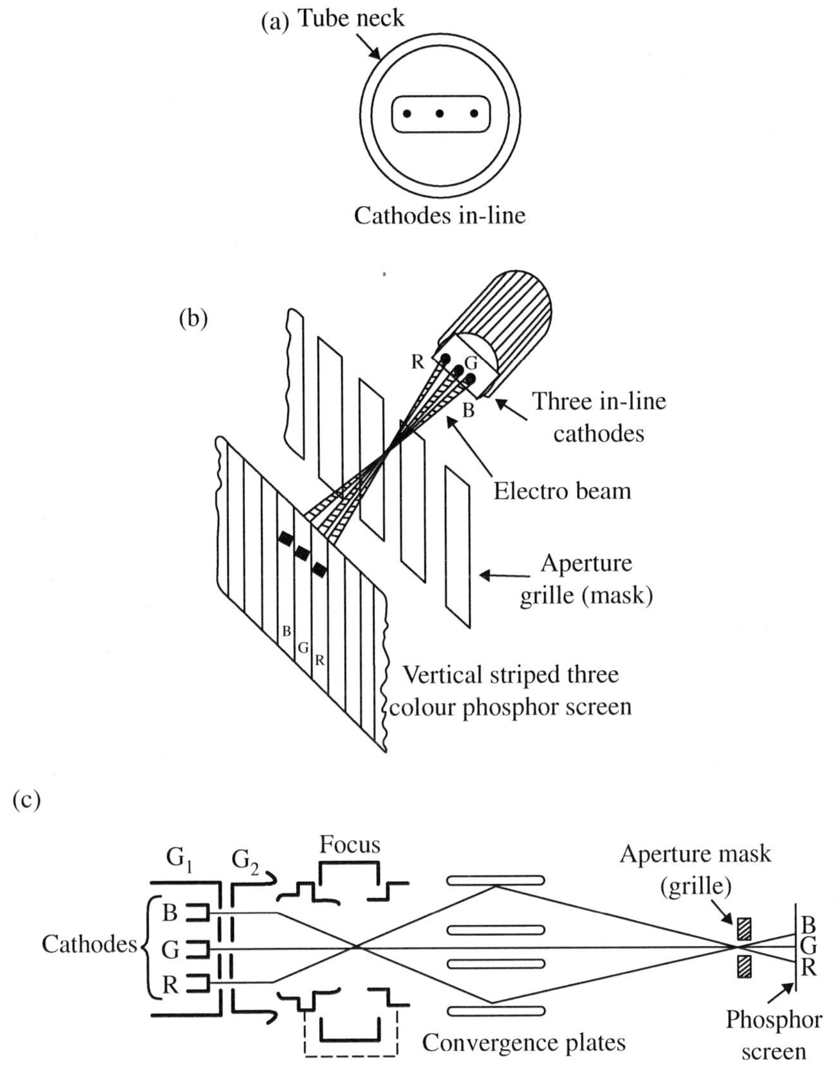

Figure 3.10 Gun and mask structure of Trintron picture tube.

of a common bigger lens. This reduces spherical abbreviations, giving a sharp picture.

3.10.3.4 Advantages of Trintron Colour Picture Tube

(a) The grill is slotted. Therefore its electron transparency is more brightness hence the efficiency of the tube increases.
(b) A convergence becomes simple because of inline beams.
(c) Brightness is more vertical misconvergence is impossible.

3.11 COLOUR PURITY

For exact reproduction of colour, the effective source of the beams must coincide exactly with the position of the light source (RGB dots in the screen). During the manufacture of tube, otherwise one or more of the beam will spill over on to the adjacent phosphors producing a wrong colour for the beam.

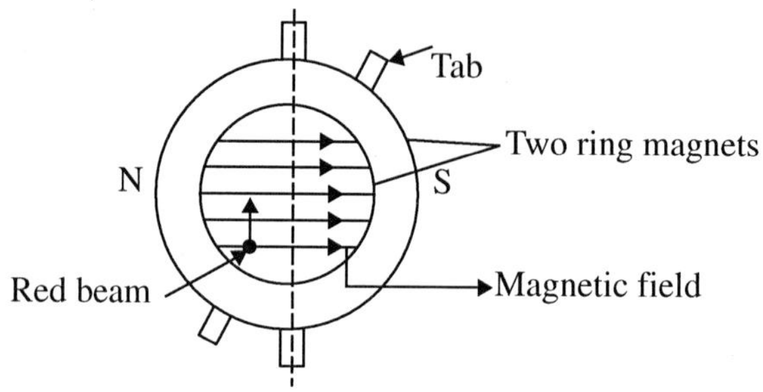

Figure 3.11 Colour purity.

This is prevented by a pair of pole magnetic ring provided with a deflection yoke on the neck of the tube. They are rotated to shift all the three beams relative to axis of the tube. The field direction and orientation can be adjusted by relative movement of two rings. The three beams can be moved in any direction together relative to the axis of the tube to give necessary purity at the centre of the screen.

The entire deflection yoke is then moved and set axially to establish coincidence between the deflection centre on the original dot fixing source position on the screen with these setting, purity is achieved over the entire screen.

3.12 COLOUR CONVERGENCE

The three electron beams, controlled independently to bring together. It is necessary to maintain three colour images precisely, so that they overlay all the points on the phosphor screen. The technique of bringing the beam together so that they hit the same part of the screen at the same time to produce three coincidence raster's is called as convergence. Convergence errors are caused by:

(i) Non-coincidence convergence plane.
(ii) Non-uniformity of deflection fields.
(iii) Flat surface of the picture tube.

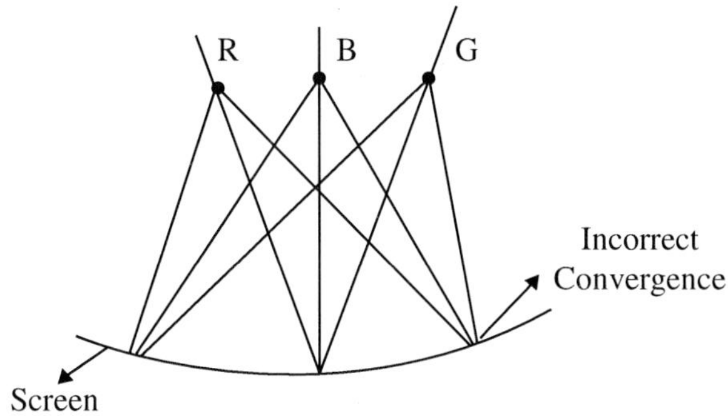

Figure 3.12 Colour convergence.

Proper convergence is achieved by positional adjustment of the individual beams it falls into two parts: (i) static convergence and (ii) dynamic convergence.

3.12.1 Static Convergence

For static convergence two sets of multi pole ring magnets are arranged on the yoke towards neck of the picture tube they produce well defined and adjustable magnetic field in the region of path of the individual electron beams. A combination of 4 pole and 6 pole ring magnets is used to set up such fields this is illustrate in Figures 3.13 and 3.14.

The 4-pole field moves the outer electron beam (red and blue) equally in opposite directions, the 6-pole field moves the outer electron beams equally in same direction. The centre (green) beam is unaffected.

3.12.1.1 Raster Symmetry

Raster symmetry is adjusted by varying the field strength of 2-pole magnet located at the near of the multi-pole unit. All the three beams are equally moved in vertical direction as shown in Figure 3.14(a).

3.12.1.2 Horizontal Colour Purity

Horizontal colour purity is obtained by varying the magnetic field strength of the 2 pole magnet situated between the 4-pole and 6-pole magnets. The three electron beams move equally in the horizontal direction as shown in Figure 3.14(b).

3.12.2 Dynamic Convergence

Dynamic convergence is necessary because of flat screen of the picture tubes. Since the outer beams are off that tube axis horizontally, the point of

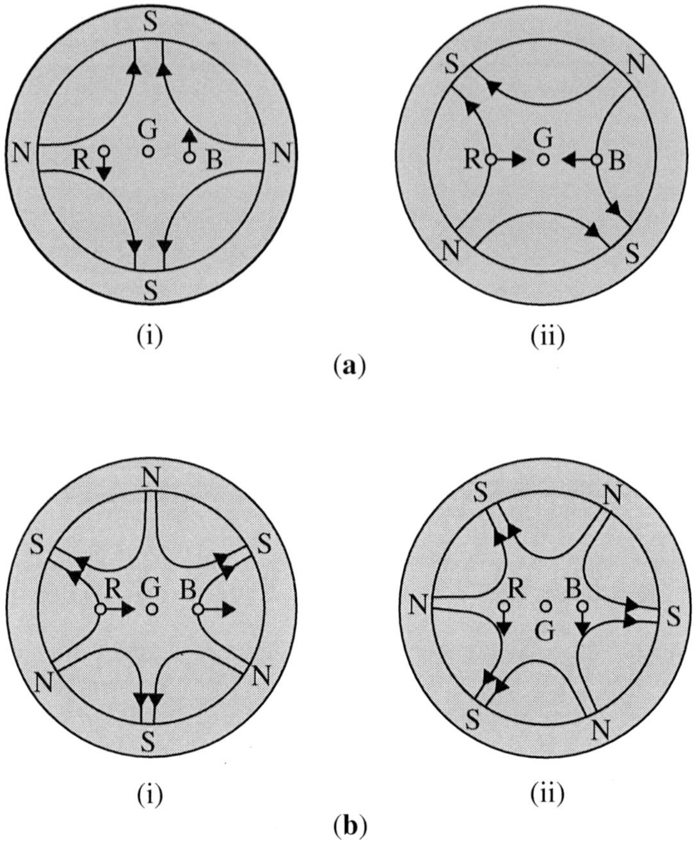

Figure 3.13 Static convergence. (a) The outer electron beams (R and B) moves equally opposite directions, (b) The outer electron beams move equally in the same direction.

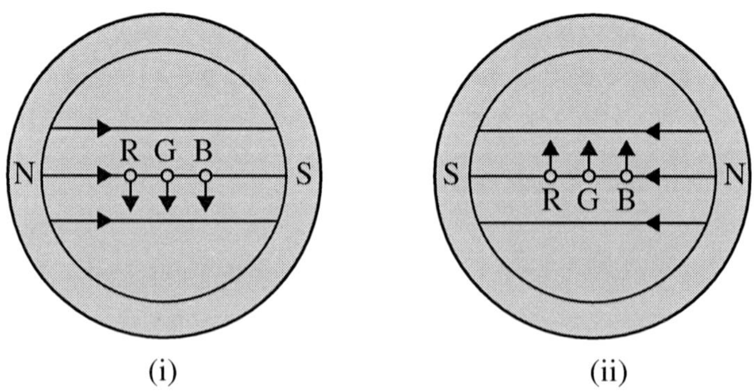

Figure 3.14 (a) Raster (horizontal axis) symmetry. All the three beams move equally in the vertical direction.

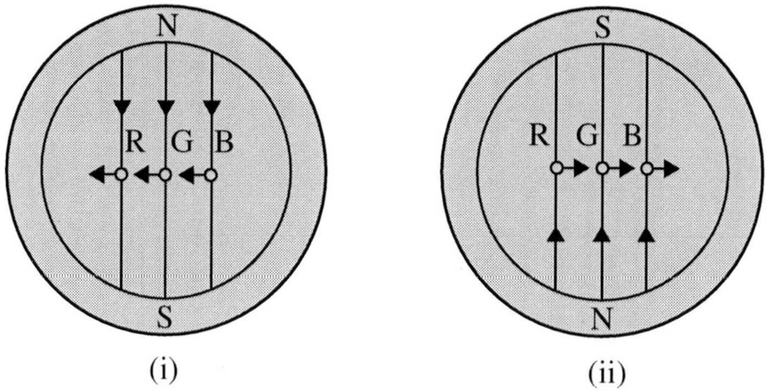

Figure 3.14 (b) Horizontal colour purity. All the three beams move equally in the horizontal direction.

convergence falls progressively short in the flat screen as the deflection is increased. Parabolic waveforms from the line time base circuits are fed to the convergence coil. Dynamic convergence is achieved by continuously varying magnetic field. The instantaneous strength of the field depends upon the position of the spot on the screen.

3.13 AUTOMATIC DEGAUSSING

Any external magnetic field can deviate the path of electron beams causing errors in beam landing and hence affects the purity of colours. This is can be due to magnetic objects toys, loudspeaker, *etc.* Even the earth magnetic field can upsets the beams to affect colour purity modern picture tubes have internal magnetic shields to prevent the external magnetic field from entering inside.

The mask and mounting frame may get magnetised by external source. To overcome this problem, the shield and mask can be provided with automatic degaussing system the degaussing means demagnetizing iron and steel parts of the picture tube mountings. It is used to remove magnetic flux from the metal parts.

3.13.1 Working Principle

By this method the picture tube is demagnetized automatically when the receiver is turned on each time. A pair of degaussing coils is built into the tube at the top, bottom and side of the screen.

Figure 3.15(a) shows the automatic degaussing circuit. It uses a thermistor and a varistor for controlling the flow of AC current through the degaussing coil. Whenever the receiver is turned on the AC voltage drop

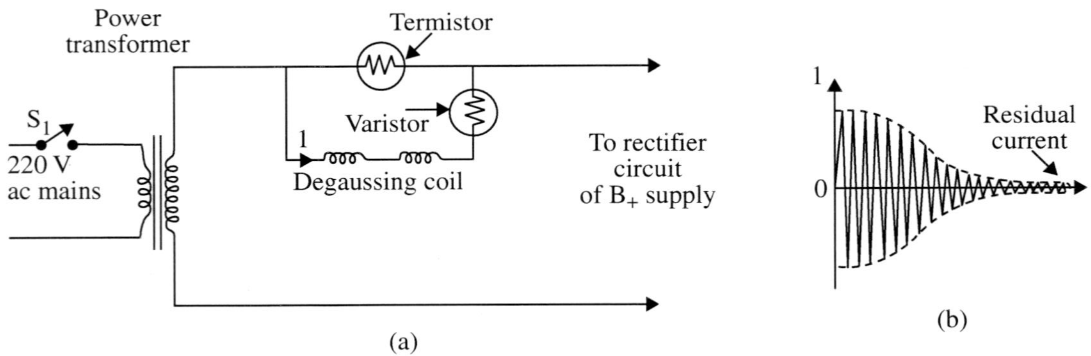

Figure 3.15 Automatic degaussing (ADG). (a) Typical circuit, (b) Variation of current in the degaussing coil when receiver is just switched on.

across the thermistor is very high (about 60 V), and this causes a large current to flow through the degaussing coil. Because of its large current thermistor heats up and its resistance decreases and the voltage drop across it decreases. As a result, voltage across the varistor decreases thereby increasing its resistance value. This will reduce the AC current through the coil to a very low value. The current components are chosen that initial current the degaussing coil is close to 4 amperes and drops to about 25 mA in less than a second as shown in Figure 3.15(b).

Once the thermistor heats up degaussing ends and normal as voltage is restored to the $B+$ rectifier circuit.

3.14 PINCUSHION DISTORTION

This is caused by the flatness of the screen and its rectangular shape. The distance between the gun and the screen goes on increasing as the beam moves towards the corners of the screen. Permanent magnet will rise purity problems in colour picture tube. A dynamic correction method is employed to eliminate the problem. This correction automatically increases the horizontal width and vertical size of the raster due to the pincushion distortion. Figure 3.16 shows the raster with and without distortion.

3.14.1 Vertical Pincushion Distortion

These magnets give the electron beam and extra horizontal or vertical deflection field. Due to this extra fields the curved side's of the raster are pulled straight. As shown in Figure 3.16 often only two magnets fitted

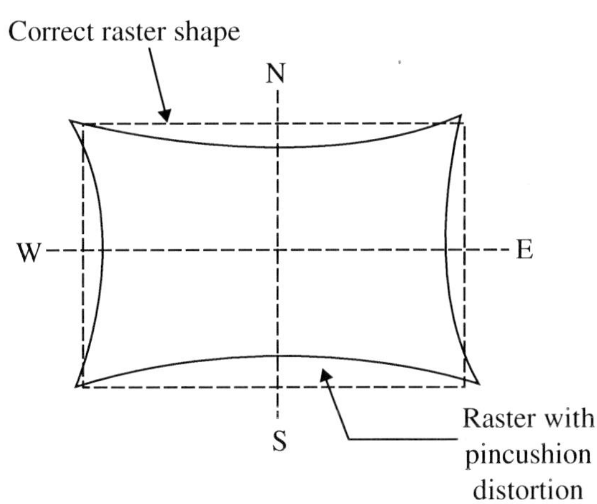

Figure 3.16 Pincushion distortion.

on the short sight of the cone of the tube are used which collect through vertical pincushion distortion.

3.14.2 Dynamic Pincushion Correction

Dynamic pincushion correction is necessary in colure picture tubes. Figure 3.16 is sketch of raster with pincushion distortion a necessary correction is achieved by introducing some cross modulation between the two deflection fields. Such a correction automatically increases the horizontal width and vertical width in those regions of the raster that are shrunken.

3.14.3 E–W Correction (Horizontal Pincushning)

To correct E–W correction (Horizontal Pincushning) in the horizontal deflection saw tooth current must be amplitude modulated at a vertical rate, so that when the electron beam is at the top or bottom of the raster, the horizontal amplitude is minimum and when its is centre of the vertical deflection interval the horizontal saw tooth current amplitude is maximum. To achieve this parabolic voltage is obtained by integrating vertical saw tooth voltage (network R1C1 is inserted as shown in Figure 3.17) is inserted in series with the DC supply to the horizontal deflection circuits.

As a result, amplitude of individual cycles of the 15,625 Hz horizontal output varies in step with a series connected 50 Hz parabolic voltage.

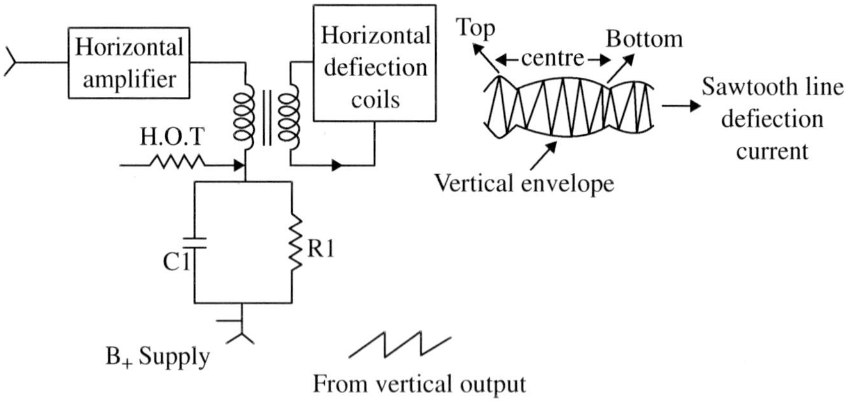

Figure 3.17 Horizontal (E–W) correction.

3.14.4 N–S Correction

To correct N–S Correction (Vertical Pincushning) in the vertical deflection saw tooth current must be pulsate in amplitude at the horizontal scanning rate so that when the electron beam is at during top or bottom scanning of the raster. A parabolic waveform at the horizontal rate is superimposed on vertical defection saw tooth current in fact these increase the vertical size during the time the beam is moving to the mid point of its horizontal scan rate. This is shown in Figure 3.18. The parabolic waveform at the top of the raster is of opposite polarity at the bottom since the raster stretch required at the top is opposite to that needed at the bottom. The amplitude of parabolic waveform required for top and bottom pincushion correction decreases to zero as vertical deflection passes through the centre of the raster.

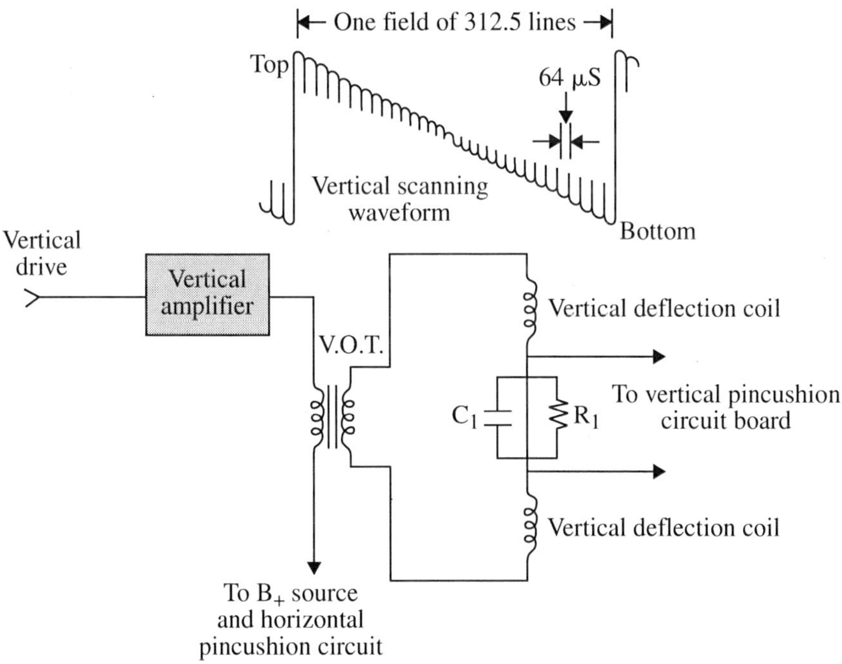

Figure 3.18 (N–S) Vertical corrections.

3.15 GRAY SCALE TRACKING

The red, green and blue lights must be combining in definite proportions to produce white light. The three phosphor have different efficiency. Also three guns may not have identical I_P/V_{gk} characteristic and cut points.

Therefore, it becomes necessary to incorporate suitable adjustments such that monochrome information is reproduced correctly (with no colour tint) for all settings of the contrast control. In practice these amounts to two distinct steps:

(i) To bring cut off points of the guns in coincidence, three cutoff control potentiometers located in the base (emitter or) leads to three video amplifier transistor are set to ensure no colour tint on the screen.

(ii) Similarly the drive control potentiometer in the input leads of green, red, blue video amplifiers are varied to obtain no colour tint and optimum reproduction of white light on the screen area.

3.16 COLOUR SIGNAL TRANSMISSION

In colour TV system we have to send the details about the colour of the scene to be transmitted so we have to send both luminance signal (Y) for monochrome and brightness, chrominance signal (C) for colour information here two colour difference signals are added together to from the chrominance signal to meet the compatibility with monochrome system we are employing two techniques called (i) Frequency interleaving and (ii) Quadrature modulation.

3.16.1 Quadrature Modulation

The two separate balanced modulators are used one for (R–Y) and other for (B–Y). The carrier frequency generated by the subcarrier oscillator is

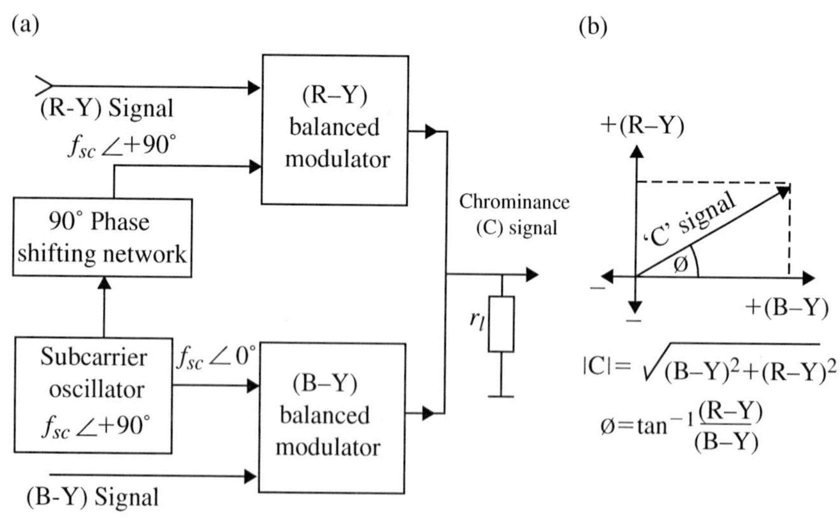

Figure 3.19 Quadrature modulation block diagram.

fed to +90° phase shift network and then it is fed to the (R–Y) balanced modulator. The carrier frequency is fed to the (B–Y) balanced modulator without any phase shift so therefore colour subcarrier fed to the (B–Y) is termed as $f_{sc}\angle 0°$ phase shift and (R–Y) modulator can be termed as $f_{sc}\angle 90°$ phase shift. These two carrier obtained are said to be quadrature (90° out of phase) and termed as 'quadrature modulation'.

Balanced modulators are used to suppress the two carrier components because their amplitudes are large when compared with the side bands and if it is not suppressed it will cause severe interference to 'Y' signal.

The two carrier suppressed modulation products are combined to get chrominance signal or chroma signal and then it is frequency interleaved with 'Y' signal to obtain colour difference signal.

3.16.2 Frequency Interleaving

The problem of transmitting colour difference signal (R–Y) and (B–Y) within the same channel bandwidth to meet the compatibility requirements has been solved by combing colour information into a single variable (C) by quadrature modulation. By employing this technique to transmit the colour information with the monochrome signal without any interference is called 'frequency interleaving'.

3.16.2.1 Operation

The camera electron beam scans the picture at the line frequency at the rate of 15,625 Hz and also the frequency components around the each line frequency harmonics are separated by field frequency of 50 Hz. Because of this frequency of the video signal increases the energy content of that frequency goes on reducing. So therefore the Y signal is kept at a bandwidth of 5 MHz and it is located in bundles or clusters with gap between them as shown in Figure 3.20(a). The unused space can be occupied by another signal without any interference.

The unused space can be accommodated with the chrominance signal (as shown in Figure 3.20(c)). The colour subcarrier frequency is chosen to be an odd multiple of half of the line frequency f_H, and then the unused space of Y signal can be occupied by the side bands of colour subcarrier. By this technique there will be no interference between the luminance (Y) and chrominance (C).

The other considerations for the location of colour subcarrier are as follows:

(a) Locating the chrominance signal (C) near the higher end of Y signal. So that the interference between the chrominance and luminance is minimum.

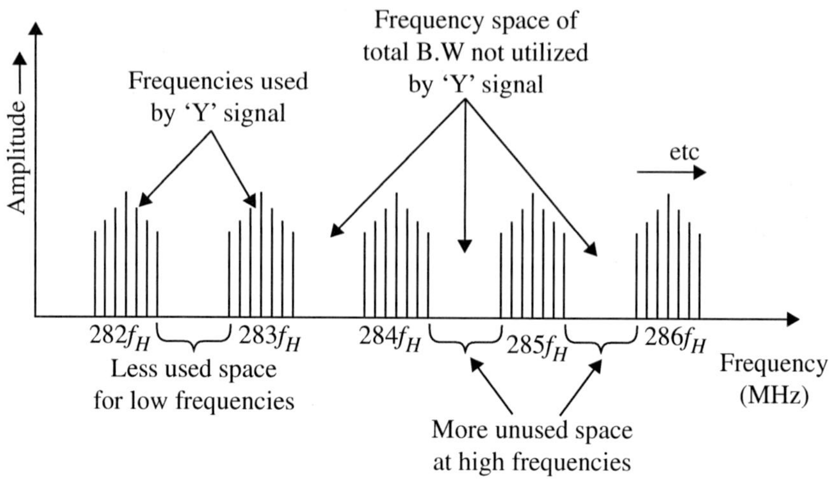

Figure 3.20 (a) Composition of video information at multiplies of line frequency.

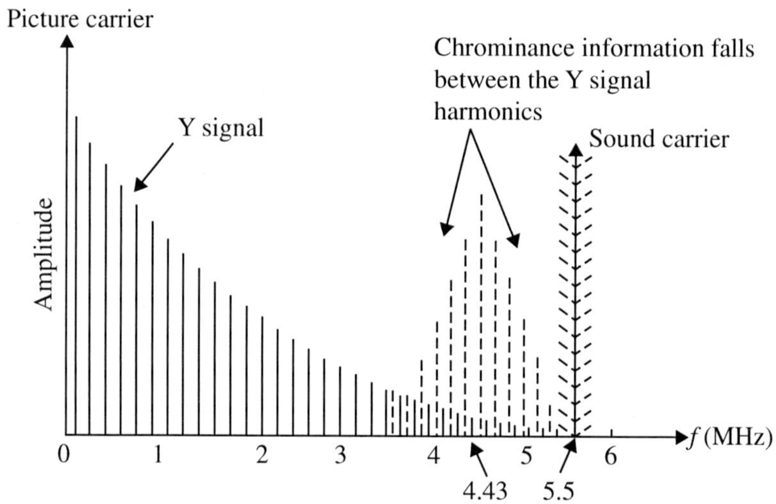

Figure 3.20 (b) Frequency interleaving.

(b) On the higher end of Y signal the energy content of luminance signal is minimum.

(c) Dot pattern suppression.

(d) For the standard PAL – subsystem the frequency of the colour – subcarrier is fixed at

$$f_{sc} = 15{,}625\,\text{Hz} \times \left(284 - \frac{1}{4}\right) \times 50\,\text{Hz} = 4.43\,\text{MHz}.$$

where $f_L = 15{,}625$ Hz and $f_V = 50$ Hz.

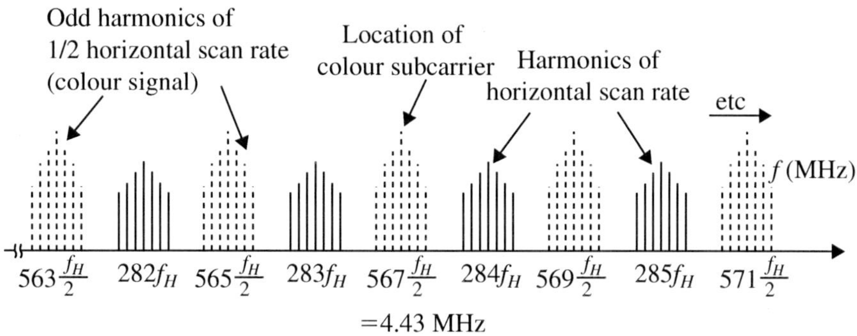

Figure 3.20(c) Interleaving of colour signal.

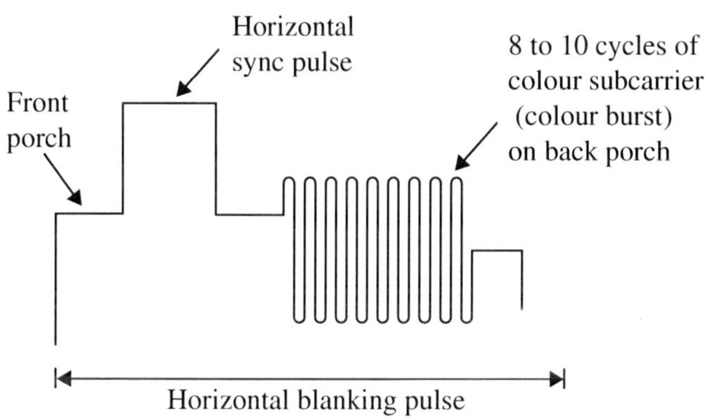

Figure 3.20(d) Location of colour burst on the back porch of each horizontal sync pulse.

This technique of combing the two distinct signals with the same bandwidth as shown in the figure is called 'frequency interleaving'. The combined signal is modulated with picture carrier and transmitted (as shown in Figure 3.20(b)).

At the receiver side, after demodulation the two signals are separated and processed to obtain Y and colour difference signals.

3.16.2.2 Colour Burst

The magnitude of 'C' signal represents the saturation of colour and phase of the 'C' signal represents the hue of the colour. Since the two carrier components are suppressed in Balanced Modulator (BM).

So therefore it is necessary to regenerate the colour—subcarrier at the receiver for demodulation (detection). The regeneration of colour subcarrier is not exactly same with that of phase relationship with the carrier at

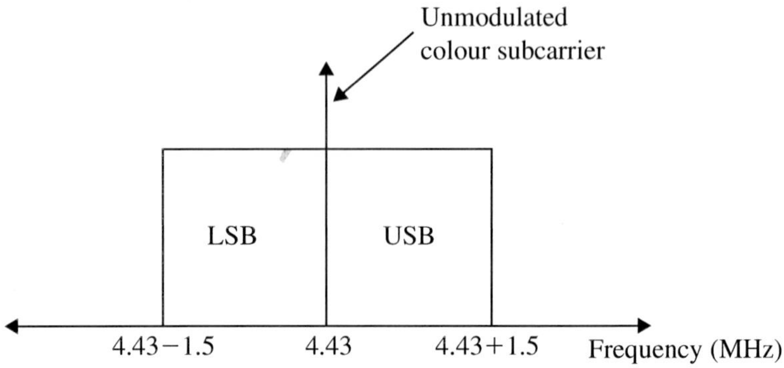

Figure 3.20(e) Frequency spectrum.

the transmitter. Therefore it results in change in the reproduction colour at the receiver side.

In order to avoid this phase difference at the receiver side, 8 to 10 cycles of the colour subcarrier oscillator output at the encoder are transmitted along with sync pulses. The sample of colour subcarrier is called 'Colour Burst', is placed at the back porch of each horizontal blanking pulse pedestal as shown in Figure 3.20(d).

At the receiver side the 'colour burst' is separated and it is used to lock the colour subcarrier oscillator with that at the transmitting end. The colour burst on the back porch of sync pulse does not interference the triggering of the horizontal oscillator.

3.17 BANDWIDTH MODULATION OF COLOUR DIFFERENCE SIGNALS

Initially monochrome transmission was used with the channel bandwidth of 7 MHz the luminance signal in monochrome transmission occupy 0 to 5 MHz produce very fine details. But for the chrominance signals this much bandwidth is not needed, because the very fine colour details in a picture is perceived by the eye only as luminance signal. This is shown in Figure 3.21. When a larger colour area is scanned, the corresponding video frequency from the camera tube output is up to 0.5 MHz.

When a medium size colour area is scanned, the video frequency at the camera tube output from (0.5–1.5 MHz). When a very small colour area is scanned, the corresponding video frequency is 1.5 to 5 MHz. But this very fine details (colour) is seen by our eye only as brightness variations not as a colour. Therefore, it is enough that the video frequencies for colour information is up to 1.5 MHz. Thus for the colour objects generating frequencies more than 1.5 MHz eye cannot perceive colour information only

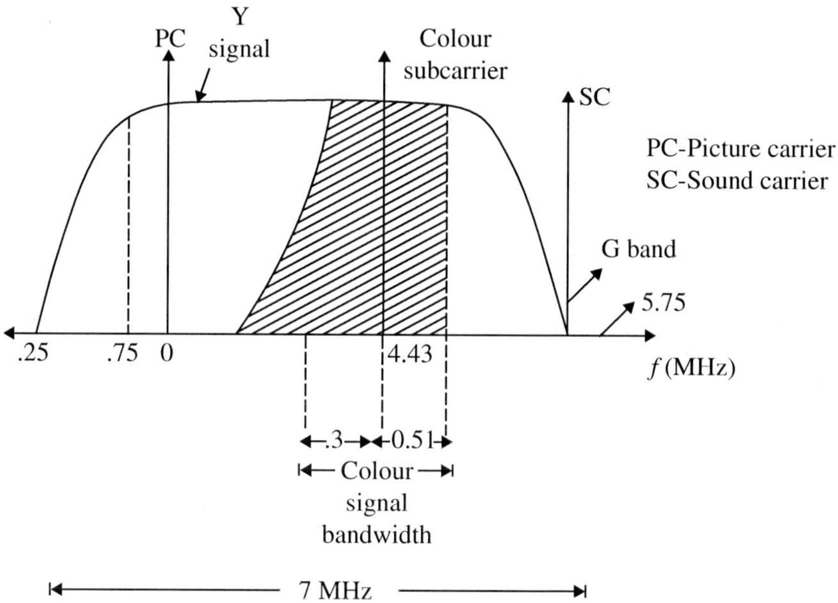

Figure 3.21 Bandwidth for colour signal transmission.

the brightness is identified. Therefore, for maximum bandwidth, necessary for colour signal transmission is around 3 MHz (±1.5 MHz).

3.18 WEIGHTING FACTOR

In the formation of Colour flexed Composite Video Signal (CCVS) the modulated signal amplitude exceed 100% modulation level this is called over modulation. This will produce distortion in the receiver while producing the colours. To prevent this weighting factors are added with (R–Y) and (B–Y) colour difference signal. (R–Y) the weighting factor is 0.877 therefore weighted (R–Y) signal is equal to 0.877× (R–Y) for (B–Y) the weighting factor is 0.493. The weighted (B–Y) signal is 0.493 × (B–Y) by doing this we can avoid over modulation in the transmitter. This is shown in Figure 3.22.

$$Y = 0.30R + 0.59G + 0.11B$$

$$(R - Y) = (1R - 0.3R) + 0.59(G) + 0.11(B)$$

$$(R - Y) = 0.7R - 0.59G - 0.11B$$

$$(B - Y) = (1B - 0.11B) + (0.59G + 0.30R)$$

$$(B - Y) = 0.89B - 0.30R - 0.59G$$

For yellow

$$(R - Y) = 0.7(R) - 0.59(G) - 0.11(B)$$
$$= 0.7(1) - 0.59(1) \qquad \text{where } R = 1,\ G = 1\ \&\ B = 0$$
$$= 0.11$$
$$(B - Y) = 0.89(B) - 0.30(R) - 0.59(G)$$
$$= 0.30(1) - 0.59(1)$$
$$= -0.89$$
$$\text{weighted } (R - Y) = 0.877 \times (R - Y)$$
$$= 0.877 \times 0.11$$
$$= 0.0964$$
$$\text{weighted } (B - Y) = 0.493 \times (B - Y)$$
$$= 0.493 \times -0.89$$
$$= -0.43$$

Therefore

$$C \text{ signal (weighted)} = \sqrt{\text{weighted } (R - Y)^2 + \text{weighted } (B - Y)^2}$$
$$= \sqrt{(0.0964)^2 + (-0.43)^2}$$
$$= 0.44$$

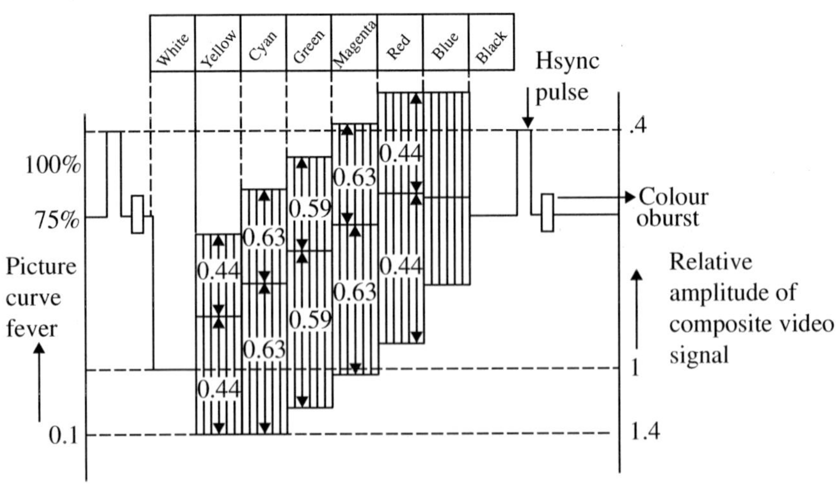

Figure 3.22 Weighting factor diagram.

For cyan

$$(R - Y) = 0.7(R) - 0.59(G) - 0.11(B)$$

$$R - Y = 0.59(1) - 0.11(1) \qquad \text{where } R = 0,\ G = 1 \ \&\ B = 1$$

$$= -0.7$$

$$(B - Y) = 0.89(B) - 0.30(R) - 0.59(G)$$

$$= 0.89 - 0.59$$

$$= 0.3$$

$$\text{weighted } (R - Y) = 0.877 \times (R - Y)$$

$$= 0.877 \times -0.7$$

$$= -0.6139$$

$$\text{weighted } (B - Y) = 0.493 \times (B - Y)$$

$$= 0.493 \times 0.3$$

$$= 0.1479$$

Therefore

$$C \text{ signal} = \sqrt{\text{weighted } (R - Y)^2 + \text{weighted } (B - Y)^2}$$

$$= \sqrt{(-0.6139)^2 + (0.1479)^2}$$

$$= 0.631$$

Table 3.1. Weighting factor for colour difference signals.

Colour	Luminance signal (Y)	Unweighted B–Y	Unweighted R–Y	Unweighted G–Y	Unweighted C_{sc}	Weighted B–Y	Weighted R–Y	Weighted C_{sc}
White	1	0	0	0	0	0	0	0
Yellow	0.89	−0.89	+0.11	+0.11	0.9	−0.4385	+0.096	0.44
Cyan	0.7	+0.3	−0.7	+0.3	0.76	+0.148	−0.614	0.63
Green	0.59	−0.59	−0.59	+0.41	0.83	−0.29	−0.517	0.59
Magenta	0.41	+0.59	+0.59	−0.41	0.83	+0.29	+0.517	0.59
Red	0.3	−0.3	+0.7	−0.3	0.76	−0.148	+0.614	0.63
Blue	0.11	+0.89	−0.11	−0.11	0.9	+0.4388	−0.096	0.44
Black	0	0	0	0	0	0	0	0

$$C_{sc} = \sqrt{(B - Y)^2 + (R - Y)^2}$$

$(B - Y)$ weighted $= 0.493(B - Y)$ unweighted

$(R - Y)$ weighted $= 0.877(R - Y)$ unweighted

Thus the weighted colour difference signal is added at the top to produce the C signal. This signal is Modulated using colour subcarrier at 4.43 MHz to produce modulated chrominance signal. This signal is added with Y signal to produce a colour plexed composite video signal as shown in Figure 3.22. From the Figure 3.22 if it is clear that there is no over modulation thus the weighting factors are used to avoid over modulation. Hence exact colour will be produced in the receiver.

3.19 FORMATION OF CHROMINANCE SIGNAL

A chrominance (C) signal gives the information about the hue and saturation of a colour signal. In colour transmission we are using the two colour difference signals (R–Y) and (B–Y) instead of using all the three primary colours and these two colour difference signals are quadrature modulated and 90° phase shift is maintained between them. By vector addition method, these difference signals are added to get the chrominance (C) signal. In this amplitude of C represent colour saturation and phase angle represent the hue.

The amplitude of 'C' its location depends on the magnitude and polarity of two colour difference signals. Maximum amplitude corresponds to maximum (100%) saturation and zero amplitude means no saturation (*i.e.*, no white). Similarly phase angle of 'C' have an angle between 0° to 360° depends on the polarity of (R–Y) and (B–Y) determines the hue of the colour.

Let's calculate the luminance level for other colours:

$$\text{Red:} \quad Y = 0.30R + 0.59G + 0.11B$$

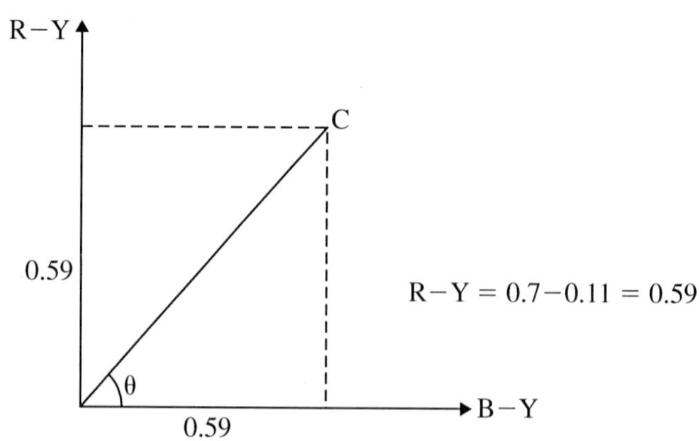

Figure 3.23 (a) Phasor diagram.

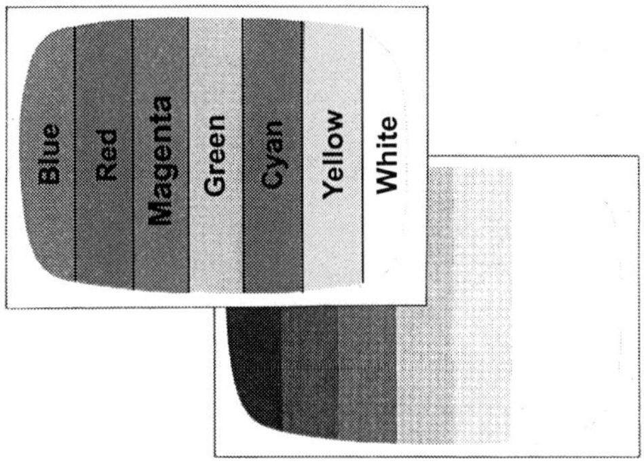

Figure 3.23(b) Colour bar pattern with colours.

For only red colour $G = B = 0$ and $R = 1$ (maximum red)

$$Y = 0.30(1) + 0.59(0) + 0.11(0)$$

$$Y = 0.30 \text{ for red}$$

Similarly $Y = 0.59$ for only green colour, and $Y = 0.11$ for only blue colour. For yellow colour is combination of red and green.

Hence $R = G = 1$ and $B = 0$

$$Y = 0.30(1) + 0.59(1) + 0.11(0)$$

$$Y = 0.3 + 0.59 + 0$$

$$= 0.89$$

For magenta is combination of red and blue.

Hence $R = B = 1$ and $G = 0$

$$Y = 0.30(1) + 0.59(0) + 0.11(1)$$

$$Y = 0.30 + 0 + 0.11$$

$$= 0.41$$

For cyan is combination of green and blue.

Hence $G = B = 1$ and $R = 0$

$$Y = 0.30(0) + 0.59(1) + 0.11(1)$$

$$Y = 0 + 0.59 + 0.11$$

$$= 0.7$$

Now let us calculate the chrominance signal separately for each colour in the bar pattern. The colour difference signals are formed by the equation represented in Table 3.1.

3.19.1 Red Colour

For red, green and blue are absent $R = 1$ and $G = B = 0$, $(R - Y) = 0.7$ and $(B - Y) = -0.3$

$$\begin{aligned} C(t) &= \sqrt{(R-Y)^2 + (B-Y)^2} \\ &= \sqrt{(0.7)^2 + (-0.3)^2} \\ &= 0.76 \end{aligned}$$

and

$$\theta = \tan^{-1}\left(\frac{0.76}{-0.3}\right) = -66^\circ$$

After modulation of subcarrier the modulated subcarrier will be in bipolar (envelope in amplitude). This signal is equal to + or −0.76 Figure 3.24(a) shows the chrominance diagram.

3.19.2 Green colour

For green colour red and blue are absent $G = 1$ and $R = B = 0$, $(R - Y) = -0.59$ and $(B - Y) = -0.59$

$$\begin{aligned} C(t) &= \sqrt{(R-Y)^2 + (B-Y)^2} \\ &= \sqrt{(-0.59)^2 + (-0.59)^2} \\ &= 0.83 \end{aligned}$$

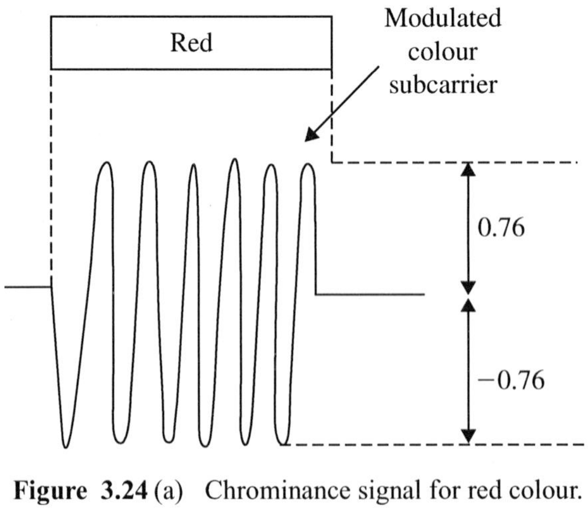

Figure 3.24 (a) Chrominance signal for red colour.

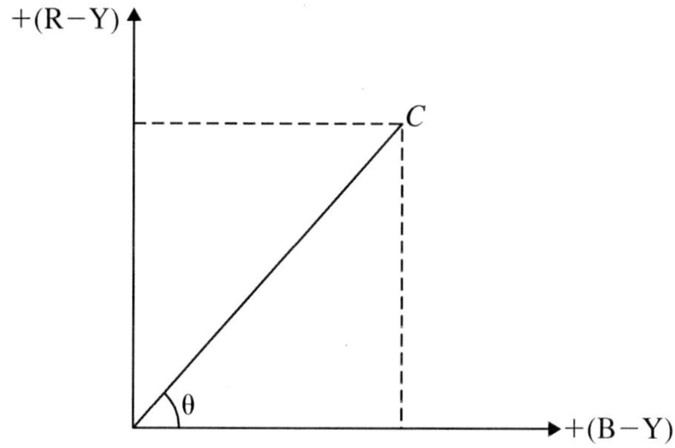

Figure 3.24(b) Phasor diagram for $C(t)$.

and

$$\theta = \tan^{-1}\left(\frac{-0.59}{-0.59}\right) = 45^\circ$$

The chrominance signal shown in Figure 3.24(b) may be little attenuated to avoid over modulation of channel carrier. The phase change in modulated colour subcarrier is not visible for each hue because of very small change.

3.20 DEGAUSSING OF PICTURE TUBES

Localised magnetic fields affect the electron beam (deviation). Due to deviation colour purity is affected. Therefore magnetic shielding is provided. When the location of the TV receiver is changed stray field changes affect the purity. To overcome these drawbacks degaussing is provided. Terrestrial magnetism affects the electron beam. This also necessary the degaussing.

KEY TERMS AND CONCEPTS

Chrominance	The combination Hue and saturation is known as 'chrominance'.
Luminance	It is the amount of light intensity as perceived by eye regardness of the colour.
Hue	It is the predominant spectral colour of the received light. The colour of any object is distinguished by its hue or tint.

Saturation	It is the spectral purity of the colour light.
Primary colours	The basic red, green, blue colours are called 'primary colours'.
Grassman's law	It is used to find out the proportion of each primary colour that combines with the other colour to create the additive effect of new colour.
Additive mixing	Two or three primary colour lights sources is combined together to generate a new different colour. In this method the principle of reflection does not exist.
Subtractive mixing	In subtractive mixing two colours are mixed. The reflecting properties of colour pigments are used.
Chromacity diagram	It implies a graphical representation of all the spectral colours and their mixtures based on the tristimulus values of the primary colour.
Rods and cones	The rods are sensitive to brightness sensation, thus perceives the objects in black and white the cones are used to identify the colour information.
ADG	Automatic degaussing circuit is used to remove the effect of magnetic field due to earth magnetic field and by other strong magnetic fields.
Triads	Three phosphor dots are arranged in a groups are known as triads.
PIL	Precision-in-Line colour picture tube. In this picture tube it has three guns which are aligned precisely in a horizontal line.
Delta gun picture tube	Radio Corporation of America was developed this tube this tube employs, three separate guns, one for each colour, in this type the colour phosphor dots are arranged in delta (Δ) shape.
Trinitron picture tube	The Trinitron or three in line cathodes colour picture tube was developed Sony Corporation of Japan.
Colour convergence	Convergence is a technique used to bring the three beams together.
Static convergence	Static convergence involves the movement of beam by permanent magnetic field.

Dynamic convergence	Dynamic convergence which involves the movement of beam by continuously varying, magnetic field.
Diachronic mirror	Allows one wavelength and rejects other wavelength.
Frequency interleaving	This technique is used to combine two distinct signals with the same bandwidth without any interference.
Quadrature modulation	This technique is used to combine the colour difference signal (R–Y) and (B–Y) into a single variable (C) Chrominance.

REVIEW QUESTIONS

Part A

1. What do you understand by hue or tint?
2. What is mean by additive mixing?
3. Write short notes on subtractive mixing.
4. Define brightness.
5. Define chrominance.
6. What do you mean by saturation?
7. Sketch the camera output voltage waveforms for a colour bar pattern consisting of red, green, blue, yellow and white in cascade.
8. How pincushion distortion is adjusted?
9. What do you understand by chromaticity diagram?
10. State Grassman's law.
11. Why the (G–Y) difference signal is not chosen for transmission?
12. What is the significance of Y signal in colour television transmission and reception?
13. What do you mean by grey scale tracking?
14. What is frequency interleaving technique?
15. Why the weighting factors are used for (R–Y) and (B–Y) signals?
16. Why the colour signal bandwidth requirement is much less than of Y signal?
17. What is gamma correction?
18. State the disadvantages of delta gun picture tube.
19. State the advantages and disadvantages of Trinitron picture tube.
20. Why Trinitron picture tube is superior to PIL picture tube?

Part B

1. Explain how by frequency interleaving the colour information is accommodated within the same channel bandwidth of 7 MHz.
2. Explain the need for pincushion distortion correction and method employed to obtain a distortion free raster.
3. What are the weighting factors and why are these applied to colour difference signals before quadrature modulation draw chrominance signal wave shape for a colour bar pattern and justify the scaling down of (R–Y) and (B–Y)?
4. Describe the gun structure and focus details of Trinitron picture tube with a neat signal. State its merits and merits.
5. Illustrate the constructional and focusing techniques employed in PIL colour picture tube. Explain the need for gray scale tracking.
6. What do you understand by degaussing and why is it necessary? Describe with a typical circuit how degaussing is affected each time the receiver is switched on.
7. What is mean by gray scale tracking? Explain the need and method of obtains a tint free associated while receiving B&W transmission?
8. Enumerate the common faults that can occur in a picture tube due to picture tube in each case list possible causes and methods of rectifying the fault.

QUESTIONS AND ANSWERS

1. **Mention the important characteristics of human eye.**
   Visual acuity, persistence of vision, brightness and colour sensation are some of the important characteristics of human eye.
2. **What are rods and cones?**
   The retina on the back side of human eye consists of two light sensitive organs. They are used to sense the variations in light. They are rods and cones. The rods are used to sense the brightness variations that perceives the object only in various shades of grey from black to white. The cones are mainly responsible for colour perception. There are 65 lakh cones and about 10 crore rods connected to the brain through 8 lakh optic nerve fibers.
3. **Give the relationship between eye parameters.**
   The eye parameters are related by the following equation

$$B = \frac{\text{constant}}{c^2\alpha_0^2}$$

where c is the constant ratio, b is the scene brightness and α_0^2 is the minimum resolvable angle.

4. **Define colour compatibility or reverse compatibility.**
   The colour compatibility implies that the colour TV signal must produce a normal black and white picture on monochrome receiver without any modification of the receiver circuit.
5. **List the requirements to be satisfied for colour compatibility in TV systems.**
   The requirements to be satisfied for colour compatibility in TV systems are:
   (a) It should occupy the same bandwidth as the corresponding monochrome signal.
   (b) The colour signal should have the same brightness information as that of monochrome signal.
   (c) The location and spacing of the picture and sound carrier frequencies should be maintained as 5.5 MHz as in monochrome system.
   (d) The colour TV signal must use the same deflection frequency and sync signals as used in monochrome signal.
6. **Define luminance or brightness.**
   The luminance can be defined as the amount of light intensity as perceived by the eye regardless of the colour. In B & W picture, bright regions have more luminance than the dark regions.
7. **What do you understand by Hue?**
   The predominant spectral colour of the received light is known as hue or tint. The colour of any object is distinguished by its hue or tint for example green leaves green Hue and red leaves have red hue.
8. **What do you mean by saturation?**
   The spectral purity of the colour light. The saturation is an indication of how little the colour is diluted by white.
9. **State Grassman's law.**
   The brightness impression produced by the three primary light fluxes that constitute the single light flux. This property of the eye of generating a response which depends on the algebraic sum of percentage blue, red and green light fluxes. This forms the basis of colour signal generation and it is known as 'Grassman's law'.
10. **Explain the significance of generating colour difference signals.**
    The significance of generating colour difference signals are as follows:
    (a) Colour difference signals are generated to avoid the separate transmission of R, G, B signals.
    (b) The amplitude of (R–Y) and (B–Y) is small so it becomes necessary to use the amplifiers in the matrix network.

11. Why is (G–Y) not suitable for transmission?
The proportion of G (59%) in Y (100%) is relatively large in luminance signal, hence magnitude of (G–Y) is relatively small in comparison with (R–Y) and (B–Y) so it causes the signal to noise ratio at the transmitting end.

12. What is gamma correction?
A colour camera is used develop three voltages proportional to red, green and blue colour contents of the picture. These voltages are represented as R, G and B. A correction is applied to these voltages to compensate for any nonlinearity of the system and that of the picture tube. This is called 'gamma correction'.

13. What is additive mixing?
The two or three primary colour lights sources is combined together to generate a new different colour. With the help of primary colours red green and blue it is possible to generate wide range.

14. What is subtractive mixing?
When two colour pigments are mixed they reflect wavelengths which are common to both. This type of mixing is called subtractive mixing and it is used for colour painting and printing. By mixing of this primary colour we can get black colour.

15. What is the significance of diachronic mirror?
The significance of diachronic mirror are as follows:

a) It splits the light signal into three basic primary colours red, green, and blue.
b) It reflect a specific colour and allow other spectral frequencies to pass through.

16. Define colour convergence.
Convergence is the process used to bring the three electron beams together. So that they hit the same part of the screen at the same time to produce three co-incidence raster's on the screen.

17. How convergence errors occur in colour picture tube?
Convergence errors occur in colour picture tube due to:

a) Non-uniformity in the deflection field.
b) Flat surface of the screen.
c) Non-coincident convergence plane.

18. Define static convergence.
It involves the movement of beam by a permanent magnetic field. This is used to converge the three electron in the central area of the screen.

19. Define dynamic convergence.
It involves the movement of beam by a continuously varying magnetic field. The proper positioning of the beam this type of problem is avoided.

20. What do you mean by pincushion distortion?
This is caused by the flatness of the screen and its rectangular shape. The distance between the gun and the screen goes on increasing as the beams move towards the corners of the screen.

21. What is the use of degaussing circuit in TV receivers?
Degaussing means demagnetizing. It is the process of removal of magnetic flux from the magnetized parts in the TV. By this method the picture tube is demagnetized automatically. When the receiver is turned on each time.

22. Comparison between the electrostatic convergence and electromagnetic convergence.
The comparison between electrostatic convergence and electromagnetic convergence is given in the table below:

	Electromagnetic convergence	**Electrostatic convergence**
1.	This is used in shadow mask picture.	This is used in Trinitron picture.
2.	Electromagnetic field is used to obtain convergence.	Electrostatic field is used to obtain convergence.
3.	Electromagnetic field is set-up by the electromagnetic and which carry currents at horizontal and vertical frequencies.	The electrostatic field is set by connecting the inner two plates to aquadag (final anode) coating and connecting the outer two plates to voltage which is less than the inner plate voltage.

23. Comparison between the shadow mask picture tube and the Trinitron picture tube.
The comparison between shadow mask picture tube and the Trinitron picture tube is given in the table below:

	Shadow mask tube	**Trinitron picture tube**
1.	The phosphor dots on the inner surface screen in the form of triads.	The phosphor dots are arranged on the inner surface of the screen in the form of vertical strips. Each vertical strip is meant for each colour.
2.	In front of the screen a shadow mask is available.	In front of the screen an aperture grille with vertical slots is provided.
3.	Electromagnetic convergence is used. It is a tri-gun, tri-colour picture tube.	Electrostatic convergence is used it is a single gun with three cathodes. Tri-colour picture tube.
4.	Brightness is less due to the opposition offered by the shadow mask.	Brightness is more due to the provision of vertical slots in aperture grill which causes less opposition.
5.	Cost is less.	Cost is more.

CHAPTER 4

COLOUR TELEVISION SYSTEM

KEY TOPICS

- ✓ NTSC colour TV system
- ✓ NTSC coder and decoder
- ✓ NTSC colour receiver and limitations of NTSC system
- ✓ PAL colour TV system
- ✓ Cancellation of phase errors
- ✓ PAL-D colour system
- ✓ PAL coder and decoder
- ✓ PAL receiver circuit operations and functions
- ✓ Merits and demerits of the PAL system
- ✓ SECAM system
- ✓ SECAM coder and decoder
- ✓ Merits and demerits of SECAM system

LEARNING OBJECTIVES

After reading this chapter, the learner should be able to:

- ♦ In olden days there are three different standards for monochrome television systems they are: (i) 525 line system; (ii) 625 line system; (iii) 819 line French system each systems are associated with different bandwidth.
- ♦ After a prolonged research and experimentation they have developed a three different systems of colour television compatible with existing monochrome TV system they are:
  - – The American NTSC (National Television System Committee) System.
  - – The German PAL (Phase Alternation by Line) System.
  - – The French SECAM (Sequential Colour A Memorie) System.
- ♦ The system of PAL, put forward by Prof. Walter Brunch, Telefunken Laboratories in the Federal Republic of Germany.
- ♦ The differential phase error that cause a phase distortion in the reproduced colours, can be overcome by shifting one of the colour difference signal phase by 180° on the alternate lines of scanning.
- ♦ Phase error problem can be solved by using a one line delay period for preceding lines signal and after that combining two colour signals (U, V) to obtain the average. This averaging technique is called as PAL-Delay (PAL-D) system.
- ♦ The PAL receiver circuits operations were discussed briefly.
- ♦ In order to avoid the problem of phase error on account of the subcarrier carrying both the chrominance signals simultaneous. H.de France proposed the SECAM system in 1958.

4.1 INTRODUCTION

In olden days there were three different standards for monochrome television systems:

(i) 525 line system.
(ii) 625 line European system.
(iii) 819 line French system.

Each system was associated with different bandwidth.

After a prolonged research and experimentation a three different systems of colour television compatible with existing monochrome TV system were developed:

(i) The American NTSC (National Television System Committee) System.
(ii) The German PAL (Phase alternation by line) System.
(iii) The French SECAM (Sequential Colour a Memoire) System.

4.2 NTSC COLOUR TV SYSTEM

The national television system committee has developed the NTSC colour TV system. It is compatible with existing American 525 line system. In order to save the bandwidth, and to maintain the compatibility with existing monochrome system two difference colour difference signals are generated they are I and Q signals having different bandwidth. They are obtained by the two colour difference signals they are:

$$I = 0.74(R - Y) - 0.27(B - Y)$$
$$Q = 0.48(R - Y) + 0.41(B - Y)$$

The values of I and Q can be derived from the colour difference signal (R–Y) and (B–Y) as follows:

$$\begin{aligned} I &= 0.74(R - Y) - 0.27(B - Y) \\ &= 0.74R - 0.27B - 0.47Y \\ &= 0.74R - 0.27B - 0.47(0.3R + 0.59G + 0.11B) \\ &= 0.60R - 0.28G - 0.32B \\ Q &= 0.48(R - Y) + 0.41(B - Y) \\ &= 0.48R + 0.41B - 0.89Y \\ &= 0.48R + 0.41B - 0.89(0.3R + 0.59G + 0.11B) \\ &= 0.21R - 0.52G + 0.31B \end{aligned}$$

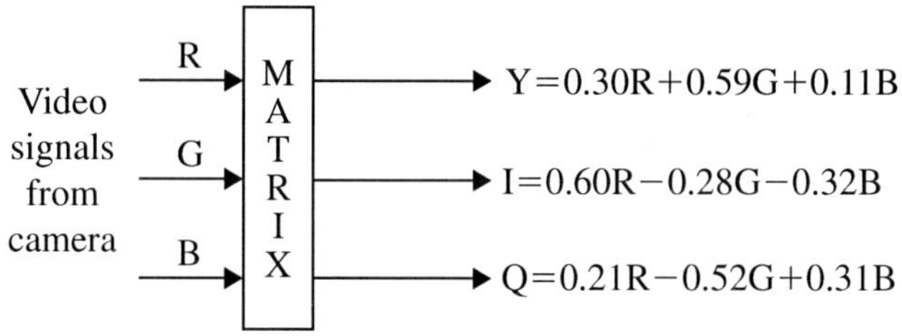

Figure 4.1 NTSC colour difference signal matrix for generating Y, I and Q signals.

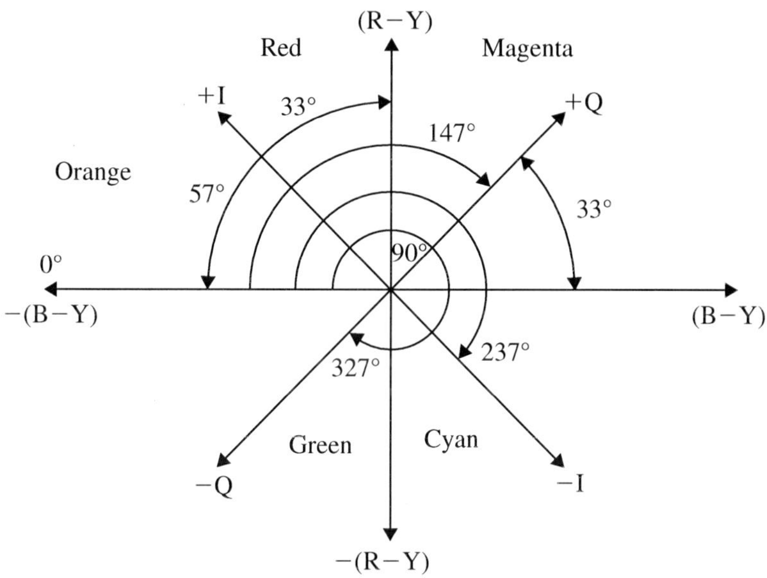

Figure 4.2 (a) Phasor diagram of I and Q signals.

The Q lies in the region of magenta. The +I is in the region of orange and −I is in the region of cyan. The eye is capable of resolving the fine details in this region. I signal allows the frequency up to 1 MHz and Q signal allows only ±0.5 MHz with respect to the subcarrier.

The matrix formation and phasor locations are shown in Figure 4.1. The I signal is equal to 0.74 (R–Y)–0.27 (B–Y) and lies in the region 33° counter clockwise to +(R–Y) where eye has maximum colour resolution.

The Q signals that is equal to 0.48(R–Y) + 0.41(B–Y) occupies a region 33° counter clockwise to + (B–Y) axis where colour resolution of the eye is least. Figure 4.2 shows the phasor diagram showing the location of I and Q signals in the colour circle.

4.2.1 Bandwidth for Colour Signal

The luminance signal Y can be accommodated with full bandwidth of 4.2 MHz and the chrominance signals I and Q are accommodated with 1.3 and 0.5 MHz respectively (< 2 dB at 1.3 MHz for I and 0.5 MHz for Q), and they are used to produce composite video signal. The bandwidth of the Y, I and Q signals are shown in Figure 4.2(b). The bandwidth are limited in order to reduce the cross talk interference between the monochrome and colour signals and hence it is convenient to accommodate these (Y, I, Q) signals in the channel bandwidth of 6 MHz.

4.2.2 Colour Subcarrier Frequency

To maintain the compatibility the subcarrier frequency f_{sc} in the NTSC system is chosen equal to:

$$f_{sc} = (2 * 227 + 1) * \frac{15750}{2}$$
$$= 3.583125\ \text{MHz}$$

Since the eye is capable of resolving fine colour details of ±I region. It is allowed to have a vestigial bandwidth up to 1.5 MHz. However, the bandwidth of Q signal is restricted to 0.5 MHz around the subcarrier because the regions around Q signal. It is interesting to note that since both I and Q signals remain active up to 0.5 MHz, they combine to produce all colours of the picture contained in the chrominance (C) signal.

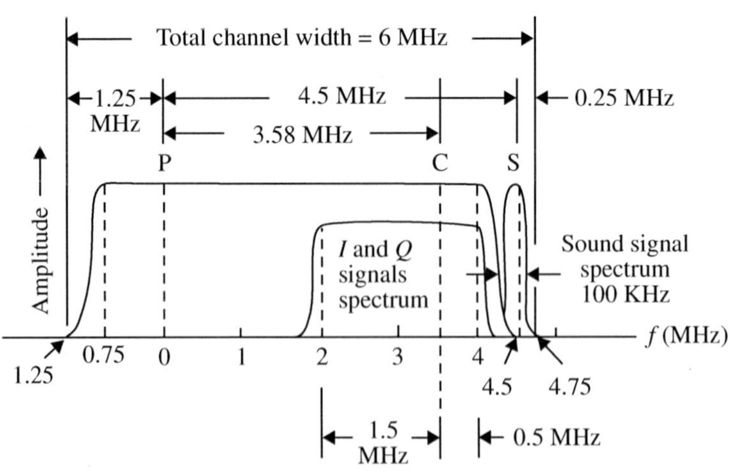

Figure 4.2 (b) Channel width details of the NTSC colour system.

4.2.3 NTSC Coder

The basic circuit arrangement of the NTSC colour shown in Figure 4.3. The weighed I and Q signals obtained from the matrix output are fed to corresponding modulators. To obtain Quadra tune modulation. The phase of colour subcarrier to the Q modulator is shifted by 90° with suitable network. Balanced modulators are used to suppress the strong carrier components. The Q modulator output is passed through a ±0.5 MHz band pass filler. Similarly the I modulation product is passed through a filter circuit to limit its bandwidth to 1.5 MHz. The band limited I and Q signals thus obtained combined with Y and C signals to form composite colour signal as shown in Figure 4.4.

It may be noted that f_{sc} at 3.58 MHz remains the same for all NTSC colour transmitters while the picture carrier frequency is different in each case and depends on channel number allotted the station.

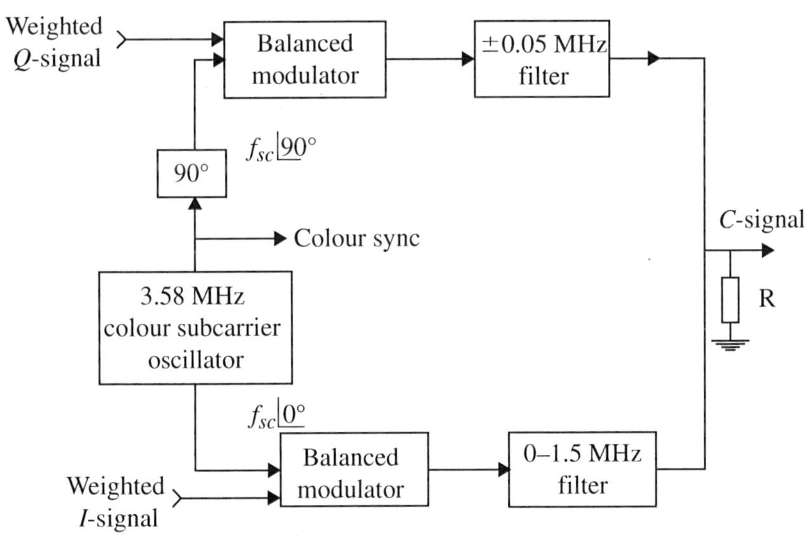

Figure 4.3 Block diagram of NTSC coder.

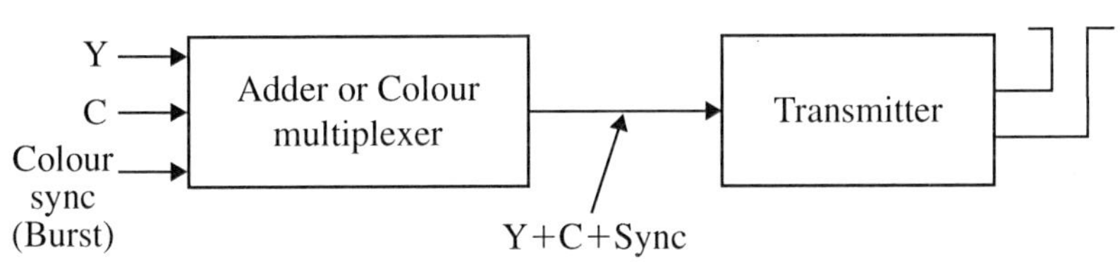

Figure 4.4 Multiplexing of Y and C signals.

4.2.4 NTSC Decoder

The major difference between a monochrome and colour receiver is that the latter has an additional circuit block called decoder or chroma sub-system. The simplified block diagram of NTSC is shown in Figure 4.5. It is meant for recovering colour difference signals from the transmitted colour information. The composite colour signal is separated by a tuned circuit connected at the output of video decoder in the receiver and fed to the decoder. The colour sync is separated from the chroma signal and fed to the colour subcarrier oscillator for keeping its output locked at 3.579545 MHz.

Two synchronous demodulators are used to demodulate the colour difference signals. The demodulators receive amplified chroma signals and colour subcarrier at constant amplified. The colour subcarrier is phase sifted by 90° before feeding it to the modulator as done at the encoder. The two outputs I and Q are then matrixes with the Y signals. For this the Y signal is obtained from the luminance channel. The three colour signals are further amplified and fed to the cathodes of R, G and B guns of the colour picture tube to obtain picture of the televised scene on the receiver screen.

4.2.5 Limitations of NTSC System

(i) The NTSC system is sensitive to transmission path differences which introduce phase errors. This results in colour changes in the picture.

(ii) At the transmitter, phase changes in the chroma signal take place when change over between programs of local and television network systems takes place and when video tape recorders are switched on.

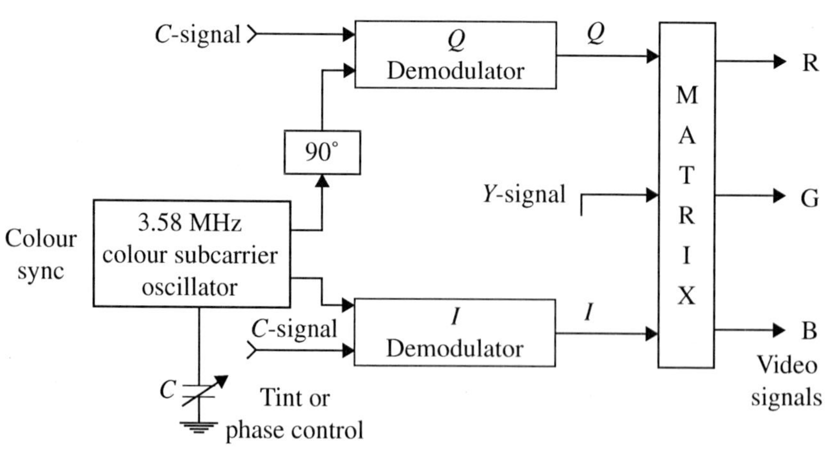

Figure 4.5 Block diagram of NTSC decoder.

(iii) The chroma phase angles is also affected by the level of the signal while passing through various circuits.

(iv) In addition crosstalk between demodulator outputs at the receiver causes colour distortion. All this requires the use of an Automatic Tint Control (ATC) circuit with the provision of manually operated tint control.

4.3 PAL-D COLOUR TV SYSTEM

The system of phase alternation by line (PAL), put forward by Prof. Walter Brunch, Telefunken Laboratories in the Federal Republic of Germany. It is a variant of the NTSC system. The differential phase error that cause a phase distortion in the reproduced colours, can be overcomed by shifting one of the colour difference signal phase by 180° on the alternate lines of scanning. Here the colour difference signal (R–Y) and (B–Y) are called '*U* and *V*'.

At the receiver side, on demodulation the same phase is shifted back by 180° to decode the transmitted information. By this process the phase error has been cancelled at the receiver side. The two colour signals (*U*, *V*) of consecutive lines are obtained at the output of the demodulator are averaged with a view to cancel the errors in the opposite direction. If a averaging is left to the viewers eye as originally conceived in the system called 'simple-PAL'.

This problem can be solved by using a one line delay period for preceding lines signal and after that combining two colour signals (*U*, *V*) to obtain the average. The averaging is done in receiver before feeding the colour signal to the picture tube. This averaging technique is called PAL-Delay (PAL-D) system.

4.3.1 PAL Encoder

The block diagram of PAL encoder is shown in Figure 4.6(a). The gamma corrected RGB signals is fed to the matrix circuit to obtain the *Y*, *U* and *V* signals. The *U* and *V* signals are fed to the quadrature modulator. The subcarrier fed to the *U* modulator has a phase shift of $f_{sc} < 0°$ while *V* modulator gets a subcarrier over a PAL switch to obtain a phase inversion, (*i.e.*, ±90° phase shift on the alternate lines).

The colour subcarrier modulated in the quadrature modulation is band limited to 1.2 MHz. The two modulated signals (*U*, *V*) are combined in the adder network. The output of the adder circuit gives the chrominance signal (C). The chrominance signal is passed to a harmonic filter to limit the bandwidth of ±1.5 MHz.

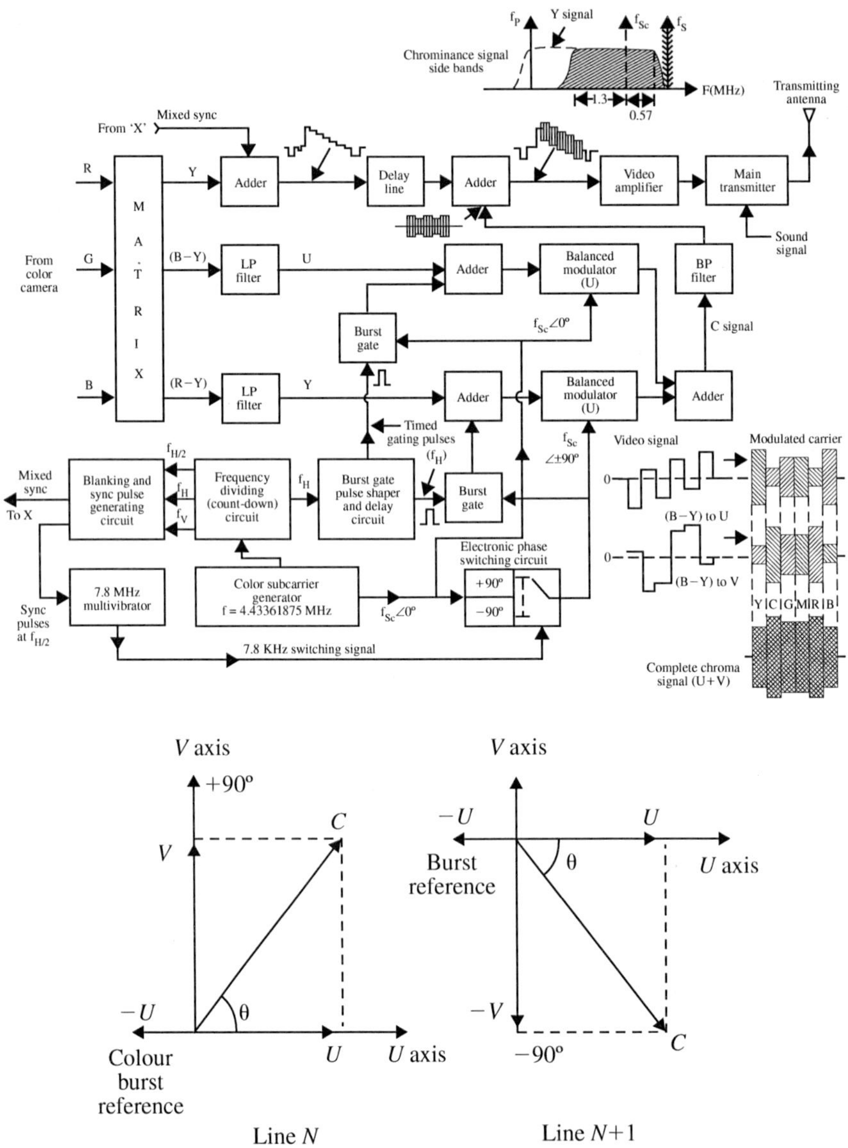

Figure 4.6 (a) Shows a basic principle of PAL-D modulation, and (b) Summation of consecutive line phasor in a PAL-D colour receivers.

The colour burst signals is also fed into the modulators along with U and V signals through adders. Before feeding this burst signal into U and V adders these are passed through separate burst gate. Each burst gate is controlled by delayed pulse at line frequency (f_H). These pulses are applied over back porch period. During the intervals 'U' modulator gets a burst with 90° phase shift, while the 'V' modulator gets the burst with ±90° phase shift on alternate lines. At the output of modulator, two burst signals are combined and produce the subcarrier burst at +45° on one line and −45° on the another line with reference to the U phasor. The luminance (Y) signal obtained at the matrix circuit is combined with line and field blanking and sync pulse. And then it is fed to the delay line circuit to maintain the time coincidence of Y and C. And then it is combined with C to form a composite video signal.

4.3.2 Weighting Factor

(i) To prevent the over modulation of colour difference (R–Y) and (B–Y) signals necessary scaling factors are employed to reduce its magnitude before the quadrature amplitude modulation.

(ii) The scaling factors chosen for this colour difference (R–Y) and (B–Y) are:

$$U = 0.439(B - Y)$$
$$V = 0.877(R - Y)$$

(iii) At the receiver after demodulation a inverse scaling factor is employed to restore U and V to their original values of (R–Y) and (B–Y).

4.3.3 Cancellation of Phase Errors

(i) The angle of C with reference to +(B–Y) determines the hue value. The C signal can suffer in phase shift both at transmitting and during transmission path cause phase shift error.

(ii) The differential phase error affect the phase shift of colour side band frequency with respect to the colour burst phase.

(iii) This can be overcome by shifting one of the colour difference signal phase by 180° on the alternate lines of scanning in the quadrature modulation. This effect leads to the inverting the error on the alternate lines.

(iv) The Figure 4.6(b) illustrates how the phase shift error gets cancelled during transmission. The location of C has 45° on N line and −45° on $N + 1$ line.

(v) Assume that a phase error of 5° occurs during transmission. Then C_N will be $(+45° + 5°) = 50°$ and C_{N+1} will be $(-45° + 5) = -40°$.

(vi) At the receiver side another phase reversal of 180° which restores the location of V and hence C_{N+1} takes a position that is 40° with respect to $+U$ axis.

(vii) Such a sequence repeats line after line at a very fast rate. The colour perceived by the eye is the average of the two signals and the location of C signal is at $(50° + 40°)/2 = 45°$.

(viii) By this the hue changes (phase error) can be cancelled.

4.3.4 PAL Colour Receiver

Figure 4.6(c) shows the block diagram of PAL colour receiver. The PAL colour system is similar to the monochrome receiver. In addition to this colour chroma section is used for colour generation.

The main blocks of PAL receiver are:

(i) Colour chroma section (PAL-D decoder).
(ii) Colour killer.
(iii) Subcarrier reference oscillator.
(iv) Phase discriminator.

In addition to these blocks the monochrome section is also used to produce luminance signal. The signal received from the antenna is fed to the tuner section. In this section, those signals are converted into IF range of frequencies. Then, the picture and sound IF produced from the tuner section is fed into the video detector.

The video detector is used to separate the video signal and it is fed to the video amplifier. After amplification, it is fed to the picture tube. The sound signal is separated using trap circuits and it is fed into FM detector for demodulation. The separation of sound signal is amplified and given to the speaker.

The PAL-D decoder consist of:

(i) Chroma band pass amplifier.
(ii) PAL delay line.
(iii) Colour signal demodulator.
(iv) U and V signal separation.

This is shown in Figure 4.6(d).

4.3.4.1 *Chroma Band Pass Amplifier*

Chroma band pass amplifier section consists of chroma filter and two stage chroma amplifiers. This is used to filter out the colour burst and subcarrier signals. These controls can be used in the front panel of the receiver.

4.3.4.2 *PAL Delay Line*

The PAL delay line is used to delay the chrominance signal by almost exactly one line period of 64 μs. Because of this delayed line $(U + V)$ signal and $(U - V)$ signals are generated.

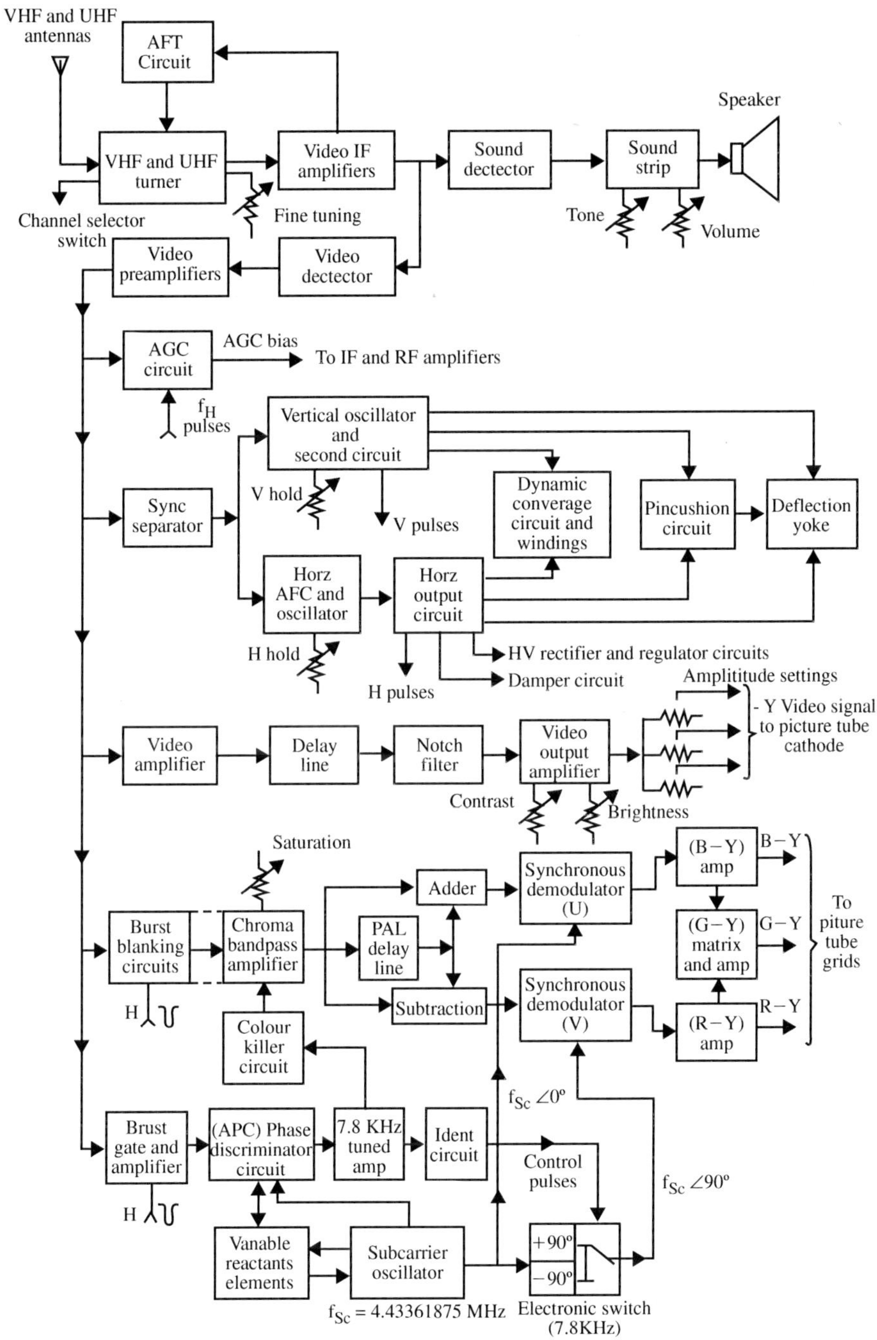

Figure 4.6 (c) PAL colour receiver.

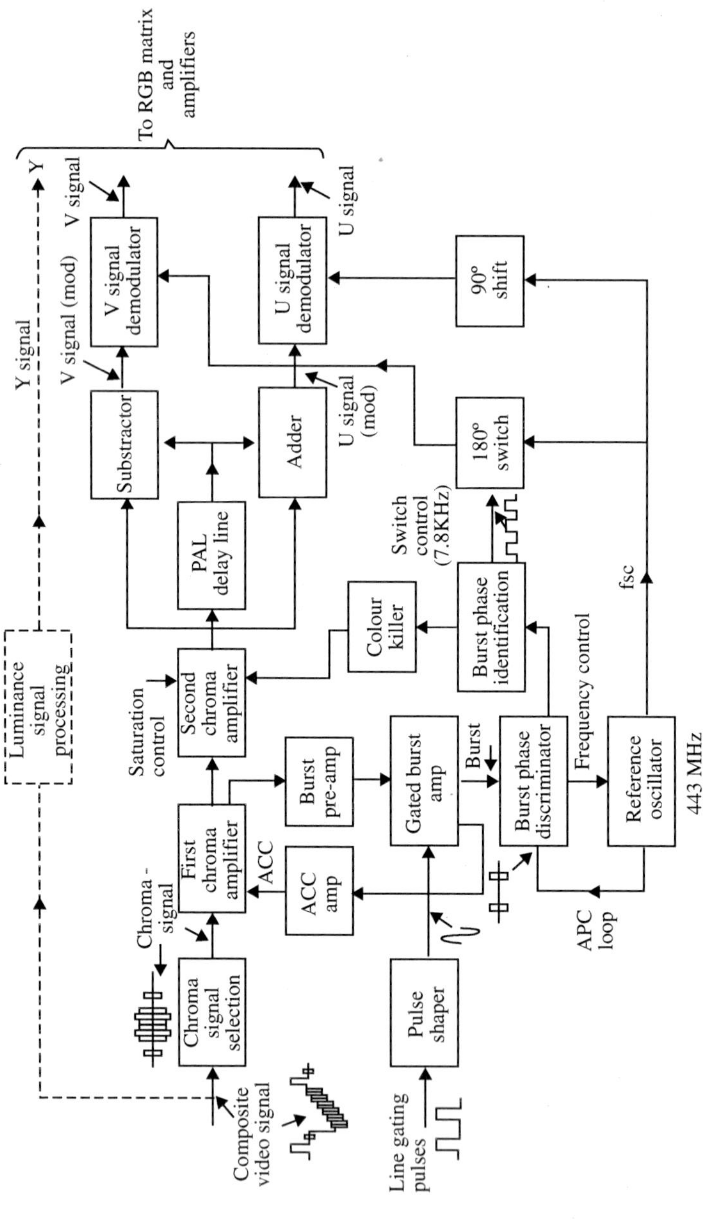

Figure 4.6 (d) Block diagram of PAL-D coder.

4.3.4.3 Colour Demodulator

To separate the U and V signals, adder and subtractor circuits are used. The colour demodulator is used to detect U and V signals. The reference carrier of oscillator is used to provide phase shift for U and V signals.

4.3.4.4 Colour Killer

If monochrome signals are received, no burst signal is present. So there is no need for chroma section and the output of chroma amplifier is zero. As there is no output at 7.8 kHz tuned amplifier, this will switch off the chroma section.

4.3.4.5 Reference Oscillator

During transmission side U&V signals are modulated by double balanced suppressed carrier. And it is necessary to regenerate the subcarrier of U&V signals at the receiver side to demodulate the colour difference signal. Here the reference oscillator is used to regenerate the subcarrier of 4.43 MHz.

4.3.4.6 Phase Discriminator

The phase discriminator is used to compare the colour burst and subcarrier frequency. If there is any difference which will produce DC error voltage. Then this is fed into the subcarrier generator.

4.3.4.7 Matrix Network

This matrix network is used to combine the U and V and luminance (Y) signal. The output of this matrix is the colour difference signal (R–Y), (B–Y) and (G–Y). Finally these signals are fed into the picture tube grids.

4.3.5 Choice of Colour Subcarrier Frequency

If the subcarrier frequency is selected on half line offset basic is carried out in the NTSC system, an unwanted line up of dots occurs on certain hues. This is because of phase reversal of the carrier at line frequency. To rectify this problem, a quarter line offset is provided and colour subcarrier frequency f_{sc} is made an odd multiple of one quarter of the line frequency.

$$
\begin{aligned}
f_{sc} &= \frac{f_h}{4(2 \times 567 \times 1)} + \frac{f_v}{2} \\
&= \frac{f_h}{4(1135)} + \frac{f_v}{2} \\
&= f_h \left(284 - \frac{1}{4}\right) + \frac{f_v}{2}
\end{aligned}
$$

substituting the values of f_h and f_v gives

$$f_{sc} = 4.43361875\ \text{MHz}$$

where f_h is horizontal line frequency (15,625 Hz) and f_v is vertical line frequency (50 Hz).

The generation of any dot pattern on the screen must be elimination by appropriate choice of subcarrier frequency. Since, the chrominance signal disappears on gray or while details in picture, the interference appears only in coloured areas of the scene. Therefore the dot effect is visible only in those coloured parts of the picture which are produced in lighter shades of grey.

4.4 COLOUR BURST SEPARATION

The transmitted signal does not contain the subcarrier frequency but it is necessary to regenerate it in the receiver for proper detection of colour subcarrier. To ensure this, a short sample of the subcarrier (8 to 10 cycles of 4.43 MHz) called colour burst. Is transmitted at the beginning of every line. Its function is to synchronize the colour oscillator for reinsertion of exact carrier into the U and V demodulator.

Before proceeding with the demodulation of U and V signals it is necessary to examine the reference oscillation, the 180° switch and 90° phase shifter because all these are connected to the demodulations. But since the logical to first examine how colour burst pre-amplifier and gated. The chroma input signal from the 1st stage of chroma amplifier is fed via C_5 (capacitor C_5) to the base as shown in Figure 4.7. Base of Q_4 transistor which function as a straight forward amplifier. The amplifier chroma signal appearing at the collector of Q_4 is fed through capacitor C_{13} to the base of Q_5, the gated burst amplifier.

4.4.1 Gated Burst Amplifier

The burst gate amplifier is used to separate the colour burst signal from the colour signal. A gated class A or C operated tuned amplifier is used for this purpose the centre frequency is set at 4.43 MHz. The bandwidth of this stage is about 0.6 MHz.

The gated burst amplifier transistor Q_5 is normally non-conducting because it is not supplied with any standing forward bias voltage as its base point is returned to ground through R_{24}. The chroma input signal is not large enough to force Q_5 in to conduction. However, it is switched on for short periods by gating pulses. These pulses are negative going and are applied at the junction of R_{27}, R_{28} and derived from the pulse sharper.

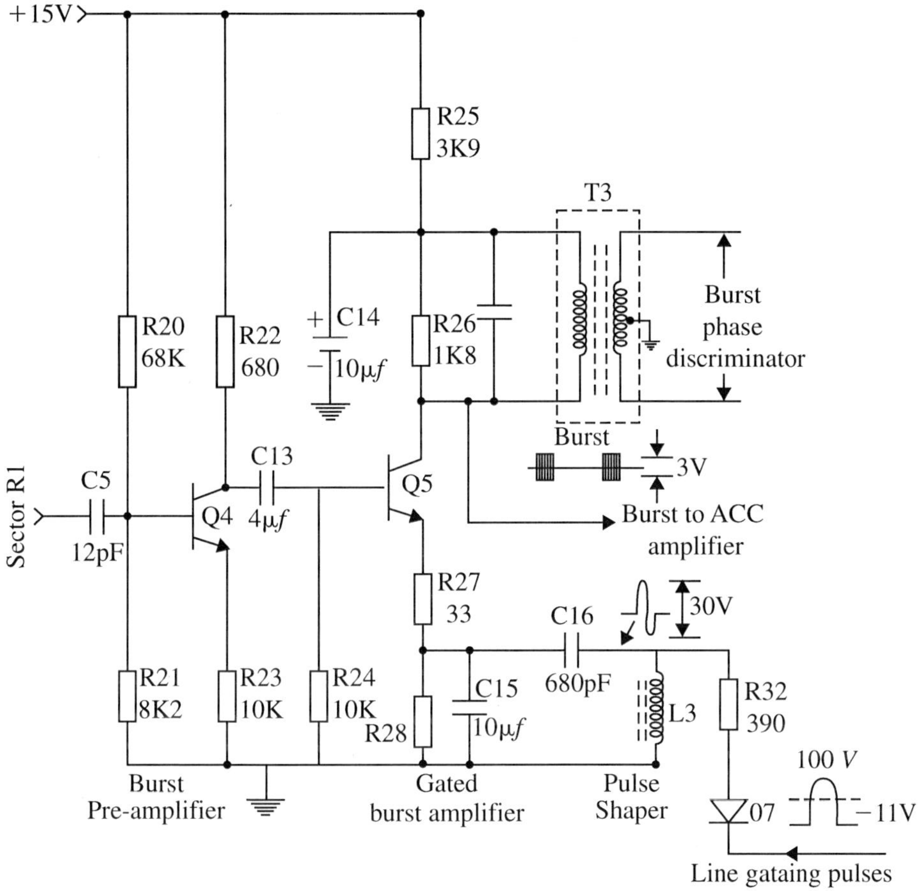

Figure 4.7 A circuit for burst phase separation from the chroma signal.

4.4.2 Pulse Sharper

The pulse shaper circuit consists of L_3, C_{15}, C_{16} and R_{28}. The capacitor C_{15}, C_{16} and resistor R_{28} is used to produce oscillations and also limit the sinusoidal amplitude necessary for forward biasing the emitter base junction of Q_5.

4.4.3 Circuit Operation

(i) The pulse shaper circuit input is sharp pulses obtained from the line time base circuit via diode D_{27} and R_{32} is shown in Figure 4.7.

(ii) During line scan periods the incoming pulse polarity is negative. Which makes the diode D_7 to be forward biased and a small current flow through the inductance L_3, with a result a small energy is stored in the inductance.

(iii) During the positive going gating pulse, which makes the diode D_7 to be cut-off instantly. This result, in a sudden interruption of current flow through the inductance L_3 and it makes self-oscillations in L_3.
(iv) The narrow frequency of self-oscillations is set at 7.8 kHz, so that one full cycle occurs during the presence of the line gating pulse and negative going half-cycle occurs coincident in time with back porch of line blanking period.
(v) This means the negative going half cycle is used to turn on the transistor Q_5 (only at the back porch period of every line). During this period only the colour burst occurs in the chroma signal, and it is amplified.
(vi) The burst output appears at the collector of transistor Q_5 is fed to the burst phase discriminator and ACC amplifier.

4.5 BURST PHASE DISCRIMINATOR

The burst phase discriminator is used to compare the phase of the colour burst and subcarrier frequency generated by the reference oscillator. If there is any phase difference, which will produce a DC control voltage (error voltage) which is fed to reference oscillator. In this way the frequency and phase are kept locked in synchronism with the colour burst.

4.5.1 Circuit Operation

(i) The discrimator circuit is formed by C_{17}, C_{18}, R_{19}, R_{30} and diodes D_5, D_6 as shown in Figure 4.8. The reference oscillator output is passed to the input of diodes D_5 and D_6 through the capacitance C_{19}.
(ii) Every time the burst arrives of equal amplitude but opposite phase appears at the two ends of the secondary transformer T_3. The primary is used to tune and pass the burst signal to the input of the discriminator circuit.
(iii) In the absence of reference oscillator input the diodes D_5, D_6 conducts to charge the capacitors to equal voltage values.
(iv) During non-conducting period the capacitor starts to discharge through the resistors R_{29} and R_{30}. The polarity of the voltage drop in the resistance is equal in amplitude and opposite in phase makes it to cancel each other. The resistance R_{29}, R_{30} and diodes D_5, D_6 forms a bridge and the potential at this junction is to be zero.

In the absence of any reference oscillator input, the diodes conduct to charge the capacitors to equal voltages across R_{29} and R_{30}. Also note that there is no voltage drop across R_{31}, the upper end of which is connected to the base of Q_6.

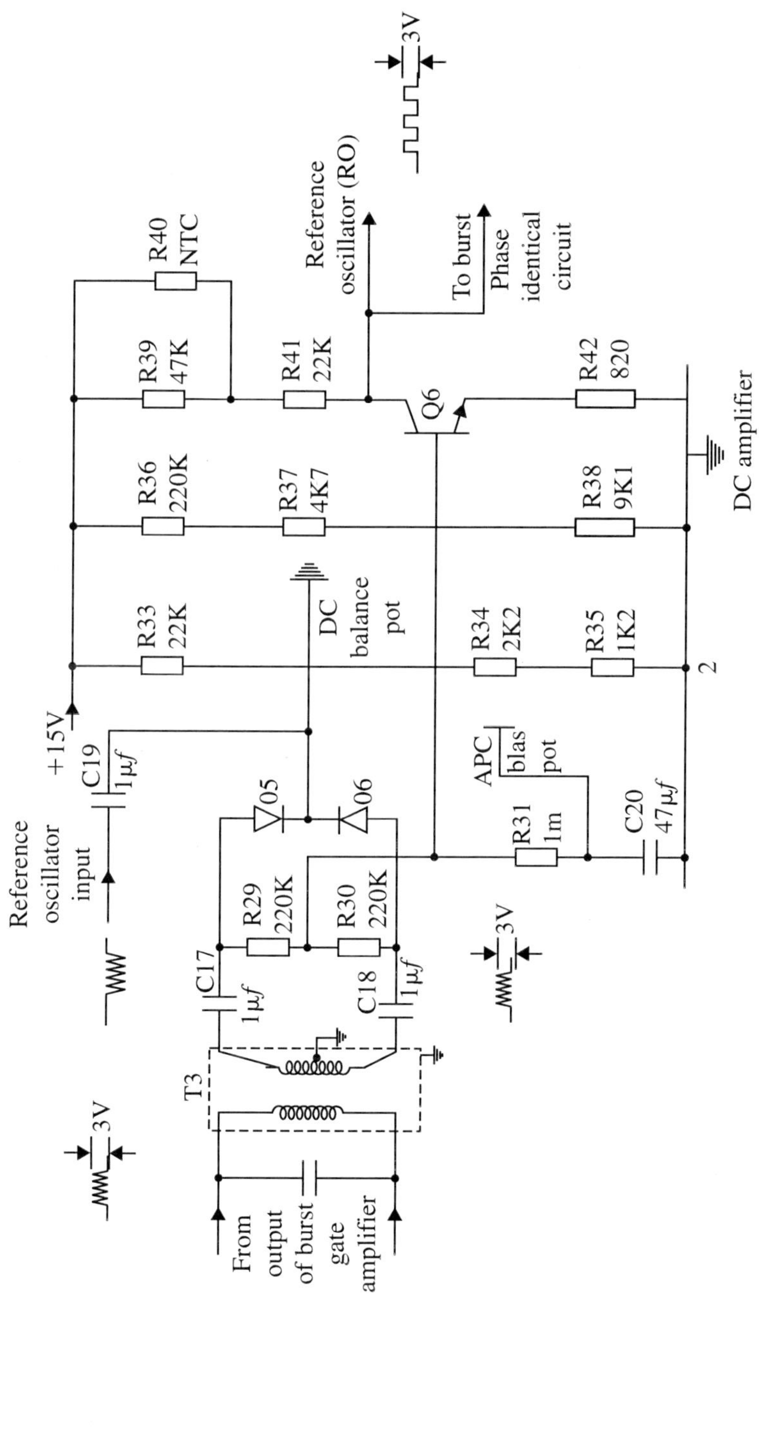

Figure 4.8 Burst phase discriminator.

(i) Next consider that both burst and reference oscillator signals are being applied to the discriminator. Note that the colour burst has about 10 cycles at a frequency of 4.43 MHz.

(ii) When the frequency of reference oscillator (RO) is exactly the same as that of anti-phase burst signals being applied at free ends of diode D_5 and D_6 and when its phase also coincides exactly with that of the un-switched TV signal, the successive 90° phase shifting of the burst which occurs from line to line will course a symmetrical, nearly square waveform to be produce at the junction of resistors R_{29} and R_{30}.

(iii) Next consider that the frequency of reference oscillator (RO) where for some reason to increase relative to the frequency of colour burst sub-carrier. Then the phase relationship with burst signals would change and the 90° phase shifting of the burst signals would cause unequal voltage drop across R_{29} and R_{30} with the result an unsymmetrical wave, though again nearly square shaped will be generated and its positive half cycle will be predominant.

(iv) When the frequency of reference oscillator (RO) falls below that of the colour burst subcarrier. This time negative half cycle of the asymmetrical wave which is predominant and the resulting negative going signal is used to increase the frequency of reference until it is restored to its correct value.

4.6 AUTOMATIC COLOUR CONTROL (ACC) AMPLIFIER

Depending upon the amplitude of colour burst, the ACC will develops a DC control voltage this DC voltage will shift the operating point of the transistor and the gain of the amplifier is automatically changed.

4.6.1 Circuit Operation

(i) The output of gated burst amplifier (Q_5) is also fed to the diode D_8 of the automatic colour control amplifier circuit as shown in Figure 4.9. The diode along with the capacitor (C_{21}, R_{43} and C_{22}) act as a rectifier-cum-low pass filter. Which is used to smooth any 4.43 MHz subcarrier frequency variations present in the rectified signal.

(ii) The output voltage at the transistor Q_7 (P-N-P) is a positive voltage which increases or decreases the signal strength of chroma signal. Under normal signal strength conditions this positive voltage is 7 V.

(iii) The resistor R_{44} provides an adjustable reverse bias for transistor Q_7 to delay its conduction, until the chroma signal exceeds the given threshold voltage.

(iv) The potentiometer formed by resistance R_{46} and R_{47} is provide a steady biase voltage for the first chrominance amplifier.

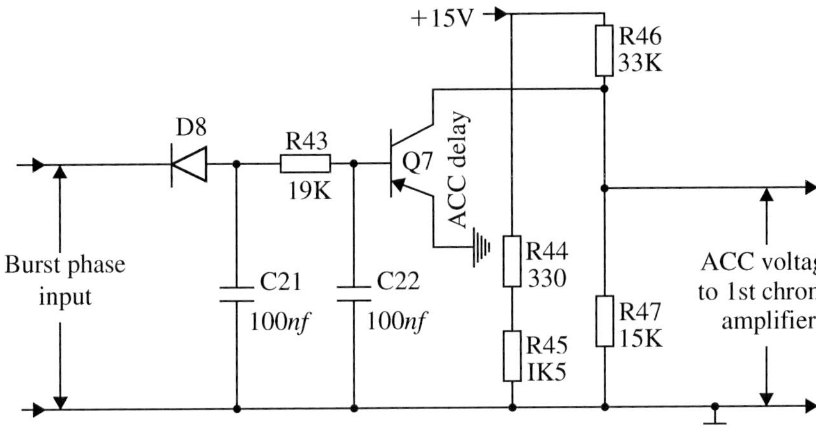

Figure 4.9 ACC amplifier circuit.

4.7 REFERENCE OSCILLATOR

During transmission side the *U* and *V* signals are modulated by double-balanced suppressed carrier. And it is necessary to regenerate the subcarrier of *U* and *V* signals at the receiver side to demodulate the colour difference signal. Here the reference oscillator is used to regenerate the subcarrier signal of 4.43 MHz. The reference oscillator circuit diagram is shown in Figure 4.10.

4.7.1 Circuit Operation

(i) The oscillator circuit is formed by transistor Q_8 and the frequency is determined by the centre—resonance frequency of the crystal (XL) which generates 4.43 MHz approximately.

(ii) Small frequency errors are corrected by trimming capacitance C_{24} and varicap diode D9. The value of capacitance C_{24} is kept large to swamp any likely variations in the input capacitance of Q_8.

(iii) The resistor R_{51} and R_{52} fix the forward bias and R_{53}, R_{54} determines the collector current. The resistance R_{55} and capacitance C_{25} forms a decoupling network that prevents the output is not to be feedback with the power supply.

(iv) The resistor R_{56} provides damping of the tuned circuit in the collector load.

(v) The output waveform is taken from the capacitive potential divider formed by C_{28} and C_{29}. This arrangement limits output voltage to about 1.3 V peak to peak.

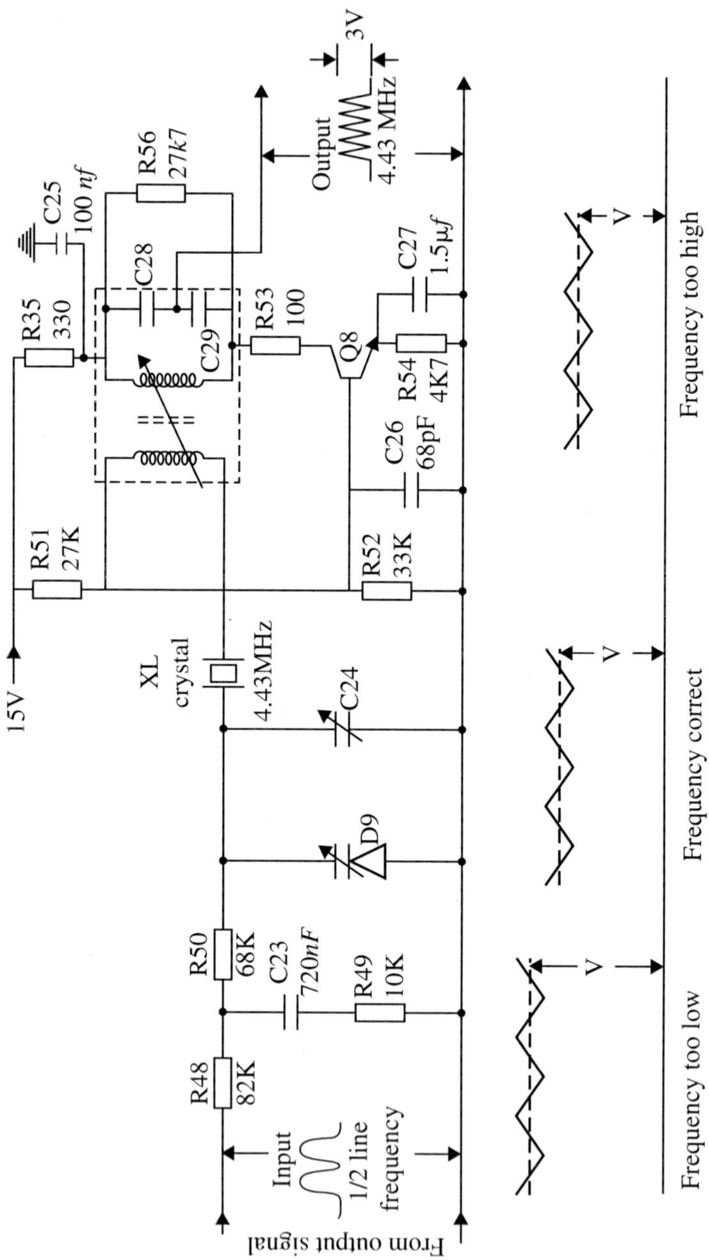

Figure 4.10 Reference oscillator.

4.8 IDENTIFICATION AND COLOUR KILLER CIRCUITS

The burst phase identification circuit accepts the near square waveform produced by the burst phase discriminator and amplifier circuit and amplifies its AC content with a high gain, narrow band amplifier and feeds it as a phase identifying signal to the 180° PAL bitable switches. The amplifier circuit is shown in Figure 4.11. This also includes the rectifier filter circuit to produce colour killer output voltage.

The 7.8 kHz signal is fed via R_{57}–C_{28} to the base of Q_9 of the high gain narrow band amplifier. While C_{28} blocks the DC content, R_{57} and C_{28} together integrate the AC content to near single wave. The resistors R_{59} and R_{60} are to provide forward bias and R_{61} determines collector current. The tuned circuit also provides desired selectivity. The waveform developed at the upper end of the tuned circuit is used as a burst phase identification reference by the 180° PAL switch. It has a peak to peak amplitude to nearly 3.5 V and is fed via R_{58} and C_{32} in series.

4.8.1 Colour Killer

During monochrome signal reception, no burst signal is present, so there is no need for chroma section and the output of the chroma amplifier is zero. As there is no output at 7.8 kHz tuned amplifier, which will switch off the chroma section.

4.8.2 Circuit Operation

(i) The amplifier circuit shown in Figure 4.11. The DC operating voltage to transistor Q_9 is supplied via a centre tap on inductance L_4. This forms a tuned autotransformer and enables a waveform of 25 V peak to peak is developed at the collector of the transistor Q_9.

(ii) This waveform is fed to the half-wave rectifier formed by the capacitance C_{30} and diode D_{10}. The components R_{62} and C_{33} forms a low pass filter which provides a steady DC voltage of 13.5 V.

(iii) During monochrome signal reception, no burst signals and there is no output from the burst discriminator and no output from the burst discriminator and no input to the burst phase ident amplifier. And the output of the 7.8 kHz tuned amplifier also falls to zero.

(iv) Under this condition the colour killer output falls to less then 2V which is not sufficient to forward bias the transistor Q_2 chroma amplifier. This in turn switch off the chroma section during monochrome signal.

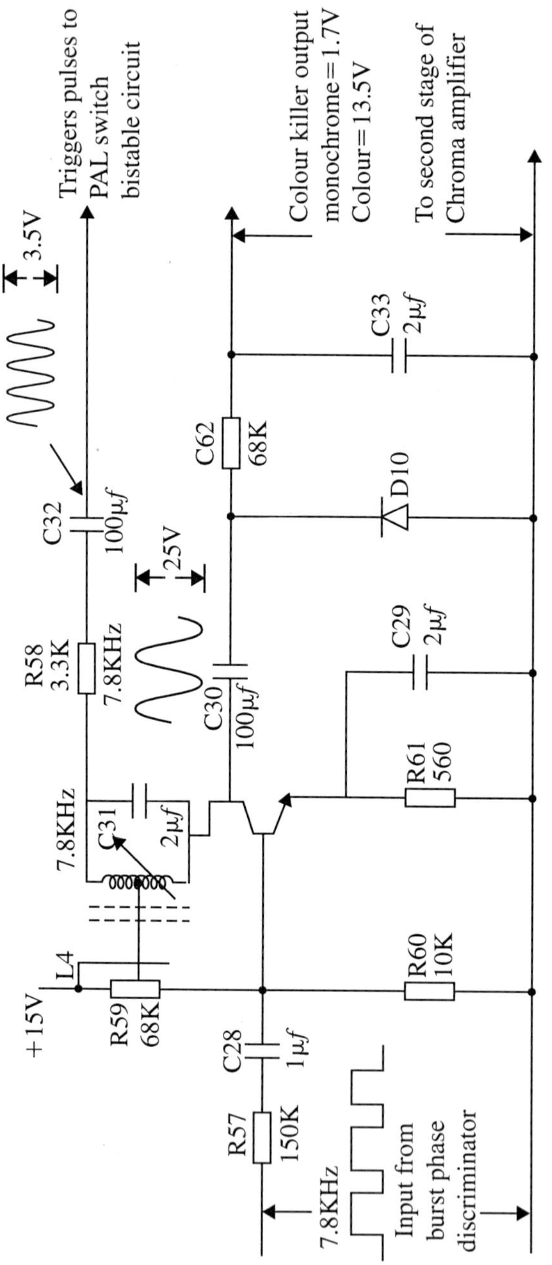

Figure 4.11 Burst phase identification amplifier and colour killer generation circuit.

4.9 SEPARATION OF *U* AND *V* SIGNALS

The technique of averaging and separation of U and V signals is based on the assumption that picture information is carried on the successive line changes. The delay line is used to do the averaging first and then present the colour to the eye. This is called PAL-D or delay line PAL method is most commonly employed in PAL colour receiver.

As shown in Figure 4.12 a delay line and adder and subtractor circuits are interposed between the chrominance amplifier and demodulator. The purpose of delay line section is used to delay the signal by $64\mu s$. Due to this arrangement the first line when the V modulator product is at 90° at the $+U$ axis, the phasor can be denoted as $(U + jV)$. This is known as 'NTSC line'. But on the next line, when the V is switched to $-90°$ the phasor becomes $(U - jV)$ and the corresponding line is called 'PAL Line'.

Line numbers	**Direct signal**
$N - 1$	$U + jV$
N	$U + jV$
$N + 1$	$U - jV$
$N + 2$	$U + jV$
$\vdots$	$\vdots$

To separate the U and V signals, the adder and subtractor circuits are used. For both the adder and subtractor, normal and delayed lines are given as input.

So, at the adder circuit:

$$(U - jV) + (U + jV) = +2U$$
$$(U + jV) + (U - jV) = +2U$$

At the subtractor circuit:

$$(U - jV) - (U + jV) = -2V$$
$$(U + jV) - (U - jV) = +2V$$

From this adder and subtractor circuit generates U and V signals, but twice the amplitude of U and V signal.

To allow the precise addition and subtraction of direct and delayed line signal. The delay line must introduce a delay which is equivalent to the exact number of half cycle of the chrominance signal.

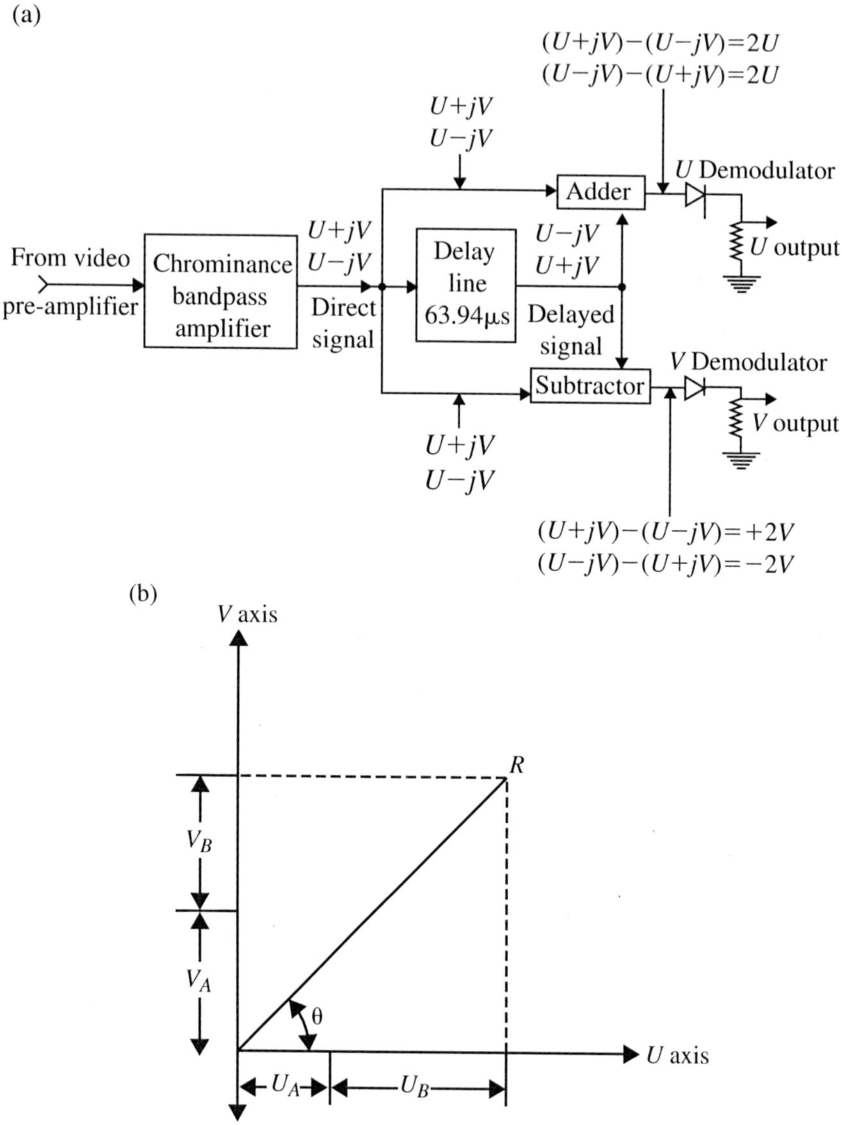

Figure 4.12 (a) Shows a basic principle of PAL-D modulation, and (b) Summation of consecutive line phasor in a PAL-D colour receivers.

At a frequency of 4.433617 MHz. The number of cycles takes place in:

$$
\begin{aligned}
64\mu s &= 4.4336187\,\text{MHz} \times 64\mu s \\
&= 283.748598\,\text{Hz} \\
&\approx 283.75\,\text{Hz}
\end{aligned}
$$

A delay line *i.e.*, to time delay $= 1/f_{sc} \times 283.5 = 63.943\mu s$.

4.9.1 *U* and *V* Demodulators

The function of *U* and *V* demodulator to detect the *U* and *V* signals and the modulated colour signal. A reference carrier frequency of 4.43 MHz is used to test the demodulators.

The reference oscillator frequency is available to *U* and *V* demodulator as:

(i) Phase shifted by 90° and given to *U* demodulator.
(ii) Phase shifted for each alternate line by 180° and locked to +ve and −ve signals.

4.9.1.1 Circuit Operation

(a) The synchronous de-modulator *U* and *V* is shown in Figure 4.13. The diodes D_{16} to D_{19} forms a detector bridge and it act as an electronic switch operating at a frequency of 4.43 MHz obtained from the reference oscillator (RO). When the diode starts to conduct it provide a low resistance path between the chroma signal input and *U* and *V* demodulator output terminals.
(b) The resistance R_{77}, R_{78} act as a current limiters for the conducting diode, and the capacitor C_{44} and resistance R_{79} forms a time constant equal to period of colour subcarrier frequency (4.43 MHz).

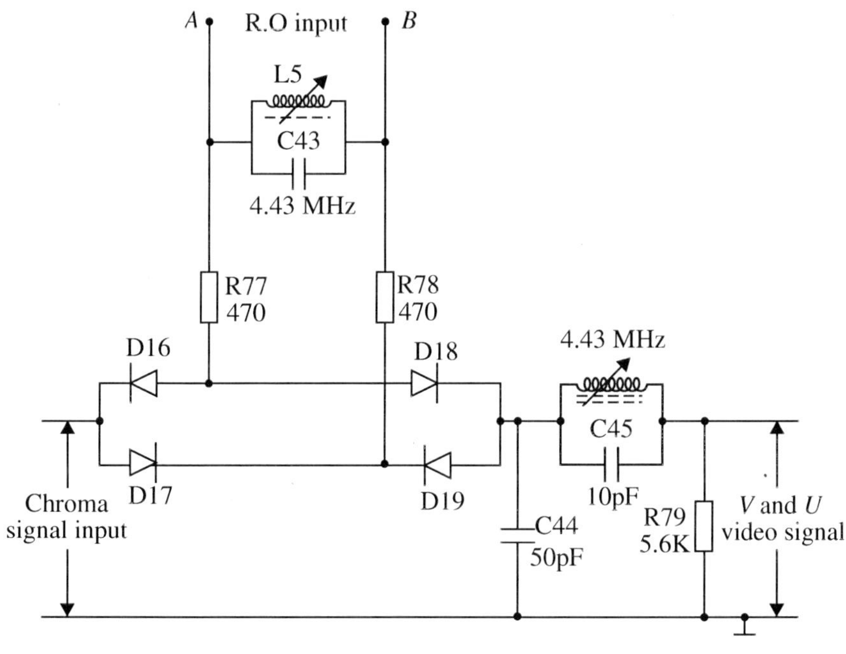

Figure 4.13 *U* and *V* demodulator.

(c) The colour subcarrier frequency is a parallel tuned circuit formed by L_6 to C_{45} and connected between C_{44} and R_{79} act as a low pass filter to the colour difference signals and rejection circuit is used to filter the content of reference oscillator.

(d) The capacitor C_{44} and resistance R_{79} have a time constant equal to the period of colours subcarrier (4.43 MHz) and is thus quite less when compared with the period of highest frequency (= 1.5 MHz) of the colour difference signals. The output is thus able to follow modulating signal variations but not to the subcarrier content. The 4.43 MHz parallel tuned circuit formed by L_6-C_{45} and connected between C_{44} and R_{79} acts as a low pass filter to the colour difference signals but a rejection circuit to the reference oscillator (RO) content of the output waveform. This circuit thus functions to reinforce filtering offered by C_{44} and R_{79}.

(e) Every time the positive half cycle of reference oscillator (RO) waveform makes point A positive with respect to point B. The four diodes become forward biased providing a low resistance path between the chroma signal input and video signal output terminals. During negative half cycle all the four diodes get reverse biased and there can be no transmission from input to output terminals. It is usual for the amplitude of reference oscillator (RO) output to be made considerably larger than that of the chroma signal so that its effect on the switching action for four diodes shall be dominant.

(f) The 4.43 MHz parallel tuned circuit formed by L_5-C_{43} and connected across the input terminals appears in the *V* signal demodulator only. Its function is to act as a low phase filter to absorb 7.8 kHz fluctuations present in the reference oscillator (RO) waveform that comes from the 180° PAL switch.

4.10 COLOUR SIGNAL MATRIXING

The function of the matrix network is used to combine the Y, *U* and *V* signal to reproduce the colour difference signal. The output of matrix network will have (R–Y), (B–Y) and (G–Y) signals, finally these signals are fed into the picture tube grids.

(i) The weighing factors employed at the transmitting end are: 0.493 for (B–Y) and 0.877 for (R–Y) signals.

(ii) To obtain (B–Y) the *U* signal must be multiplied by 2.03 (*i.e.*, 1/0.493) and *V* signal by 1.14 (*i.e.*, 1/0.877).

The transistor Q_2 is an inverting amplifier with signal input via R1 and Y signal input through R3. Based on the values of associated resistors, the *V* signal is multiplied by the ratio R7/R1 = 1.14 and the Y signal is multiplied by the ratio R7/R3 = 1. These two then add to give inverted *R* signal at the collector of Q_2, *i.e.*

$$Vr = -\frac{1.14(R - Y)}{1.14 + Y} = -(R - Y) + Y = -R$$

The transistor Q_1 operates in much the same way to produce:

$$Vb = -\frac{2.03(R - Y)}{2.03 + Y} = -(B - Y) + Y = -B$$

The transistor Q_5 and Q_6 are connected as emitter followers to avoid any mutual loading effects while combining R and B with Y to obtain G.

As shown on Figure 4.14, the matrixing of -R, -B and Y signals to obtain G takes place at the input of inverter amplifier Q_4. An (-R) signal of fixed amplitude derived from the emitter follower Q_3 and applied to Q_4 via R_{14}. Similarly (-B) signal of fixed amplitude derived from the emitter follower Q_6 is applied to Q_4 via R15. The Y signal input is derived from Q_1 via resistor R_{13}. With these inputs and appropriate choice of input and feedback resistance the transistor Q_4 multiplies (-R) signal by the ratio R_{18}/R_{14}. $(16.9k/33k) = 0.51$, (-B) signal by the ratio R18/R15. $(16.9/90k) = 0.186$ and Y signal by the ratio $R_{18}/R_{13}(16.9k/10k) = 1.69$. It then adds the three output together to yield the inverted G signal at the collector. This operation can be expressed as

$$-Vg = -(-0.15R - 0.186B + 1.69Y) = -G$$

note that $Y = 0.3R + 0.59G + 0.11B$.

The inverted R1, B and G colour signals are fed to the red, blue and green cathode electrodes of the picture tube by the way of variable resistors R_{11}, R_{12} and R_{19}. These controls are used to the level which on combining give correct grey scale tracking and therefore correct balance of reproduced coolers and black and white shades.

4.11 MERITS AND DEMERITS OF PAL SYSTEM

4.11.1 Merits

(i) The problem of differential phase errors has been successfully overcome in the PAL system. This is the main merit.

(ii) A manual hue control becomes unnecessary.

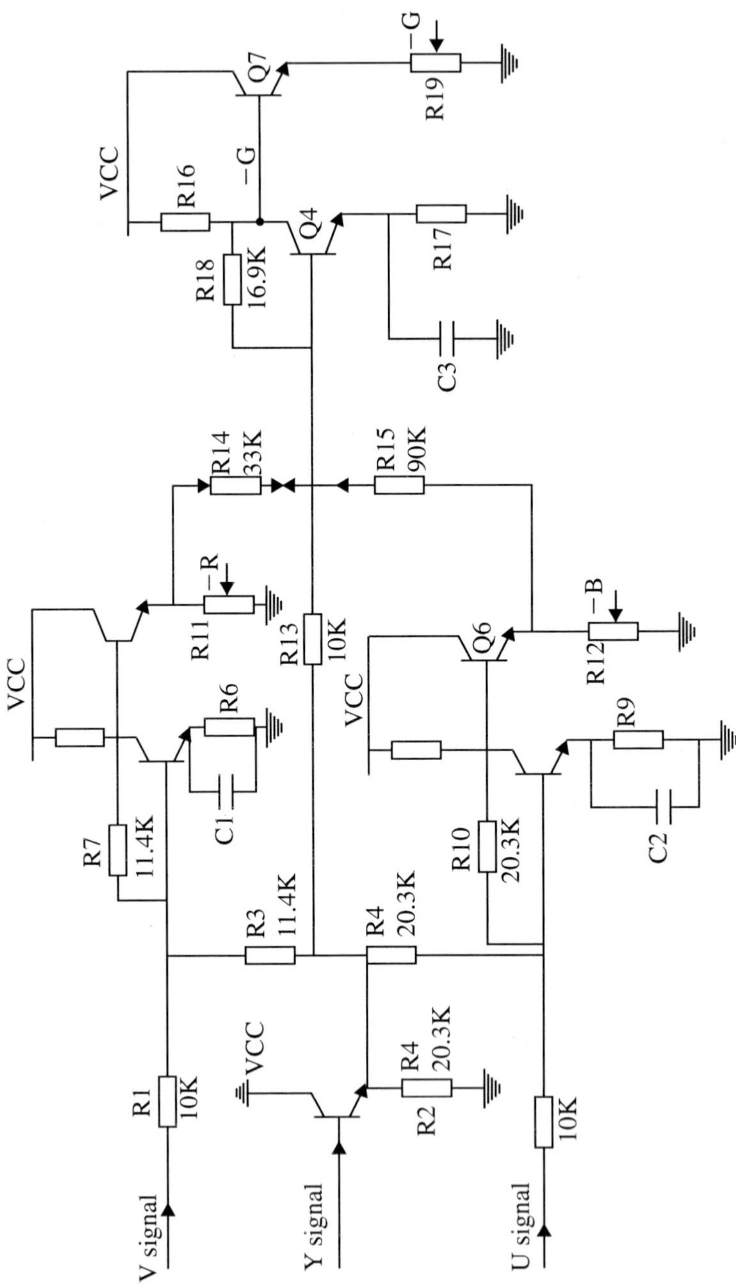

Figure 4.14 A sample circuit for RGB matrixing and driving amplifier.

4.11.2 Demerits

(i) Reduction in the vertical resolution of the chrominance signal.
(ii) The use of phase alternation by line technique and associated control circuitry together with the need of a delay line in the receiver makes the PAL system more complicated and expensive.
(iii) The receiver cost is higher for the PAL colour system.

In addition the PAL system presents problem in magnetic recording since a complete colour coding sequence requires eight field instead of four necessary in the NTSC system.

4.12 SECAM SYSTEM (SEQUENTIAL COLOUR A MEMORIE)

In order to avoid the problem of phase error on account of the subcarrier carrying both the chrominance signals simultaneous. H.de France proposed the SECAM (Sequential Colour a Memorie) System in 1958. It is compatible with 819 line black and white system. Later a 625 line 50 field 9 MHz channel bandwidth version of SECAM was developed and now it is used in France, Russia and some other countries.

4.12.1 SECAM Coder

In SECAM it can transmits and receive two chrominance signals on alternate lines. The colour difference signals are band limited to 1.5 MHz and then it is applied to the FM Modulator alternatively via an electronic switch at a horizontal line frequency rate as shown in Figure 4.15.

The subcarrier is frequency modulated by colour difference signal before transmission. For every Horizontal line only one weighted colour difference signal is transmitted either DR or DB. For first line DR is transmitted along with luminance and for second line DB is transmitted along with Y (Luminance) signal. This sequence follows up to the last line in the field. To maintain this sequence identification pulse is send during every vertical blanking period. Therefore the receiver can follow the same sequence of lines as in the transmitter.

The SECAM weighting factors and Ident signal are added in the matrix. An electronic switch which changes its mode during every line blanking interval directs DR and DB signals to the frequency modulator in a sequential manner *i.e.*, when DR is being transmitted on the line, then DB is not used and vice versa.

The band limited signals are pre-emphasized and then made to frequency modulate the subcarrier. The frequency modulator output, sync pulses and blanking pulses are passed to an adder network which also

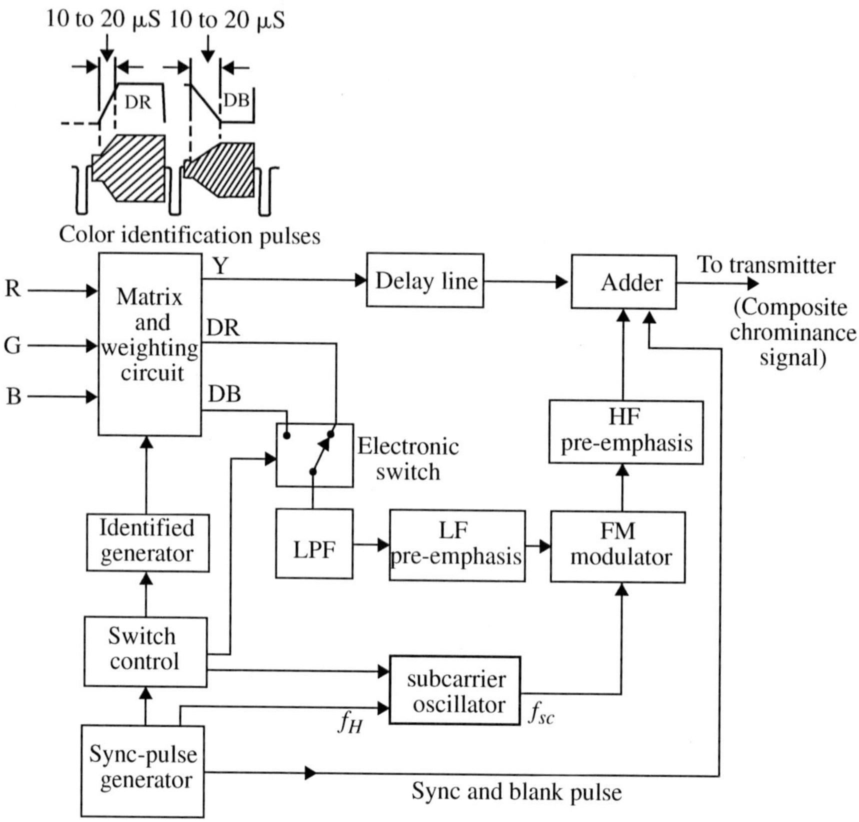

Figure 4.15 Block diagram of SECAM coder.

receives Y signal. The adder output produces a composite video signal and it is passed to the transmitter.

4.12.1.1 Weighting Factors

Different weighting factors are used and weighted colour difference signal $(R - Y)$ and $(B - Y)$ are termed as DR and DB.

$$DR = 1.9(R - Y)$$

$$DB = 1.5(B - Y)$$

To suppress the visibility of the dot pattern on monochrome reception two colour difference subcarrier are used.

$$R - Y \rightarrow 282F_L = 4.40625\,\text{MHz}$$

$$B - Y \rightarrow 272F_L = 4.250\,\text{MHz}$$

4.12.1.2 *Pre-emphasis*

The colour difference signals are band limited to 1.5 MHz. Before modulation of chrominance signals are low frequency pre-emphasis. And then it is passed to an FM Modulator. After Modulation the output passes through a high frequency pre-emphasis filter. This increases the amplitude of subcarrier and it increases the deviation and it increases the SNR ratio and interference is very much reduced.

4.12.2 SECAM Decoder

In SECAM receiver, the decoder transforms this sequential colour difference signal into simultaneous ones. A simplified block diagram of SECAM decoder is shown in Figure 4.16.

The band pass filter rejects unwanted low frequency components. The output from the band pass filter is amplified and fed to the electronic line-by-line switch via two parallel paths. The 64μs delay line is used as a one line memory device to decode the colour difference signals transmitted in alternate lines. The electronic switch is used to switch the DR and DB to the respective demodulator block. The switch is operated by line frequency pulses.

The switch is operated by line frequency pulses. If the phasing of the switch turns to be wrong means it directs DR and DB signals to wrong demodulators. In order to avoid this sensing circuit is used in the ident module.

The electronic switch directs the frequency modulated signals to limiters and frequency discriminators. The discriminators have a wider bandwidth used for detecting the FM sound broadcasts. After Demodulation the colour difference signals are de-emphasized with same time constant as employed in the pre-emphasis circuit.

In the receiver the matrix network combines the colour difference signals with the Y signal to give primary colour signals R, G and B which control the three electronic colour beam of the picture tube.

4.12.3 Conclusion

SECAM system is free from problems of hue and saturation due to colour difference signal being sent transmitted on the FM carrier. There is no need for the generation of a local reference subcarrier.

4.12.4 Merits of SECAM System

(i) SECAM system has several advantages because of frequency modulation of the subcarrier and transmission of one line at a time.

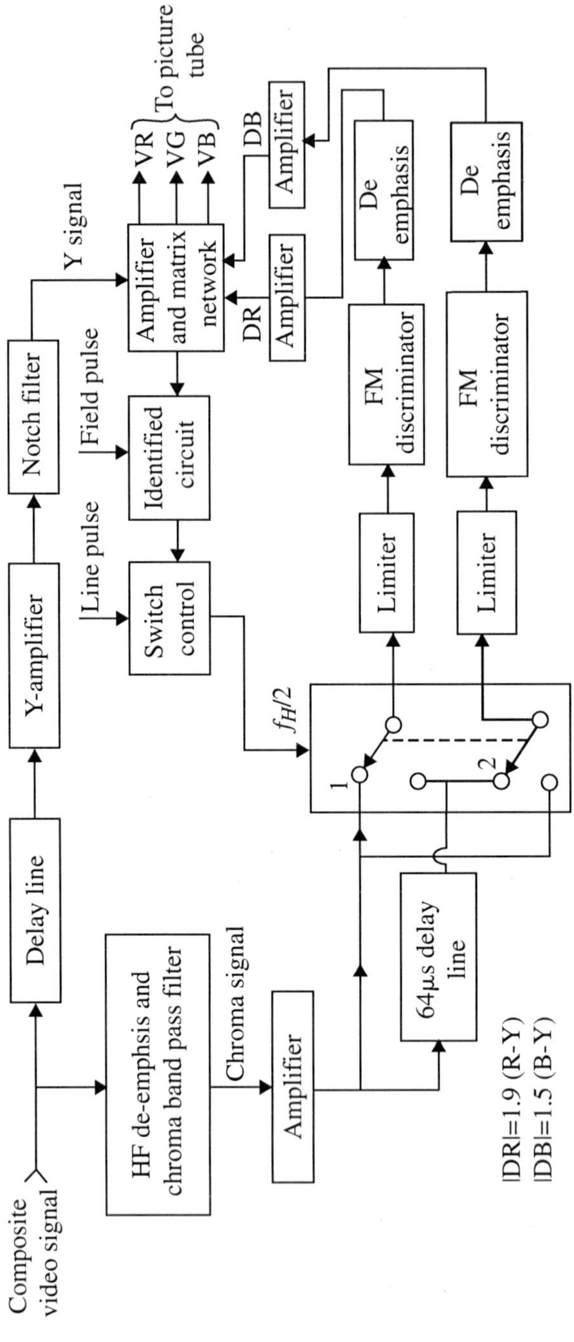

Figure 4.16 Block diagram of SECAM decoder.

(ii) SECAM receivers are immune to phase distortion. Both the luminance and chrominance signals are not present at the same time, there is no possibility of cross talk between the colour difference signals.
(iii) There is no need for the use of QAM at the transmitter and synchronous detectors at the receiver.
(iv) The receiver does not need automatic tuning control and automatic colour control.

4.12.5 Demerits of SECAM System

In SECAM system luminance is represented by the amplitude of voltage but hue and saturation are represented by deviation of the subcarrier. When a composite signal involving luminance and chrominance is faded out in studio operation, it is the luminance signal that is readily attenuated and not the chrominance. This makes the colour more saturated during fade to black.

KEY TERMS AND CONCEPTS

NTSC	The National Television System Committee has developed the NTSC colour TV system. It is compatible with existing American 525 line system.
PAL	The system of phase alternation by line (PAL), put forward by Prof. Walter Brunch, Telefunken Laboratories in the Federal Republic of Germany. The differential phase error that cause a phase distortion in the reproduced colours, can be overcome by shifting one of the colour difference signal phase by 180° on the alternate lines of scanning.
SECAM	In order to avoid the problem of phase error on account of the subcarrier carrying both the chrominance signals simultaneous. H.de France proposed the SECAM (Sequential Colour a Memorie) system in 1958.
FCC	Federal Communication Commission.
CCIR	International Radio Consultative Committee.
RCA	Radio Corporation of America.
Colour burst	The transmitted signal does not contain the subcarrier frequency but it is necessary to generate it in the receiver for proper detection of colour subcarrier. To ensure this, a short sample of the subcarrier (8 to 10 cycles subcarrier frequency) is called 'colour burst'.

ACC amplifier	Automatic Colour Control Amplifier, depending upon the amplitude of colour burst, the ACC will develops a DC control voltage this DC voltage will shift the operating point of the transistor and the gain of the amplifier is automatically changed.
Chroma band pass amplifier	It consists of chroma filter and two stages of chroma amplifier. The chroma subcarrier signals and chroma burst signals are separated by chroma filter from the incoming composite video signal. A band pass amplifier is used to amplify the chroma signals.
Colour killer circuit	During monochrome reception, no colour burst signals are present. So the output of the burst amplifier is zero. The colour killer circuit is used to turn off the chroma amplifier.
Phase discriminator	The phase discriminator compares the reference subcarrier oscillator frequency with the burst signal frequency.
Reference oscillator	This is a subcarrier oscillator generates a frequency of 4.43 MHz.
Matrix network	The matrix network is used to combine Y, U and V signal. The output of this matrix is the colour difference signal (R–Y), (B–Y) and (G–Y). Finally these signals are fed into the picture tube grids.
PAL delay line	The PAL delay line is used to delay the chrominance signal by almost exactly one line period of 64μs. Because of this delayed line $(U+V)$ signal and $(U-V)$ signals are generated.
Colour demodulator	The function of colour demodulator is used to detect the U and V signals and the modulated carrier.
Burst phase discriminator	The burst phase discriminator is used to compare the phase of the colour burst and subcarrier frequency generated by the reference oscillator. If there is any phase difference, which will produce a DC control voltage (error voltage) which is fed to reference oscillator.

Chroma band pass amplifier	Chroma band pass amplifier section consists of chroma filter and two stage chroma amplifiers. This is used to filter out the colour burst and sub-carrier signals. These controls can be used in the front panel of the receiver.
Identification pulse	To maintain this sequence identification pulse is send during every vertical blanking period. Therefore the receiver can follow the same sequence of lines as in the transmitter.
HF-pre-emphasis	This increases the amplitude of subcarrier and it increases the deviation and it increases the SNR ratio and interference is very much reduced.

REVIEW QUESTIONS

Part A

1. Write the basic principle of colour killer circuit.
2. What are the main functions of a colour decoder?
3. State the merits and demerits of PAL system.
4. Describe the PAL burst swing on alternate lines.
5. List the functions of the following of a PAL-D colour receiver mentioning their input and output (a) burst phase discriminator, and (b) colour killer circuit.
6. Justify the choice of 3.579545 MHz as the subcarrieer frequency in the NTSC system.
7. List the four major limitations of NTSC standard.
8. Draw the block diagram of the circuit that separates U and V colour signals.
9. State the merits of RGB matrix in PAL-D TV receivers.
10. Mention the major features of PAL system.
11. Write short notes on NTSC system.
12. Differentiate between NTSC, PAL and SECAM.
13. What is the main use of the identification pulses?
14. What do you mean by automatic frequency tuning?
15. What is the use of colour subcarrier oscillator?
16. How the phase error cancelled in the PAL system?
17. Give the abbreviation for NTSC, SECAM and PAL.
18. Define PAL-D colour system.
19. Write a short notes on colour subcarrier frequency of PAL-D system.

20. What is the use of chrominance band pass filter?
21. What do you mean by compatibility?
22. State the merits and demerits of the SECAM system.
23. Why it is necessary to prevent any deviations of IF frequencies in a colour receiver?

Part B

1. Explain how the differential phase error is continuously corrected in the PAL system while affecting QAM of the colour difference signals.
2. Show the colour killer circuit and explain how the chrominance signal remains interrupted during monochrome transmission show ident signal is generated.
3. Explain the delay line method of separating U and V signals in PAL receiver.
4. Describe the circuit of colour burst blanking first stage of chroma band pass amplifier. Explain the circuit of the saturation control in the chroma band passes amplifier.
5. Explain the generation and modulation details of I and Q signals in an NTYSC colour TV system. Why the colour subcarrier is fixed at 3.579545 MHz?
6. Briefly describe the characteristics of NTSC, PAL and SECAM TV standards, compare them and explain why India chose PAL-B standard?
7. With a neat block diagram explain the operation of PAL coder.
8. Draw the circuit and explain the operation and working principle of the burst phase discriminator.
9. Explain the subcarrier generation and control in the PAL colour receiver.
10. Explain in detail the NTSC colour receiver system.
11. Draw the block schematic that produces R, G, B signals and Y, U, V signals in PAL-D colour receiver. Explain its signal flow.
12. Describe the operation of the synchronous demodulator circuit to provide U and V video signals.
13. Describe briefly the basic difference between the SECAM and other colour TV systems. Draw a simple block diagram of a SECAM decoder and explain how the colour signals are recovered what is the purpose of one H delay line.

QUESTIONS AND ANSWERS

1. **Give the abbreviation for NTSC, SECAM and PAL.**
   (a) NTSC—National Television System Committee System.
   (b) PAL—Phase Alternation by Line System.
   (c) SECAM—Sequential Colour a Memory System.
2. **Why is video amplifier required?**
   The amplitude of the composite video signal at the output of video detector is not enough to drive the picture tube directly. Therefore further amplification is required. This is done by video amplifier.
3. **Write a short notes on PAL system.**
   The system of PAL, put forward by Prof. Walter Brunch, Telefunken Laboratories in the Federal Republic of Germany. It is a variant of the NTSC system. The differential phase error that cause a phase distortion in the reproduced colours, can be overcome by shifting one of the colour difference signal phase by 180° on the alternate lines of scanning. Here the colour difference signal (R–Y) and (B–Y) are called '*U* and *V*'.
4. **What do you understand by PAL-D colour system?**
   The use of eye as the averaging mechanism for the correct hue is the basic concept of simple 'PAL' system. Beyond a certain limit, the human eye see the effect of colour changes on alternate lines hence the system needs modification. Considerable improvement found in the system of a delay line is used to do the averaging first and then present the colour to the eye. This is called PAL-D or delay line PAL method and is most commonly employed in PAL receivers.
5. **What do you mean by colour burst?**
   In PAL system the two carrier components are suppressed in the balanced quadrature modulator it is necessary to regenerate at the receiver for demodulation. For this, 8 to 10 cycles of the colour subcarrier oscillator output at the encoder is transmitted along with other sync pulses. This sample of the colour subcarrier called colour burst, is placed at the back porch of each horizontal blanking pulse pedestal.
6. **What is swinging burst?**
   The PAL burst phase actually swings 45 about the $-U$ axis from line to line and indicates the same sign as that of the *V* signal, thus the switching mode information is the swinging burst. This is known as swinging burst.
7. **Mention some features of PAL system.**
   Features of PAL system are as follows:
   (a) The weighted (B–Y) and (R–Y) signals are modulated without being given a phase shift of 33 as is done in the NTSC system.

(b) On modulation both the colour difference signals are allowed the same bandwidth of about 1.3 MHz.
(c) The colour subcarrier frequency is chosen to be 4.43 MHz.
(d) The weighted colour difference signals are quadrature modulated with the subcarrier.

8. What is the use of chrominance band pass amplifier?

The chroma band pass amplifier selects the chrominance signal and rejects other unwanted components of the composite signal.

9. What do you mean by Automatic Color Control?

The ACC circuit is similar to the AGC circuit used for automatic gain control of RF and IF stages of the receiver. It develops a dc control voltage that is proportional to the amplitude of the colour burst.

10. Write short notes on colour killer circuit.

When a monochrome transmission is received there is no input to the colour killer and no positive voltage is developed. Therefore no input is given to the second chroma amplifier from the colour killer circuit, it blocks the second chroma amplifier. Thus it prevents the colour noise on black and white picture.

11. Mention merits of PAL system.

The problem of differential phase errors has been successfully over come in the PAL system and a manual hue control becomes unnecessary.

12. Mention demerits of PAL system.

The demerits of PAL system are:

(a) Reduction in the vertical resolution of the chrominance signal.
(b) Use of phase alternation by line technique and associated control circuitry together with the need of a delay line in the receiver makes the PAL system more complicated and expensive.
(c) Receiver cost is higher for the PAL colour system.

13. Why the scaling factors are chosen for colour difference signal in PAL System?

The scaling factors are chosen for colour difference signal in PAL System for:

(a) Preventing the over modulation of colour difference (R–Y) and (B–Y) signals necessary scaling factors are employed to reduce its magnitude before the quadrature amplitude modulation.
(b) Scaling factors chosen for this colour difference (R–Y) and (B–Y) are:

$$U = 0.439(B - Y) \qquad \text{and} \qquad V = 0.877(R - Y)$$

(c) At the receiver after demodulation an inverse scaling factor is employed to restore U and V to their original values of (R–Y) and (B–Y).

14. Why 64 μs delay line is used in PAL-D decoder section?

The PAL delay line is used to delay the chrominance signal by almost exactly one line period of 64μs. Because of this delayed line $(U + V)$ signal and $(U - V)$ signals are generated.

15. What do you mean by Automatic Frequency Tuning?

AFT is used to improve the stability of the oscillator circuit, some drift does occur on account of ambient temperature changes, component aging, power supply voltage fluctuation and so on. The fine tuning control is adjusted to get a sharp picture.

16. How the *U* and *V* signals are demodulated in PAL-D system?

To separate the U and V signals, adder and subtractor circuits are used. The colour demodulator is used to detect U and V signals. The reference carrier of oscillator is used to provide phase shift for U and V signals. And its frequency is available to U and V demodulators are as follows:

(a) Phase shifted by 90° and given to U demodulator.

(b) Phase shifted for each alternate line by 180° and locked to +ve and −ve signals.

17. What is the use of the matrix network?

This matrix network is used to combine the U and V and luminance (Y) signal. The output of this matrix is the colour difference signal (R–Y), (B–Y) and (G–Y). Finally these signals are fed into the picture tube grids.

18. Write short notes on burst separator.

The burst separator circuit has the function of extracting 8 to 10 cycles of reference colour burst which are transmitted on the back porch of every horizontal pulse. The circuit is tuned to the subcarrier frequency and is keyed on during the fly back time by pulses derived from the horizontal output stage.

19. Write short note on burst phase discriminator.

The burst phase discriminator is used to compare the phase of the colour burst and subcarrier frequency generated by the reference oscillator. If there is any phase difference, which will produce a DC control voltage (error voltage) which is fed to reference oscillator. In this way the frequency and phase are kept locked in synchronism with the colour burst.

20. What is the use of reference oscillator?

During transmission side the U and V signals are modulated by double balanced-suppressed carrier. And it is necessary to regenerate the subcarrier of U and V signals at the receiver side to demodulate the colour difference signal. Here the reference oscillator is used to regenerate the subcarrier signal of 4.43 MHz.

21. How the phase error is cancelled in the PAL system?

In PAL system phase error can be overcome by shifting one of the colour difference signal phase by 180° on the alternate lines of scanning in the quadrature modulation. This effect leads to the inverting the error on the alternate lines.

22. Write short note on colour subcarrier frequency of PAL-D system.

The colour subcarrier frequency of 4.43 MHz is produced with a crystal controlled oscillator. To accomplish minimum raster disturbance through the colour subcarrier it is important to maintain correct frequency relationship between the scanning frequencies and subcarrier frequency. Therefore, it is usual to count down from the subcarrier frequency to twice the line frequency pulses.

23. Write notes on NTSC system.

NTSC system is compatible with 525 line American system. In NTSC system, the I and Q singles carry the colour information. The I signal is equal to 0.74 (R–Y)-0.27(B–Y) and lies in the region 33° counter clockwise to +(R–Y) where eye has maximum colour resolution. The Q singles that is equal to 0.48 (R–Y) + 0.41 (B–Y) occupies a region 33° counter clockwise to +(B–Y) axis where colour resolution of the eye is least, it is allowed to have a maximum bandwidth of 1.5 MHz. The bandwidth of Q signal is restricted to 0.5 MHz.

24. What is the use of colour subcarrier oscillator in NTSC system?

The function of subcarrier oscillator is to generate a carrier wave output at 3.57 MHz and feed it to the demodulators. The subcarrier frequency is maintained at its correct value and phase by the APC circuit.

25. State of limitations of the NTSC system.

The limitations of the NTSC system are:

(a) The NTSC system is sensitive to transmission path differences which introduce phase errors. The phase errors results in colour change in the picture.

(b) All the transmitter, phase change in the chroma signal take place when changeover between programs of local and television network systems takes place and when video tape recorders are switched on.

(c) The chroma phase angles are also affected by the level of the signal while passing through various circuits.

(d) In addition crosstalk between demodulator outputs at the receiver causes colour distortion. All this requires the use of an Automatic Tint Control (ATC) circuit with the provision of manually operator tint control.

26. State the bandwidth required for NTSC system.

The luminance signal Y can be accommodated with full bandwidth of 4.2 MHz and the chrominance signal are accommodated with 1.3 and 0.5 MHz respectively (< 2 dB at 1.3 MHz for I and 0.5 MHz for Q). And they are used to produce composite video signal. The bandwidth are limited in order to reduce the cross talk interference between the monochrome and colour signals and hence it is convenient to accommodate these (Y, I and Q) signals in the channel bandwidth of 6 MHz.

27. What is the difference between NTSC, PAL and SECAM?

The difference between the SECAM system on one hand and NTSC and PAL on the other is that the later transmit and receive two chrominance signals simultaneously while in the SECAM system only one of the two colour difference signal is transmitted at a time.

28. Write short note on SECAM system.

In order to avoid the problem of phase error on account of the subcarrier carrying both the chrominance signals simultaneous. H.de France proposed the SECAM (Sequential Colour a Memorie) System in 1958. It is compatible with 819 line black and white system and it can transmits and receive two chrominance signals on alternate lines. The colour difference signals are band limited to 1.5 MHz and then it is applied to the FM Modulator alternatively via a electronic switch at a horizontal line frequency rate.

29. State the merits of SECAM system.

The merits of SECAM system are:

(a) SECAM system has several advantages because of frequency modulation of the subcarrier and transmission of one line at a time.

(b) SECAM receivers are immune to phase distortion. Both the luminance and chrominance signals are not present at the same time, there is no possibility of cross talk between the colour difference signals.

(c) There is no need for the use of QAM at the transmitter and synchronous detectors at the receiver.

(d) The receiver does not need automatic tuning control and automatic colour control.

30. State the demerits of SECAM system.

In SECAM system luminance is represented by the amplitude of voltage but hue and saturation are represented by deviation of the subcarrier. When a composite signal involving luminance and chrominance is faded out in studio operation, it is the luminance signal that is readily attenuated and note the chrominance. This makes the colour more saturated during fade to black.

31. What do you mean by high frequency pre-emphasis?

In SECAM system, the chrominance signals are pre-emphasized before modulation. After modulating the carrier with the pre-emphasized and weighted colour difference signals, another form of pre-emphasis is carried out on the signals. This increases the amplitude of subcarrier and it increases the deviation and it increases the SNR ratio and interference is very much reduced.

32. What is the use of line identification pulses?

In SECAM system, the switching of DR and DB signals line by line takes place during the line sync pulse period. The sequence of switching continues without interruption from one field to the next and is maintained through the field blanking interval. However it is necessary for the receiver to be able to deduce as to which line is being transmitted. Such an identification of the proper sequence of colour lines in each field is accomplished by identification pulses.

33. Write note on luminance channel.

The video amplifier in the luminance channel is DC coupled and has the same bandwidth as in the monochrome receiver. It is followed by a delay line to compensate for the additional delay the colour signal suffers because of limited band pass of the chrominance amplifier. This ensures time coincidence of the luminance and chrominance signals. The channel also includes a notch filter which attenuates the subcarrier by about 10 dB. This helps to suppress the appearance of any dot structure on the screen along with the colour picture.

CHAPTER 5

ADVANCED TELEVISION SYSTEM

KEY TOPICS

- ✓ Satellite TV technology
- ✓ Cable TV
- ✓ Video disc recording and playback
- ✓ Stereo sound in TV
- ✓ Teletext broadcast receiver
- ✓ Digital television receiver
- ✓ Flat panel display TV receiver
- ✓ Stereo sound in TV
- ✓ 3D TV
- ✓ EDTV
- ✓ Digital equipments for TV studios

LEARNING OBJECTIVES

After reading this chapter, the learner should be able to:

- ♦ During television signal transmission, the difficulty of line of sight restriction is overcome by satellites TV technology.
- ♦ CATV was initially developed for the benefit of communities in rural areas on shadow zones, which distribute a good quality television signal to a very large number of television receivers.
- ♦ The video disc is essentially a gramophone record with pictures on it. The first video disc was the laser optical and the second video disc system is based on the capacitance principle.
- ♦ Digital television means digital processing of TV signals both at the transmitter and in the receiver.
- ♦ The stereo sound transmission in television uses the basic FM–FM stereo multiplex system, but left and right channel signals are accommodated in different way.
- ♦ A projection television uses a projector to create a small image from a video signal and magnify this image onto a viewable screen.
- ♦ A flat panel television receiver is one in which the conventional picture tube is replaced by flat panel display.
- ♦ EDTV is mainly intended for direct broadcasting satellite system. To improve the performance of TV systems it requires different transmission standards but it maintain the present line numbers and field rates.

5.1 SATELLITE TV TECHNOLOGY

The first satellite was launched in 1957 and then the satellite communication opened up area for communication. The line of sight range was limited by curvature of the earth, there fore long distance communication was required, to adopt some other solutions. The greatest advantage of satellite communication is that there is direct line of sight communication directly towards earth. This advantage of satellite communication solved many problems in the field of communication. The long distance communication requires the cables and repeaters. But satellite communication does not require, any cable or repeaters stations. With the satellite, it is possible to cover a specific limited area for transmission. This area is called foot print area of the satellite or coverage area. Signals can be relayed via a satellite to earth in two ways:

(i) Using an artificial earth satellite as a passive reflector for radio waves.
(ii) Using an artificial earth satellite as an active repeater.

A satellite is placed in a elliptical orbit with an pogge (the point in the orbit farthest from the earth) of 39,957 km and a perigee (the point in the orbit closest to the earth) of 548 km and inclined at an angle of 65 to the equator as shown in the Figure 5.1(a) travels at a varying speed which is a maximum at perigee and minimum at a pogge. In other words, the satellite traverses the region around perigee faster than the region around a pogge. Because of this contact between the satellite and the ground based stations is maintained for about 9 hours.

5.1.1 Satellite Frequency Plans and Allocations

The frequencies to be allocated to various satellites are established by ITU. This provides the international regulation of frequencies, the maximum radiation levels from space, coordination with terrestrial systems and use of satellite locations in the space *etc.* The special frequency bands are provided for military and non-military applications. At present, C and X bands are widely used by satellites. The band of higher frequencies (k9, k6 and *V*-band) are made available with higher bandwidth. The bandwidth is roughly 10% of the carrier frequency of that band.

5.1.2 Satellite Communication System

During television signal transmission, the difficulty of line of sight restriction is overcome by satellites. Here the transmitting station sends the information signal to the satellite and in turn it receives and transmits

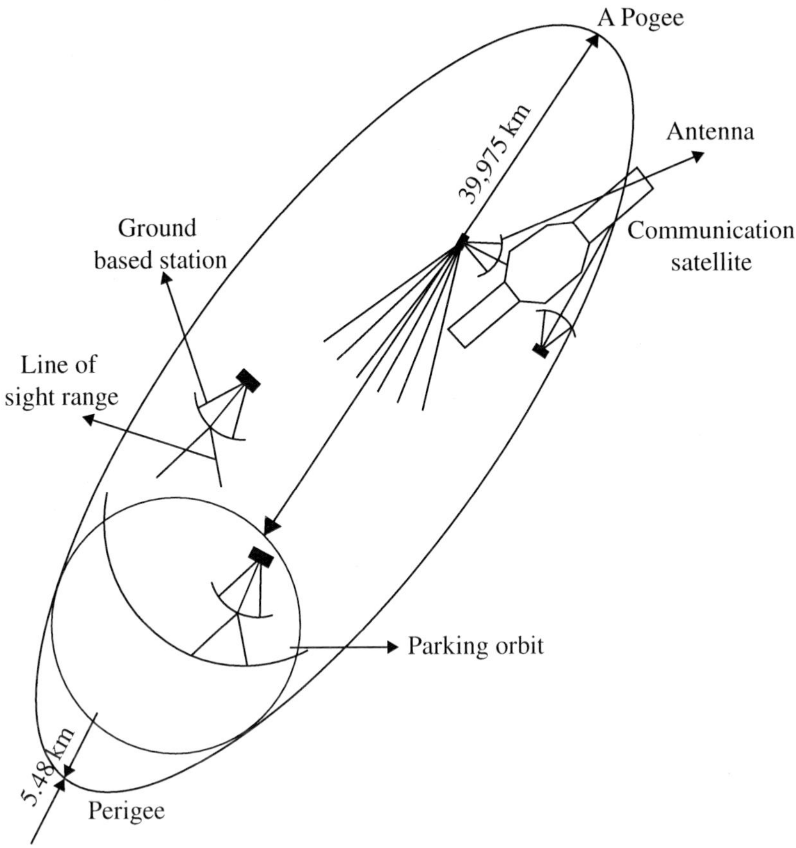

Figure 5.1 (a) General satellite diagram.

the signal to the receiving station. The frequencies used for this purpose are in microwave frequency range from (3 to 30 GHz). During UHF, the atmosphere does not act as a barrier and signals can able to travel out into space and return back without any absorption or deflection. Figure 5.1(b) shows the basic principle of a communication satellite.

5.1.2.1 Up-Link Frequency

The frequency at which the information signals are transmitted from earth station to the satellite is called 'uplink frequency'. Most the communication satellite uses 'C' band frequencies.

$$\text{Uplink frequency range} = 5.925 \text{ to } 6.425\,\text{GHz}$$

$$\text{Bandwidth} = 500\,\text{MHz}$$

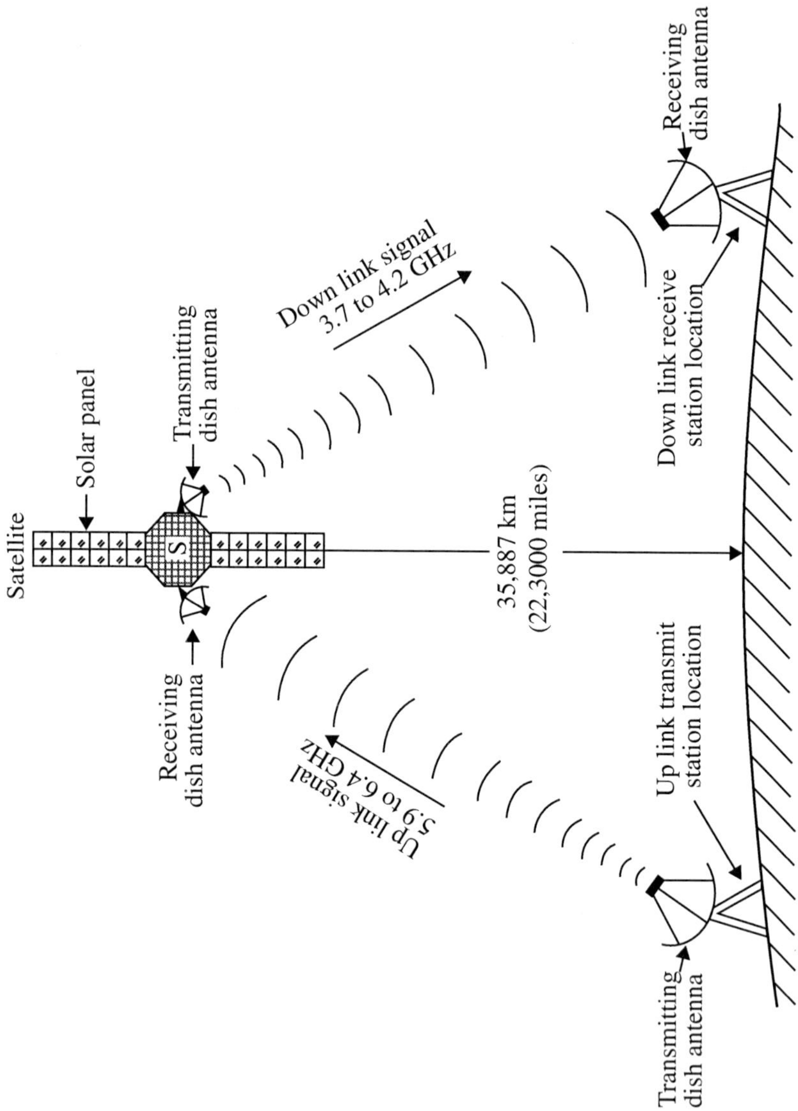

Figure 5.1 (b) Signal path in a satellite communication system.

5.1.2.2 Down-Link Frequency

The frequency at which the information signals are transmitted from satellite to the earth receiving station is called 'down-link frequency'.

$$\text{Down-link frequency range} = 3.7 \text{ to } 4.2\,\text{GHz}$$
$$\text{Bandwidth} = 500\,\text{MHz}$$

During transmission, the transmitting dish antenna transmits the signal in the direction of unseen satellite. And the satellite is equipped with own antenna and frequency converters. It receives uplink frequency signal and fed to its receiver circuit. The receiver is designed to amplify and converts the uplink frequency into down link frequency signal and then it is fed to the satellite transmitter and it transmits back to earth station through another antenna. This is shown in Figure 5.1(b).

5.1.2.3 Foot Print

The satellite communication does not require cable or repeater stations. Within the satellite itself it is possible to cover a specific limited area for transmission. This area is called foot print area of satellite or coverage area.

The satellite antenna is designed for a wide angle beam called 'global-beam' which extends an angle of 18°, and covers maximum possible surface area of the earth. The coverage area is about $1/3^{rd}$ of surface area of earth.

Points to Remember

- One degree (1°) of satellite antenna beam-width covers an earth surface area of about 300,000 square kilometres.
- Four degree (4°) of satellite antenna beam width will cover the whole India.

5.1.2.4 Satellite Frequency Plans and Allocation

S.No.	Band	Uplink Frequency	Downlink Frequency
1.	C Band	5.925 to 6.425 GHz Bandwidth = 500 MHz	3.7 to 4.2 GHz Bandwidth = 500 MHz
2.	KU band	14 to 14.5 GHz Bandwidth = 500 MHz	10.95 to 12.75 GHz

5.1.3 Satellite Transponders

A simplified block diagram of a transponder is shown in Figure 5.2. Attenuation of radio waves that propagate from earth to satellite is around 200 dB. There will be an equal attenuation from the satellite to earth will be 200 dB. It makes the total loss between earth transmitters to earth receiver to about 400 dB.

$$\begin{aligned}\text{Total loss} &= 200\text{ dB} + 200\text{ dB}\\ &= 400\text{ dB}\end{aligned}$$

For a satisfactory operation an earth station receiver needs an RF input power of about 10^{-12} watts. And then the total power radiated by the earth transmitter should be above 10^{-12} watts and 400 dB. *i.e.*, $[(-12+40) = 10^{-28}]$ watts. Thus it makes impossible to achieve.

This problem can be solved by providing a transmitter–receiver combination in satellite called transponder. The satellite transponder receives uplink frequency and provides high amplification and frequency translation. And retransmits the signal to down link frequency to earth stations.

5.1.3.1 Reason for Frequency Translation

(a) By using different frequency for uplink and downlink, frequency interference can be avoided.

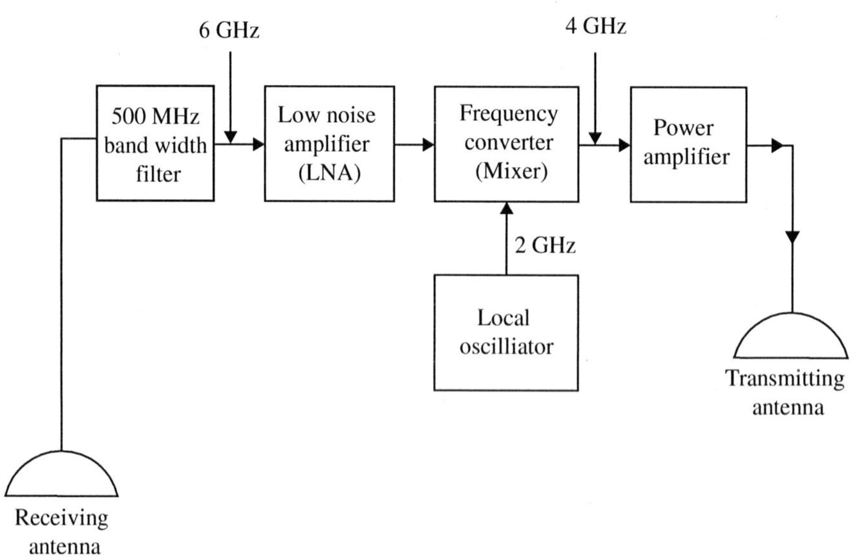

Figure 5.2 Block diagram of transponder.

(b) The transponder cannot transmit and receive on same frequency, because the transmitter signal can overload the receiver and block the small uplink frequency signal and thereby prohibiting the communication.

5.1.3.2 Transponder Channels

Frequencies designed for 'C' band uplink and down link are 5.925–6.4 GHz and 3.7–4.2 GHz respectively. Thus the available bandwidth for the both up and down transmission is 500 MHz. Such a 500 MHz wide bandwidth is capable of accommodating of a 12 number of RF channels. Each channel has a bandwidth of 36 MHz. A 4 MHz guard band is allowed between channels to minimize adjacent channel. Interference as shown in Figure 5.3. The 36 MHz bandwidth of each transponder is wide enough and can handle up to 1,000 telephone conversations and full colour TV channel.

5.1.3.3 Transmitted Earth Station

A simplified block diagram of a transmitting earth station is shown in Figure 5.4.

The base band video and associated signals are frequency modulated around a centre IF frequency of 70 MHz. FM modulation preferred over AM modulation because it has high signal to noise ratio.

The FM modulated signals are up-converted to 6 GHz frequency range before high power amplification and transmission.

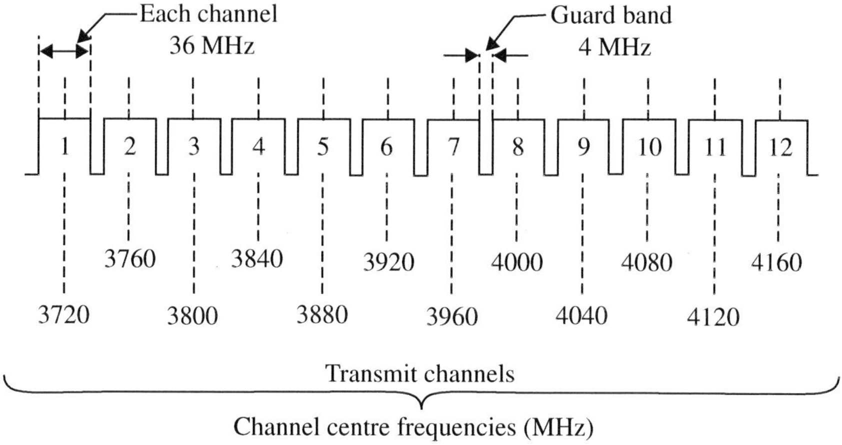

Figure 5.3 Channel bandwidth (36 MHz + 4 MHz) = 40 MHz and transmit centre frequency in a 'C' band communication satellite.

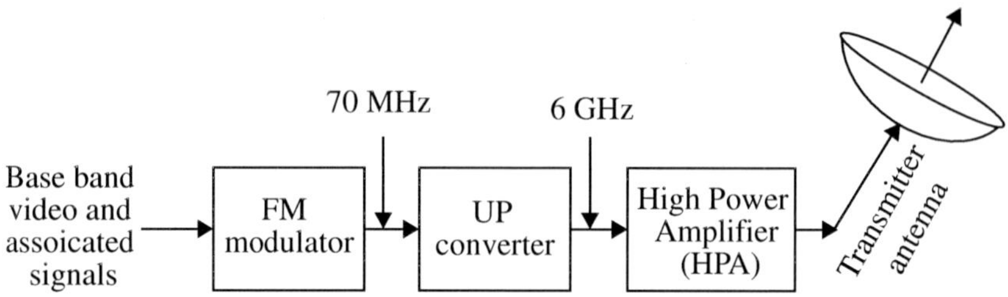

Figure 5.4 Simplified block diagram of a transmitting earth station.

5.1.3.4 Number of Transponders

The number of transponders provided is equal to the number of channels. Thus, normally, there would be 12 transponders, one for each particular channel. The 12 channel transponder is shown in Figure 5.5.

The uplink signal is picked up by the receiving antenna and channelized by a band-pass filter before feeding it to the corresponding transponder. Here the (LNA) Low noise amplifiers provides a sufficient amplification and it is frequency translated. This is done by the mixer circuit, where the original modulation is retained during this process. The mixer output is amplified again and fed to a band pass filter (BPF), the purpose of BPF is to remove all the uplink frequency.

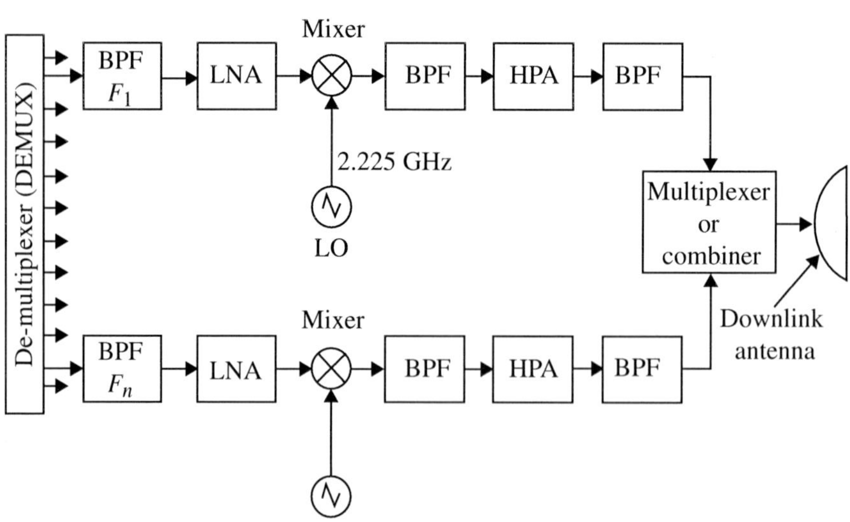

Figure 5.5 Channel transponder architecture.

Finally the downlink signal as amplified by (HPA) high power amplifier and the amplified signal is fed to BPF to further filter to retain only desired channel in between.

The resulting output is combined with output of other transponders and fed to downlink antenna. Note that each channel has its own mixer driven by local oscillator (LO). Whose frequency is 2.225 GHz on 'C' band.

5.1.4 Frequency Re-Use

In order to increase bandwidth and signal carrying capacity. The most commonly used technique is frequency re-use. In this system, a communication satellite is provided with two identical sets of 12 transponders. The two sets through operate at the same channel frequencies. The radiation pattern of one set is choose with vertical polarization (V) and other with horizontal polarization (H).

A signal that is transmitted with vertical polarization will not be received on an antenna that is horizontally polarised. As such that, if the antenna is set suitably there will be no interference between the two signals. Thus a 24 channel capability is achieved with the same bandwidth of 500 MHz.

5.1.5 Earth Station

The communication is established to the satellite through the earth station. The earth station can be located on the top at the sea or it can be located on the space craft or actually on the earth. The location of the earth station is decided depending upon the ease of control of satellite and function of the satellite. The type of earth station depends upon:

(i) Function of the station.
(ii) Type of the service.
(iii) Frequency bands used.
(iv) Transmitter.
(v) Receiver.
(vi) Antenna characters.

Figure 5.6 shows the block diagram of earth station.

5.1.5.1 TVRO Earth Station

The receive subsystem of a TVRO earth station is designed to receive a down link frequency from two transponders of (INSAT), each channel is 36MHz with 4 MHz guard band, *i.e.*, 40 MHz (36 MHz $\pm$ 4 MHz) it occupy 2.55 to 2.635 GHz with centre frequency of 2.575 and 2.615 GHz respectively.

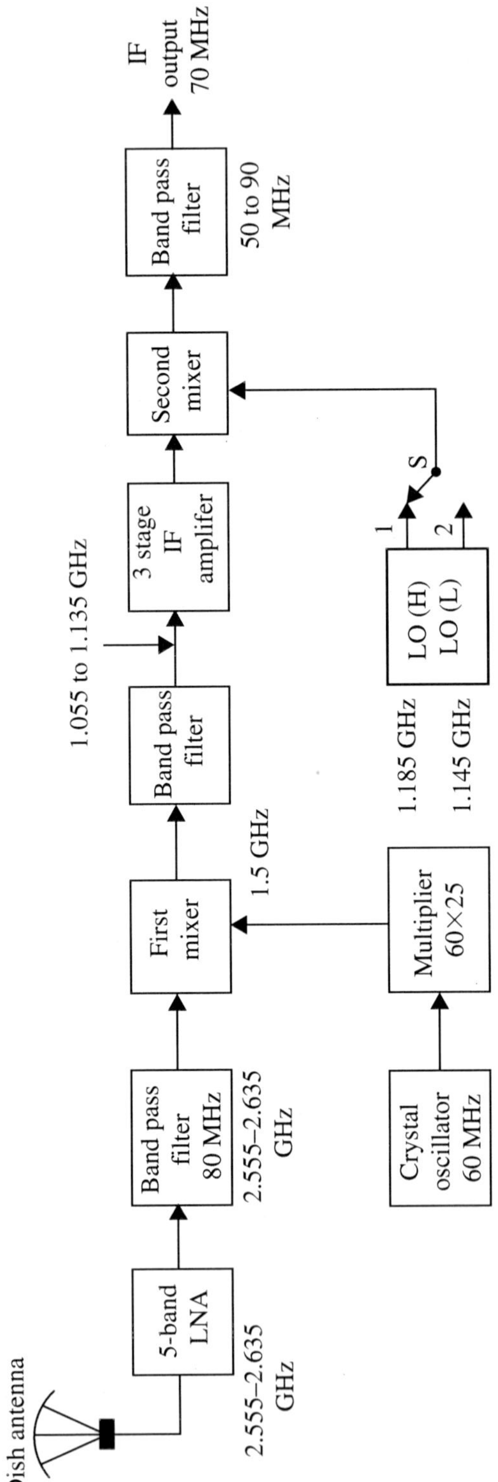

Figure 5.6 Block diagram of receiver subsystem of through earth station.

A double conversion circuits is provided to translate downlink signals to an IF of 70 MHz after passing through the S-band LNA (Low Noise Amplifier). The band pass filter (BPF) is used to select the desired frequency of 80 MHz (*i.e.*, $2.635 - 2.555 = 80$ MHz). This feeds the signal to the first mixer with local oscillator frequency (LO) of 1.5 GHz. A crystal oscillator is used to generate 60 MHz sinusoidal signal and its output signal is passed to an multi-harmonic generator, which produces an Continuous Wave (CW) output at 1.5 GHz.

The BPF is used to pass the frequency band between 1.055 to 1.135 GHz followed by the mixer and the BPF output signal is amplified by 3 stage IF amplifier to increase the gain of about 40 dB.

The amplified signal is passed to an 2nd mixer to obtain an IF output of either of the two incoming channels. For this one VCO is tuned to 1.180 GHz other is at 1.145 GHz. A change over switch enables the selection of one of these depending upon the downlink channel to be selected. The difference IF output is ($1.145 - 1.095 = 50$ MHz) to ($1.145 - 1.055 = 90$ MHz) with centre IF frequency $= 70$ MHz. similarly the output is (50 to 90 MHz) with centre frequency of 70 MHz, but with the modulation of second channel. The BPF after the second mixer selects the IF signal (50 to 90 MHz) for further processing.

5.1.6 Demodulation to Base Band Signal

Figure 5.7 shows the block diagram of the earth station receiver's demodulation equipment that converts the FM modulated. IF signal at (70 MHz + 18 MHz) to the base band range of video and audio signals. A two stage band width limited IF amplifier followed by a BPF enabled IV (p–p) output at 70 MHz + 18 MHz, this signal is necessary for effective detection by the FM demodulator. A limiter process the detector to ensure equal amplified of all the sideband of FM signal.

A PLL type FM demodulator is commonly used to obtain video and sound IF base band range.

A de-emphasis circuit is provided at the demodulator to obtain an original amplitude of the modulating signal. The video and inter carrier sound output are separated by suitable band pass filter.

The satellite has provision for accommodating two audio channels, one centered around 5.5 MHz and other at 6 MHz. one of these is used for carrying audio in a regional language and other is used for Hindi or English. Depending on the desired output, the FM demodulator oscillator is set either 5.5 MHz or 6 MHz.

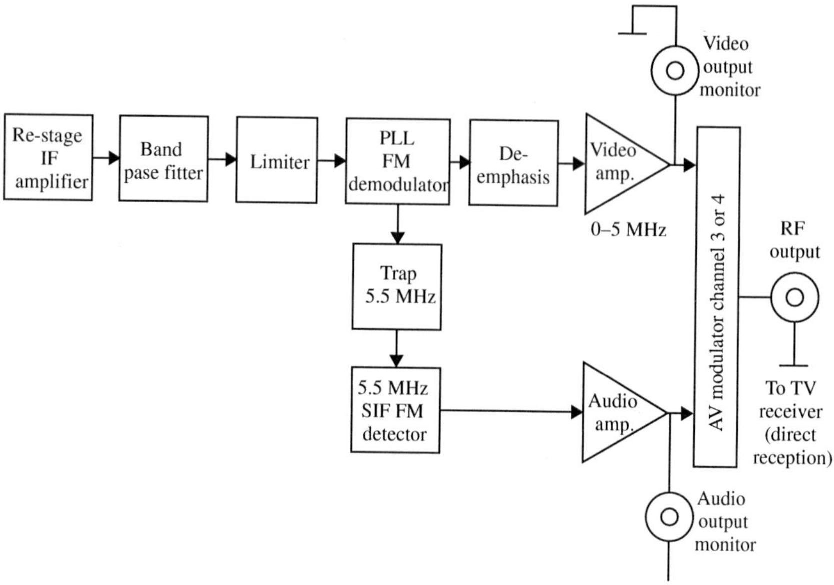

Figure 5.7 Block diagram of the demodulation section of the earth station receiver.

Video and audio amplifiers provide necessary amplification before feeding the signal to AV modulator, and then it modulates the audio and video signal with a carrier frequency of channel 3 or 4.

5.1.7 Geo-stationary Satellite

We have seen that the satellites orbits in the geo-stationary orbit complete one revolution per day and they appear stationary to the point on earth. The geo-stationary satellites are quite above the high intensity inner radiation belt. The stationary orbit is sunlit for about 99% of the time. Hence it can get solar energy from the sun. If therefore does not need heavy storage of energy. The batteries of the satellite always remain charged and provide power at the time of eclipses. Also the strength of earth magnetic field is very small (negligible) at the height of geo-stationary orbits.

Therefore the effects due to earth magnetic fields are neglected. The geo-stationary satellite is visible from around 42% of the earth surface. Therefore continues international and national communication can be covered by these satellites. The earth stations to transmit and receive information to and from the geo-stationary satellite are relatively simple and low cost. The earth stations in the same country or other countries can communicate with each other through these geo-stationary satellites.

5.1.8 Communication Satellites

Geo-stationary satellites used for communication are called 'communication satellites'. These satellites can offer services like point to point communication. Broadcasting of radio and TV signals, commercial, military, amateur, mobile communication, maritime, aeronautical or fixed points *etc.*, are depends on the coverage area. Communication satellites are classified into global, regional or domestic satellites. The international telecommunication satellites (INTELSAT) are also communication satellites shown in Figure 5.8. The communications like telephony, data and TV signal is carried out by these INTELSAT systems. Over 100 nations are inter-connected through 200 earth stations via INTELSAT systems. If the INTELSAT systems are placed at 120 to each other in geo-stationary orbit then they can cover complete earth surface. The earth station transmitting signals, to one of these satellites can be communicated from one satellite to another satellite. All these three satellites then transmit down the signals to earth. Thus any TV programme or information can be relayed at the same time through out the world.

5.1.9 Direct Broadcast Satellites

The block diagram of TV broad system form an Indian Nation System is shown in Figure 5.9. To avoid the crowed of satellites in the space and mixing of frequencies, an international committee regulates the frequencies to be used for particular applications. And also the distance between the two satellites. Permission for launching *etc.*, activities are carried out by this committee few channels of these satellite or complete satellite can be used for TV signal broadcasting. Since these satellite broadcast directly to the TV receivers, they are also called direct broadcast satellites, all these national and international channels which we observe today on the TV and cable network are brought to us by these DBS system. Depending on the orbit of the satellite it is called polar, synchros or geo-stationary satellite. For effective utilization of the frequency band provided to the satellite normally multiplexing is used to accommodate maximum channels.

5.1.10 Domestic Broadcast Satellite

Domestic broadcasting from geo-stationary satellite is another way that is used to supplement the microwave links and co-axial cable system for TV signal distribution all over the country. This system has much smaller coverage area as compared to an INTELSAT type of satellite which may cover approximately one third of the globe. Thus the domestic satellite can

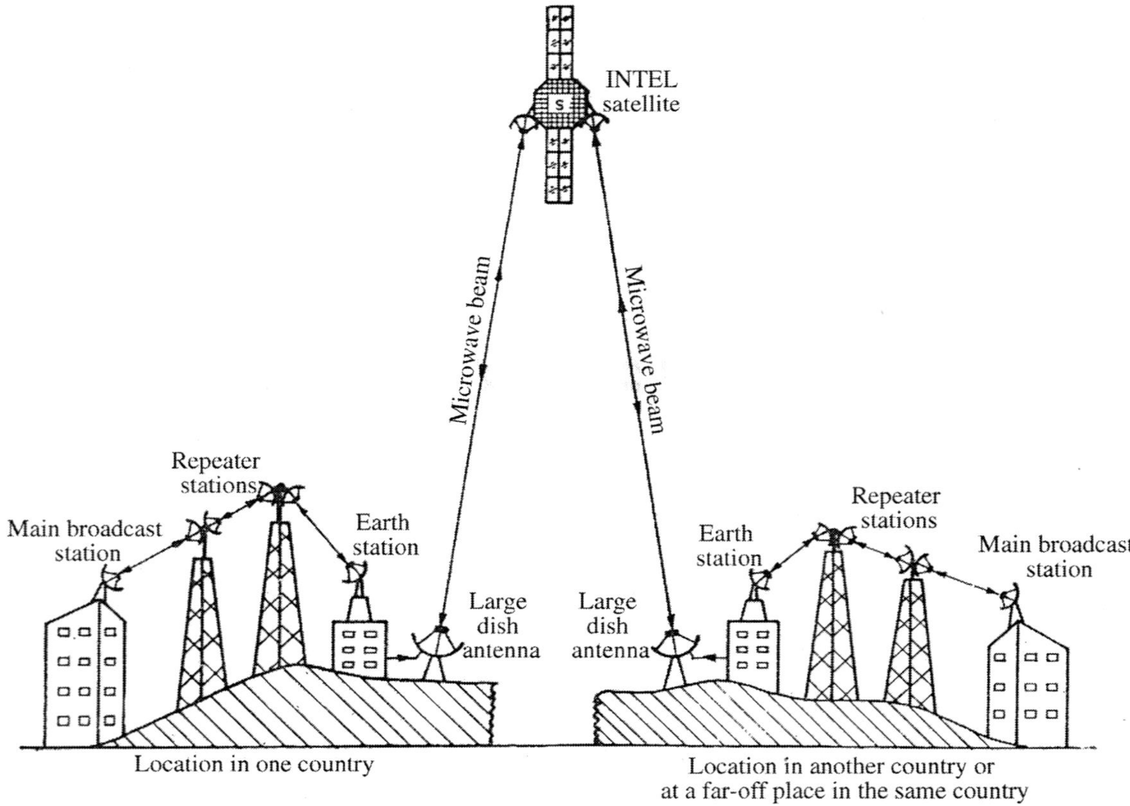

Figure 5.8 Typical set-up for television broadcast via an International communication satellite.

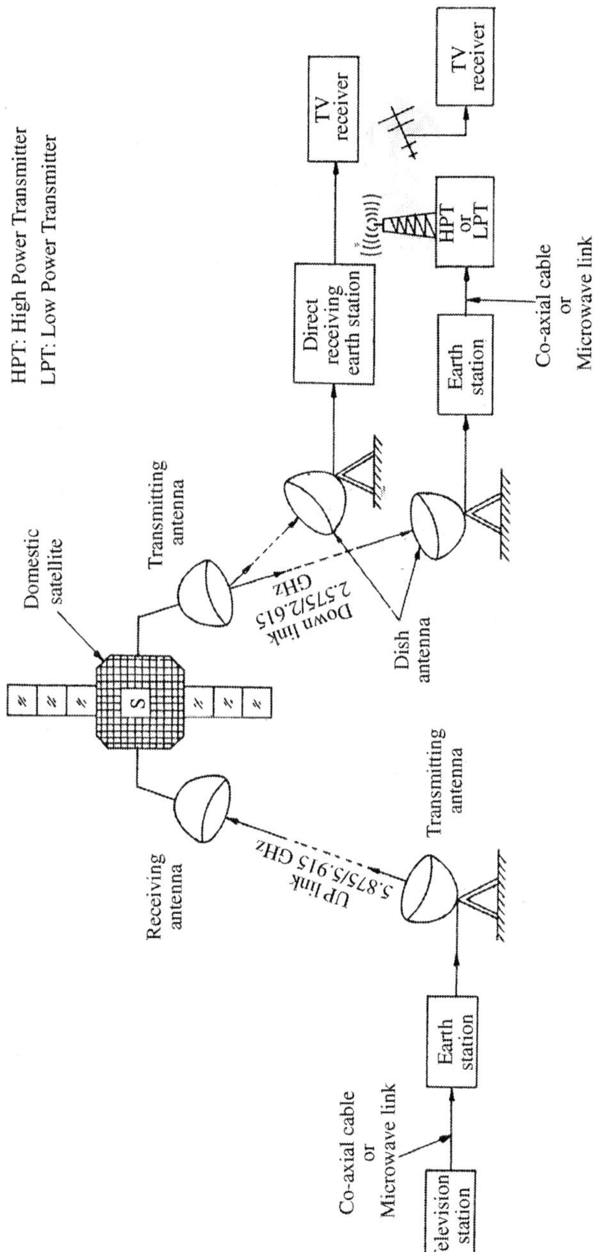

Figure 5.9 Block diagram of the TV broadcast system from an Indian National System.

be of medium power and can get supply fairly strong signal by restricting its foot print to a limited area. Consequently smaller receiving dish antenna and simpler earth station receivers can handle these signals usefully. In this typical setup the signal is beamed up the satellite from an origination station. The satellite can then re-broadcast the signal to a large number of fairly small and simple earth receiving stations as shown in Figure 5.10. Located all over the country. The signals on suitable conversion and processing are broadcast by low power transmitter (LPT) operating at different channels in the VHF band. In India there are above 800 such small stations that cover practically its entire populated area.

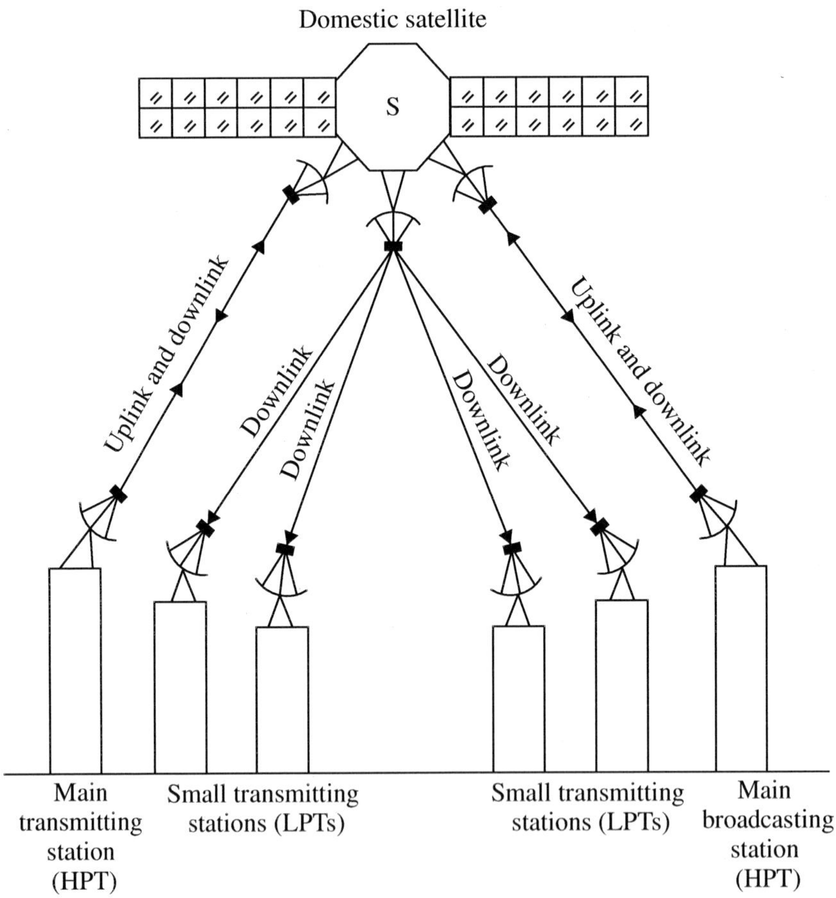

Figure 5.10 TV reception by domestic broadcast system where small LPT station distribute the signal over a large area.

5.2 CABLE TELEVISION

Cable television (CATV) was initially developed for the benefit of communities in rural areas on shadow zones, which distribute a good quality television signal to a very large number of television receivers. This shown in Figure 5.11. This system provides may different channels to subscribers who pay the fee for this service. With this CATV developed into a complex multi-channel system expanded in a big way in all the countries.

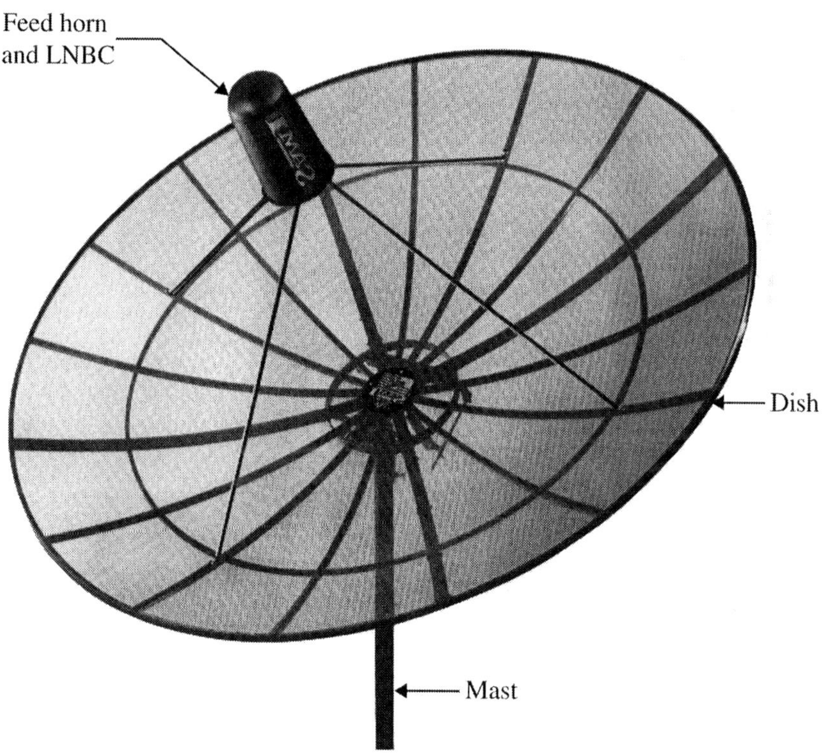

Figure 5.11 Dish antenna assembly with feed-horn and LNBC.

5.2.1 Cable Signal Sources

Figure 5.12 shows block diagram of low noise block converter where:

(i) A horn type parabolic dish antenna of large diameter is used to collect the signal arriving from various satellites.

(ii) It collects the signal arriving from it and refract them to a common point called focal point which is located in front and above the centre of antenna dish.

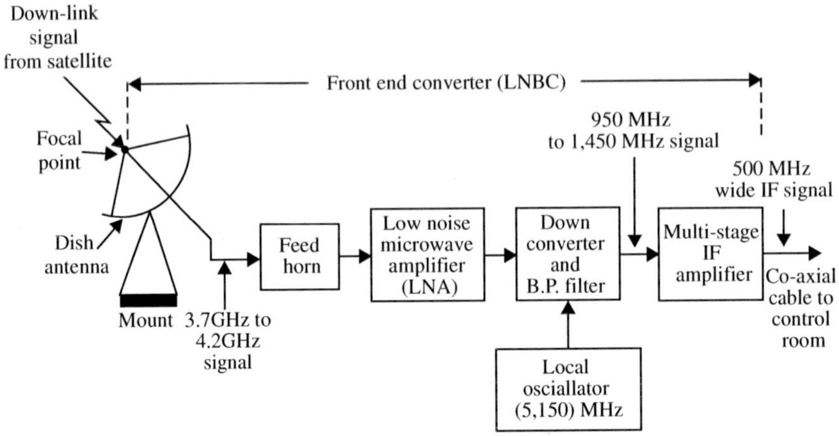

Figure 5.12 Block diagram of LNBC.

(iii) A feed horn which is a small wave guide section is mounted at the focal point and its function is to receive the signals reflected towards by the dish and deliver these to the low noise block converter.

(iv) In order to minimize the losses in the co-axial cable carry the dish antenna signals to control room, and the collected signals are first translated to a lower frequency range. This is done by LNBC.

(v) The composite signal collected by the feed horn is fed to low noise amplifier which is specially designed to provide enough gain which maintain a maximum possible Signal to Noise Ratio (SNR).

(vi) The LNA output is fed to mixer which translates the incoming microwave signals to a lower frequency range of 950 to 1,450 MHz. This is achieved by fixed down frequency converter and select only the difference products form its output.

(vii) The difference products will have the desired range of 950 to 1,450 MHz, a BPF at the output of mixer separates the desired IF signals from other signals and it is amplified by multistage IF amplifier and then sent through a high grade co-axial cable to the cable station.

(viii) The LNA is provided in the middle of co-axial cable. This LNA is often called as bullet amplifier. The feed horn and LNBC unit is called Front End Converter (FEC) and also referred as outdoor equipment because it is located in the open and close to the dish antenna where it is mounted. It is often necessary to install 2, 4, 6, or even 8 dish antenna units associated with feed horn and LNBC to collect signal from different satellites.

5.2.2 Cable Signal Processing

The cable signal processing unit also called 'Head-End' equipment consists of

(i) Satellite receivers.
(ii) Terrestrial TV signals.
(iii) VCR/CD-player outputs.
(iv) Video camera.
(v) Camcorder.
(vi) Channel modulators.
(vii) Monitors and combining network.

5.2.2.1 Satellite Receivers

(a) The satellite receiver accepts input from one of the output port on the power divider. It is desired to select any one desired channel present on the 500 MHz IF signal either manually or by remote control.
(b) The receiver function like a TV receiver and it demodulates the received frequency modulated signal of the selected channel to provide a base band signal.
(c) The base band is used to indicate combined audio and video (0 to 5 MHz) signal around 5.5 MHz.

The two outputs (V and A) as obtained at the output of satellite receiver are fed to channel modulators. This is shown in Figure 5.13.

5.2.2.2 Terrestrial TV Signal

The directional antenna is used to collect the TV signal and the low loss co-axial cable are used to carry them to the cable station. If the received signal are picked up signal is weak, a low noise booster amplifier is connected near the antenna in the cable length.

Signal Processor: In the cable system, it is not possible to assign the same channel number to programmes on which on they are received. In order to rectify this problem the equipment used to do this transformation is called 'signal processor'. The terrestrial signal is to first down convert the receiver RF signal to the IF band maintaining video IF at 38.9 MHz and audio IF at 33.4 MHz. This 7 MHz wide signal is then up converter to the assigned cable channel spectrum.

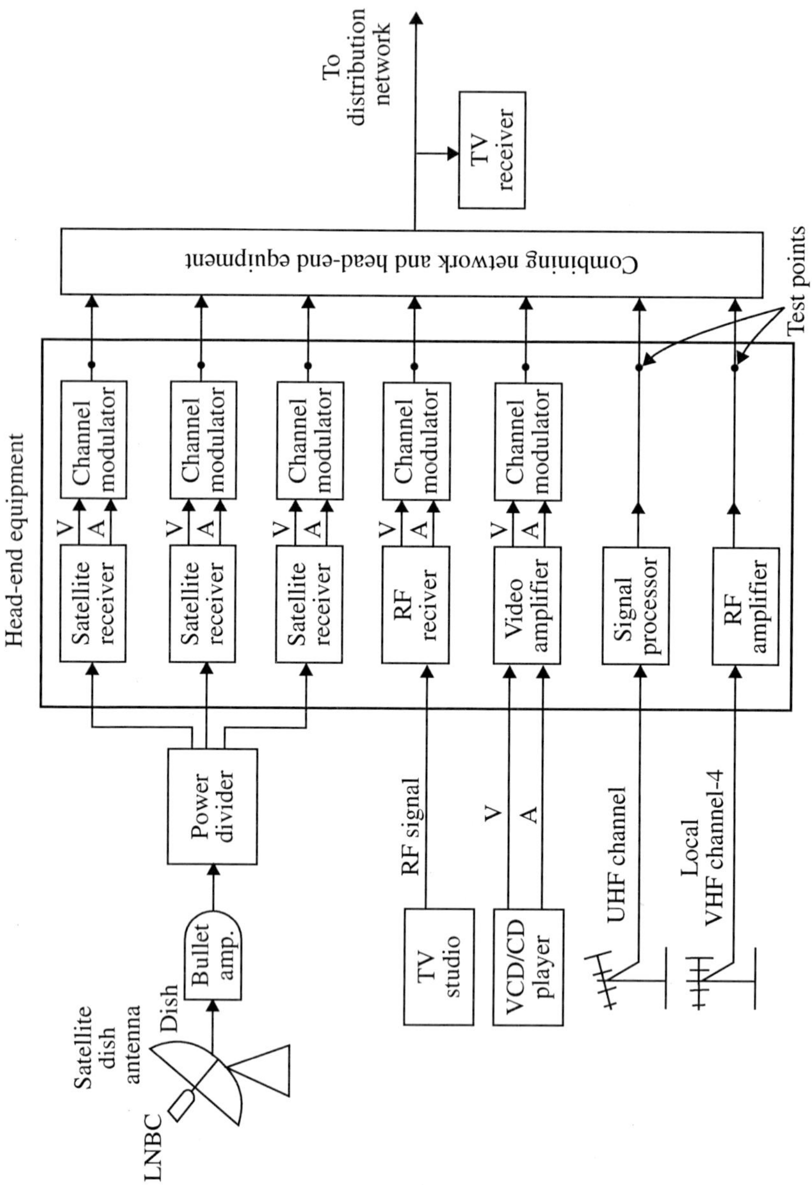

Figure 5.13 Block diagram of head-end installation.

5.2.2.3 VCR/CD Player Output

The VCR/CD player produces two types of output they are:

(a) Separate video (V) and audio (A) signals.
(b) Standard RF of 7 MHz channel width on VHF channel.

The signal processor is used to transfer these outputs to the allotted channel number.

5.2.2.4 Video Camera

During direct transmission through cable station, the video Camera is used to produce a base band video signal. Professional TV cameras are used for Live programs and the outputs are continuously monitored in the control room and selected signal is transmitted directly to cable station through the co-axial cable or a mini-transmitter.

5.2.2.5 CAM Corder

Many big cable companies collect local news and events during the day and show these in evening on their cable system. A cam corder (CAMERA RECORDER) is essentially a video-cum-sound camera and a VCR combination. The recorder which becomes available on the tape are edited and transferred on a standard tape (VHS) for playing on a VCR for distribution to subscribers over the cable system.

5.2.2.6 Channel Modulator

Channel modulator are two types: (a) fixed channel carrier frequency and (b) agile fixed frequency modulators are preferred at cable stations because no need to change their carrier frequency there for a particular channel number.

5.2.2.7 Combining Network

The combining of several modulator and signal process or output is essential to multiplex of two or more signals for simultaneous transmission, Over a single communication channel. Where the multiplexer is a linear mixer and all the signals are simply added together algebraically. The resultant output signal is a composite signal of all the channel carriers and their modulation product as shown in Figure 5.13.

5.2.3 Satellite Signals

5.2.3.1 LNBC Output

In order to minimize the losses in the co-axial cables that carry the dish antenna signal to control room and the collected signals are translated to a lower frequency range this is done by LNBC.

5.2.3.2 Power Divider

The signal from the LNBC is delivered to a signal splitter. It divides the signal into independent paths. The signal splitter is commonly called as 'power divider' because it enables equal division of signal power at its output ports.

5.2.3.3 Character Generator

It is often necessary to super impose spot news, messages or advertisement on pictures via satellites or from local transmissions, this is done at the output of the IF section. This information to be inserted is obtained through equipment called character generation.

5.2.3.4 Monitor Receiver

In the control room of a cable station, a remote controlled colour receiver is connected at the output of the combiner network to enable the cable station. Supervisor to periodically scan all the channels to monitor on the performance.

5.2.4 Cable signal distribution

Figure 5.14 shows the cable signal distribution in a cable TV system.

5.2.4.1 Combiner and Head End Amplifier

The signals received from the different antennas and local programs are fed into the combining network. The combining network is a broad band amplifier, which is used to increase the signal level from 120 dB μv to 60 dB mV (*i.e.*, 1V peak to peak). After amplification it feeds the signal to several trunk amplifiers on cable line, which carry the signal to long distances.

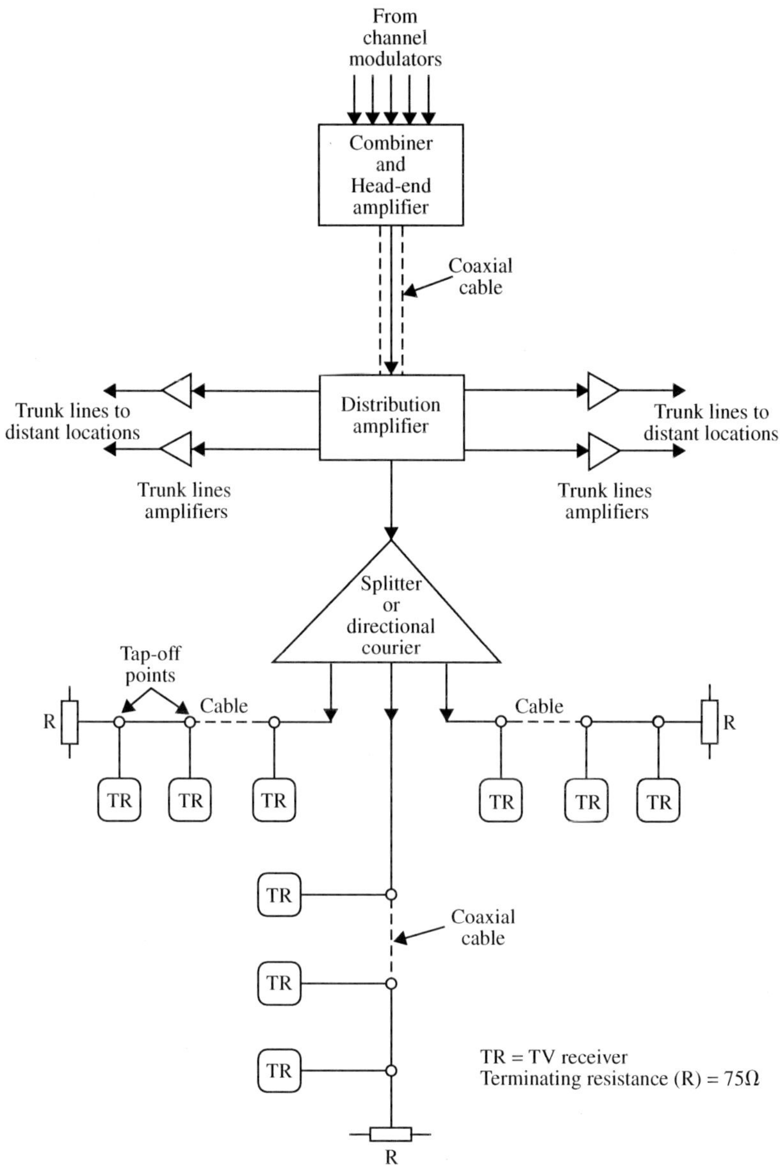

Figure 5.14 Signal distribution in cable TV system.

5.2.4.2 Distribution and Trunk Line Amplifiers

(a) In order to overcome the signal attenuation due to cable losses, several Trunk and distribution amplifiers are located at several points along the cables.

(b) On trunk lines the signal amplitude is not allowed to fall much below 100 dBμv to avoid this several trunk line amplifiers are placed at different points.

5.2.4.3 Splitter and Directional Couplers

(a) The signal distribution is done through the splitters and directional couplers. It divides the signal power equally in all the branches and also provides the impedance matching and keeps the losses to be minimum.
(b) Splitters are made up by line amplifiers which maintain a signal amplitude level close to 10 mV (peak to peak) at output terminals.

5.2.4.4 Tap-Off Point and Terminating Resistance

(a) The tap off networks divide the signal equally and also maintains the continuity of signal flow along the distribution cable. It has 2, 3 or 4 output ports for signal distribution to an equal number of subscribers.
(b) The improperly terminated lines will develop standing waves, in order to avoid this at the end of each distribution cable is terminated with a 75Ω resistor called terminator.
(c) A signal level of about 1–1.7 mV is available to the input of each customer TV receivers, this is enough to produce a good quality picture.

5.2.5 Bi-directional Cable System

The two-way cable system provides communication with subscribers for shopping at home, market surveys and pay per view for special programmes.

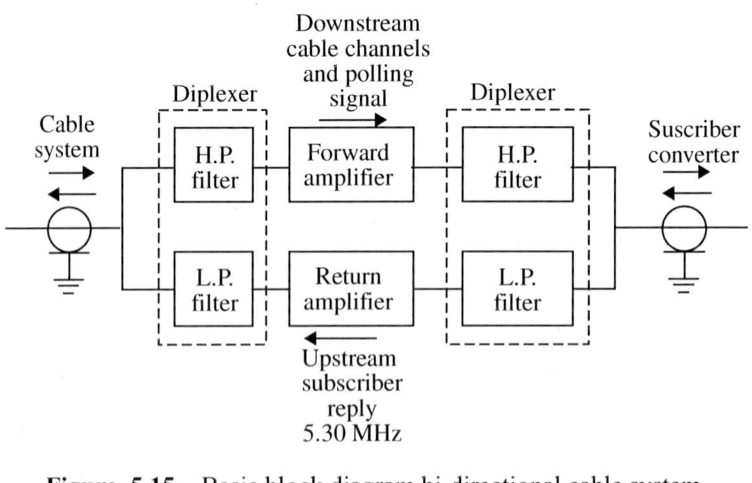

Figure 5.15 Basic block diagram bi-directional cable system.

In this service, the same cable can be used for both the directions. Here the separate amplifiers and diplexers are used for upstream and down stream signals as shown in Figure 5.15. The down transmission to subscriber is around 50 MHz or higher. And the upstream transmission towards the head end communication is in the frequency range of 5–30 MHz.

The methods used to reduce the cable losses for long distances are:

(i) Super trunks.
(ii) Fiber optics.
(iii) Microwave links.

5.2.5.1 Super Trunks

In this method, large cables having less I^2R losses are used. And the cable channels are heterodyned down to lower frequencies to keep radiation and skin effect losses to be minimum.

5.2.5.2 Fiber Optics

The latest type of communication link uses a fiber optic cable consist of glass fibers and uses natural light source to transmit a signal over a long distance with minimum loss. The major advantage of using fiber-optic cable over a long distance has a minimum loss, light weight and easy to install.

5.2.5.3 Microwave Links

Satellite communication uses a microwave links. It turns to be the best solution because the signal frequencies are very high, for their transmission over long distances.

5.2.6 Scrambling and Descrambling

5.2.6.1 Methods of Scrambling

Scrambling means, making the picture unintelligible on the screen. Some cable companies offer programs for a minimum charge, but they collect fee for telecasting the special channels. So, that particular signal is scrambled at the transmitter side and it should be descrambled at the receiver restored to its normal form with a signal supplied by the cable operator at the subscriber's request with additional payment.

There are five different types of scrambling:

(a) Trap method.
(b) SYNC suppression scrambler.

(c) Base band scrambling.
(d) Encryption method.
(e) Addressable converter.

Trap Method: The signal blocked by interfering carrier in the pay channel and it is filtered out by a suitable filter at the subscribers end.

SYNC Suppression Scrambler: The most commonly used method of scrambling the signal is known as 'SYNC suppression'. The SYNC pulse is suppressed in the RF modulation envelope of the signal to be scrambled. If any one tries to receive the signal at receiver side, picture rolls continuously with horizontal tearing of its details.

Base Band Signaling: Here the polarity of the randomly selected line is inverted. This technique provides greater security. This requires demodulation and re modulation after separating the code.

Encryption Method: Here the video and audio signals are digital processed. Then the digital processed signal is modified before transmitting using encryption technique.

Addressable Converter: Addressable converter includes a tunable converter and reconstruction circuits. Each converter can be addressed and controlled from the head-end by a computer.

5.2.6.2 Descrambling

Descrambling means, to restore to its normal form with a signal supplied by the cable operator at the subscribers request with additional payment. It is the reverse process of scrambling.

Figure 5.16 shows the block diagram of descrambler. The descrambler is used to restore the sync signal to RF signal.

In this method shown in Figure 5.16 a keyed attenuator is bypassed with a diode switch. The attenuator is represented by resistor RA and a diode switch by 's'. The pulses needed for operating the attenuator switch are sent to the descrambler unit by a separate route through the pulse shaping circuit. For this purpose a pilot carrier signal having a frequency lower that of the channel is used.

The descrambler contains a narrow band amplifier receiver is tuned to the pilot carrier frequency. It has amplitude detector and pulse shaping

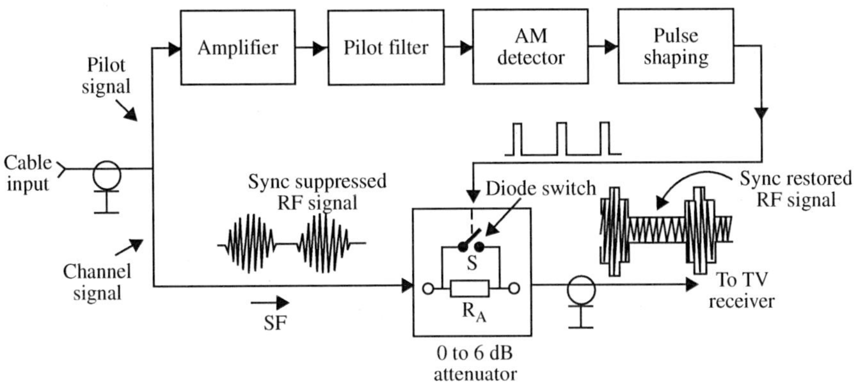

Figure 5.16 Block diagram of sync suppression type of descrambler.

circuits to drive the diode switch. The decoded pulse in the pilot carrier is used for descrambling. As a result, the SYNC is restored in the RF signal for the TV receiver enabling normal viewing in TV screen.

5.2.7 CATV-Community Antenna TV System

The block diagram of CATV system is shown in Figure 5.17.

In previous days, CATV system was used only in fringe areas or in valleys surrounded by mountains where the reception of signals was difficult.

But, now days it is used in big cities also because due to the very tall buildings it produces ghost images due to multi-path reflection and signal strength becomes weak.

A several high gain antenna are employed to pick up signals from one direction. Most cable television systems provide additional services like household, business and educational besides commercial TV and FM broadcast programmes. To provide the above options cable operators can charge fees from the subscribers.

The signals received from Various TV channels are processed, both VHF and UHF signals are received, to process the UHF channel, frequency translator is used to convert the UHF frequency into VHF. Because the CATV uses lengthy cables and loss through the cable is much greater than at UHF than VHF. This entire set-up is called ‘head-end’.

The output of the combining network is fed into the distribution amplifier through the trunk cables. Pre-amplifier and equalizer are used before the distribution amplifier to avoid the signal attenuation. Then the output of amplifier is given to the splitters then through the tap-off points the signals are passed to different receiver.

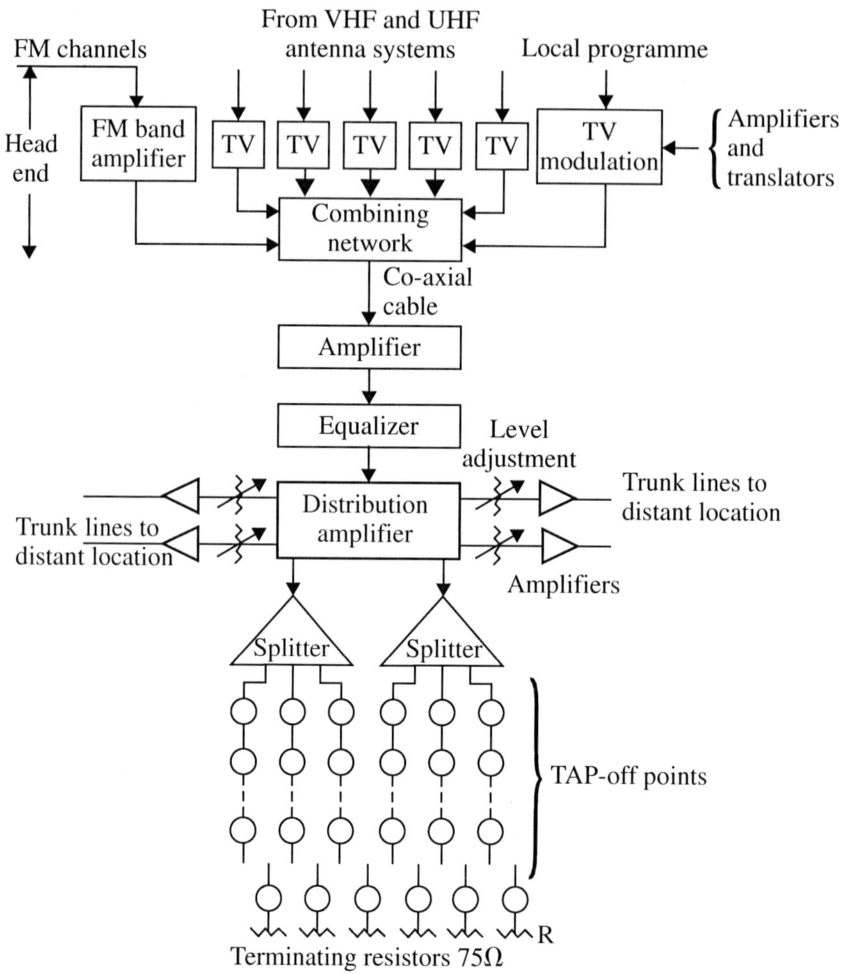

Figure 5.17 Block diagram of CATV system.

5.3 DIGITAL TV TRANSMISSION AND RECEPTION

5.3.1 Digital Colour Television

Digital television means digital processing of TV signals both at the transmitter and in the receiver. In transmitter side analog signals are first converted to digital form and similarly digital to analog converters are used at the final stages of the receiver to change video and audio signals from digital to analog form.

Because, due to technical problems and cost of the equipments digital pictures cannot be televised, the existing TV transmitters are continuing to process the video and audio signals in the analog form. These radiated signals are picked up by the receiver antenna is fed to the IF amplifier and after amplification process, all the signals are converted in digital form. The chrominance and luminance signals are processed in digital form. The micro-computer used in digital receiver is used to control all the circuits operations in receiver. Due to this additional options such as digital zooming are possible.

At the final stage of processing these signals are again converted into analog form and fed to picture tube and speakers this is shown in Figure 5.18.

5.3.2 Video Signal Quantization

For digital processing the analog TV signal from the camera is first converted to digital form as illustrated in Figure 5.18(a), this is called qunatising of analog signal, where it is sampled at a very fast rate to obtain discrete levels with the magnitude of each quantum representing the amplitude of analog at the instant of time at which it is sampled. It has been practically verified that sampling of the reproduced signal to be almost undetectable from the original analog signal. Each sample as shown in Figure 5.18(a) is represented by a digital code (0,1).

In Figure 5.18(a) illustrates a black level in the signal at 64th level (in code 01000000) and peak white to 204 (in code 11001100).

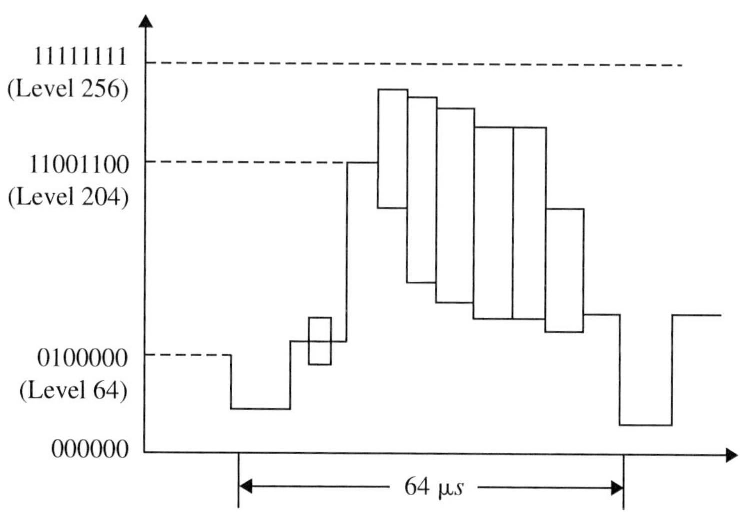

Figure 5.18 (a) Quantization of analog signals into discrete digital levels.

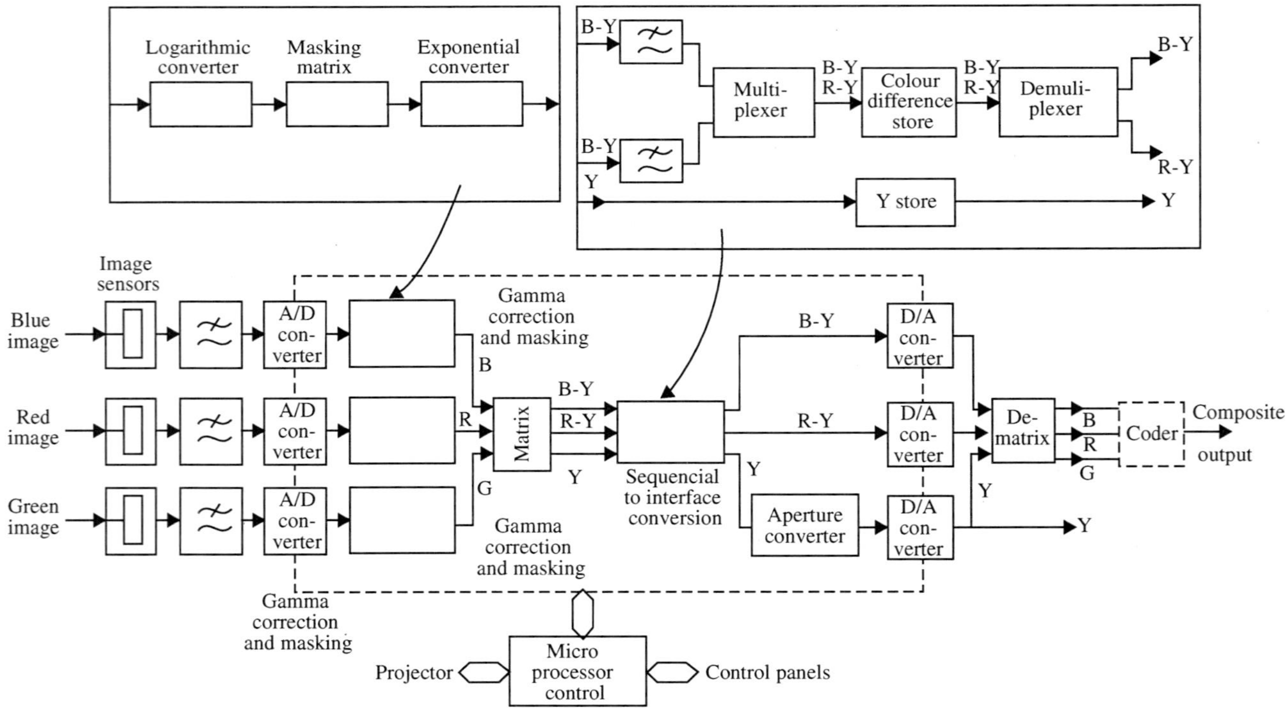

Figure 5.18 (b) Block diagram of digital colour receiver.

5.3.3 Audio Quantization

For the quantization of analog audio, the sampling levels can range from 64 to 128 depending on desired sound output quality. For best sound output for CD's, 128 sampling levels can be used. For this 7-bit code can be used $2^7 = 128$.

5.3.3.1 *Encoder*

The quantization of analog signals and generation of corresponding binary code with an A/D converter is shown in Figure 5.18(a). For sampling and quantization PAM.

5.3.4 Digital Colour TV Receiver Block Diagram

A simplified block diagram of such a colour receiver is shown in Figure 5.18(b). The RF signal from the antenna is demodulated and amplified in the conventional way to obtain IF output.

The demodulated signal is digitized and presented to digital signal processing circuits with the aid of a micro computer. Before digitization, the video and audio IF signals are separated to centre to different bandwidths and processing requirements. The clock frequency for the digital circuits is derived from the colour subcarrier burst present in the video sync.

The central control units (CCUs), a micro computer. This controls and coordinates all the circuits and signal processing functions and provides a user interface. The functions of various IC chips used *I* the receiver is described briefly.

The main blocks used in this digital receiver are:

(i) Video codec
(ii) Video processor
(iii) Deflection processing unit
(iv) Audio codec
(v) Audio processor
(vi) Central control unit

5.3.4.1 *Video Codec*

The video codec which takes signals form the IF stage converts the analog composite video signal by its high flash A/D converter into an 8-bit digital signal and feeds this to the video processor.

5.3.4.2 Video Processor

The signals received from the codec are fed to the video processor for separating the luminance and chrominance signals.

Luminance signals are separated and fed into multiplier stage to produce required contrast and brightness.

The comb filter, colour decoder is used to decode the relative weight age of red, green and blue signals. The delay line averaging technique is used to demodulate the *U* and *V* signals. After demodulation these R, G, B signal are amplified and fed to the R, G, B matrix formation and it passes the corresponding signals to the guns of the picture tube.

5.3.4.3 Deflection Processing Unit

In this section, the sync pulse from composite video signal is separated and it is fed to integrator and differentiator for separation of vertical and horizontal sync pulse respectively. And it is used to trigger the horizontal and vertical oscillators. This is done by counting the colour subcarrier and locking to it.

5.3.4.4 Audio Codec

The audio codec *i.e.*, A/D converter samples the input signal to produce a 1-bit data stream and then converts this into a 16-bits resolution stream at 35 kHz.

The digital identification filter is used the identification signal to identify whether the broadcast sound signal is mono or stereo. A parallel to series converter enables one stream output which is made input to the audio processor.

5.3.4.5 Audio Processor

Audio processor is used to demodulate the frequency modulated signal and this signal is passed to filter for controlling the stereo, mono, tone and other functions and splits it into two channels. It then sends each signal through a series of filters that control stereo balance, tone, loudness and other necessary functions. The co-efficient of filters are controlled by signals from the computer control and are based on user's settings.

5.3.4.6 Central Control Unit (CCU)

The final main chip is the central control unit. It is in fact, a computer control that controls other chips and translates the user's instructions from an entry keyboard or from remote transmitter. The control computer also stores factory alignments programmed during manufacturing and feeds these two other chips, each time the set is tuned on. It has 8-bit processor with EPROM, timer and other necessary circuits to decode the user commands.

5.3.5 Merits of Digital Processing

However complete digital receivers when available will have the following merits in terms of overall pictures quality and sound reproduction:

(i) **Reduced Ghosts:** By employing digital processing techniques, partial or near total filtering out of ghost signals is possible. Similarly digital means enable better synchronization and hence, interference from electrical appliances is suppressed.

(ii) **Reduction of 50 Hz Flicker:** The capability of digital TV sets to store picture signals can allow faster re-scanning of each picture on the screen and thus flicker is nearly eliminated.

(iii) **High Resolution Pictures:** The picture storage capability of the digital system makes it possible to scan a second set of lines in between the usual scan lines thus enabling a high resolution picture with the same low resolution in coming signal.

(iv) **Picture-In-Picture (PIP):** With digital signal storage facility it is possible to see two programmes from different stations at the same time. The second channel picture is usually one-forth of the main picture and can be shifted in anyone corner of the screen. The two pictures can be interchanged. Such a provision is commonly known as 'P-I-P' facility.

(v) **Slow Motion Picture:** When a digital TV set is coupled to a VCR, the ability to store images allows the user to view the action in slow motion. Each frame can be scanned onto the screen at several times to freeze a frame.

(vi) **Easy Adoption to Additional Displays:** The digital system adapts more easily to additional video display functions and teletext, TV games and computer facility can be easily provided in digital TV receivers.

(vii) **Reduced Operational in Stability:** The effect of component aging on alignment tuning and consequent instability problems in analog receivers is eliminated in digital receivers.

(viii) **Improved Reception in Fringe Areas:** Noise is considerably reduced with digital signal processing. Thus the qualities of picture in poor reception areas are improved at additional cost.

5.4 STEREO SOUND SYSTEM

In order to stimulate the characteristics of hearing with two ears that enable perception of both direction and depth of sound besides its loudness, it is

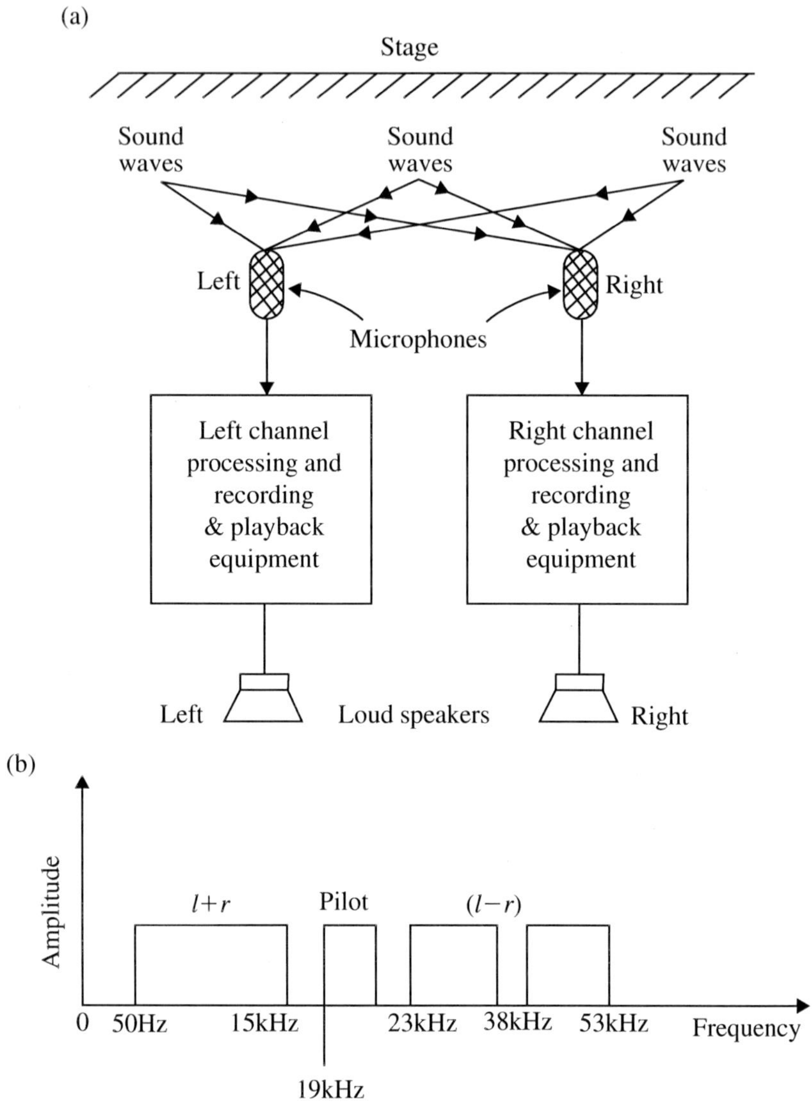

Figure 5.19 (a) Method of stereophonic sound recording and reproduction, and (b) FM multiplex stereo signal spectrum.

necessary to have two completely separate channels at the transmitter and independent detection and amplifying path in the receiver.

5.4.1 FM–FM Stereo Multiplex System

The stereo sound broadcasting is done by FM–FM stereo multiplex system, which uses only one FM transmitter to broadcast audio signals on both (left and right) channels. This is shown in Figure 5.19(a).

As shown in Figure 5.19(b), two channels signals are combined in a predetermined manner in the FM multiplex transmitter before transmission. At the transmitter side the FM station transmits two signals they are $(L+R)$ and $(L-R)$.

$(L+R) \Rightarrow$ Left channel is added with right channel.

$(L-R) \Rightarrow$ Left channel is subtracted with right channel.

5.4.2 Stereo Sound In Television

The stereo sound transmission in television uses the basic FM–FM stereo multiplex system, but left and right channel signals are accommodated in different way. In a television system the IF spectrum consist of:

(i) Vision carrier.
(ii) First sound IF carrier at 5.5 MHz.
(iii) Second sound IF carrier at 5.7421875 MHz.

To minimize the interference, the two sound carriers are separated by a Frequency of

$$(15.5 \times F_H)$$

where F_H horizontal line frequency of 15625 Hz.

$$\text{Sound IF carrier is at} = 5.5\,\text{MHz}.$$

To minimize interference $= 5.5\,\text{MHz} + (15.5 \times 15{,}625\,\text{Hz}) = 5.7421875.$

During stereo transmission, the first sound IF is FM modulated with $(L+R)$ information, so that it produces a mono sound. Similarly the second IF carrier is FM modulated with 2R information.

In order to provide synchronization at receiver side, a pilot carrier is transmitted along with two channel signals. At the receiver side FM multiplex system is used to separate the left and right channel components by using a adder and subtractor.

To obtain left channel component

$$(L+R) + (L-R) = 2L$$

To obtain right channel component

$$(L+R) - (L-R) = 2R$$

Then the separated left and right channel sound signals are amplified and reproduced by the corresponding loud speakers. To identify the transmitted

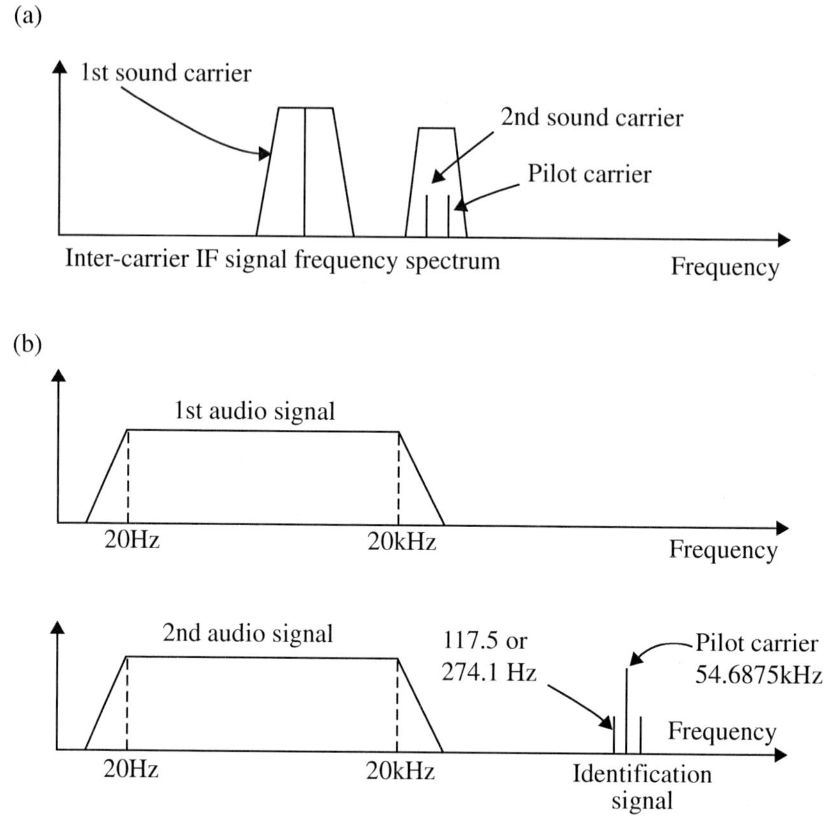

Figure 5.20 (a) Inter-carrier IF signal frequency spectrum and (b) Audio signal frequency spectrum identification.

signal is mono or stereo or dual sound. The 2nd IF carrier has the identification signal called pilot carrier (54.68 kHz),which is amplitude modulated with 117.5 MHz for stereo or 274.1Hz for dual transmission.

The frequency spectrum of the inter-carrier IF signal is shown in Figure 5.20(a). The frequency spectrum after demodulation as shown in Figure 5.20(b). The stereo information can be separated by doubling the amplitude of 1st sound signal and subtracting the demodulated second sound signal from it *i.e.*, $2(L + R) - 2R = 2L$. After the identification signal from the pilot carrier is used to activate the logic circuits to route the required information to the sound outputs, under the control of the viewer.

5.5 PROJECTION TELEVISION

A projection television uses a projector to create a small image from a video signal and magnify this image onto a viewable screen. This is shown in

Figure 5.21(a). The projector uses a bright beam of light and a lens system to project the image to a much larger size. It can be classified into two types they are: (i) Front-projection Television system and (ii) Rear-projection television system.

Figure 5.21 (a) Picture view of TV projector.

5.5.1 Front-Projection Television

Front-projection television uses a projector that is separate from the screen which could be a suitably-prepared wall, and the projector is placed in front of the screen. This is shown in Figure 5.21(b). front projection television functions can be classified into two general approaches they are:

(i) Transmittive projectors.
(ii) Reflective projection.

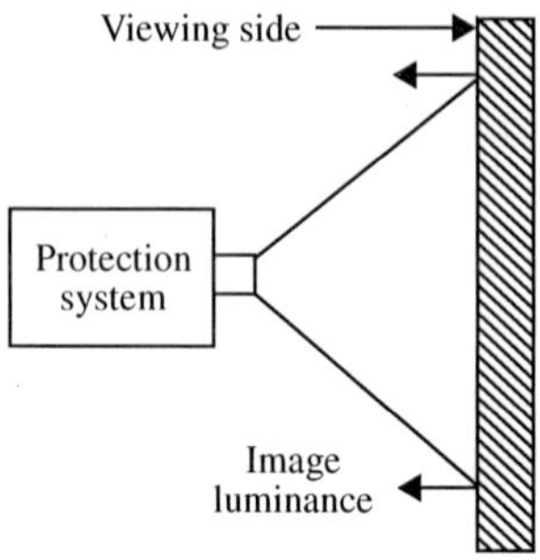

Figure 5.21 (b) Front projection.

5.5.1.1 *Transmittive Projectors*

Based on the image forming elements it can be classified into:

(a) **CRT transmittive projections:**
In this diagonal CRTs, are used to emit a light. A lens is placed in front of the CRT tubes and magnifies the image and projects it on to the screen. It has three configurations used in many applications.
- **Single colour CRT tube:** In this a single lens is placed in front of the CRT tube and magnifies the image and projects it on to the screen located at a distance.
- **Single B&W CRT with colour filters:** It has a fast rotating colour filter wheel with red, green and blue segments is placed between CRT and projection lens. It results in continues changing of three colour images and the ability of the human eye is to integrate them.
- **Three CRTs:** In this it uses a three lenses one for each CRTs. The lenses are aligned, so that a single coloured image appears on the screen.

(b) **LCD transmittive projectors:**
To increase the resolution of the image LCD technologies have been developed. This is shown in Figure 5.21(c). A transmittive LCD projector uses a bright light to illuminate on the LCD panel. The image formed on the LCD is projected by a lens system on the screen located at a suitable distance.

Advantages

- Cost is less.
- Front-projection picture quality approaches that of movie theater.
- It take's very little space because a projector screen is extremely slim, and even a suitably-prepared wall can be used.
- Display size can be extremely large, up to hundreds of inches.

Figure 5.21 (c) LCD projection television.

Disadvantages

- Front-projection more difficult to set up because projector is separate and must be placed in front of the screen, typically on the ceiling.
- Lamp may need to be replaced after heavy usage.

5.5.2 Rear-Projection Television System

It is similar to a traditional television, the projector is placed inside the television box and projects the image from behind the screen. this is shown in Figure 5.21(d). The screen used in this method should have high transmission of brightness and low reflectance.

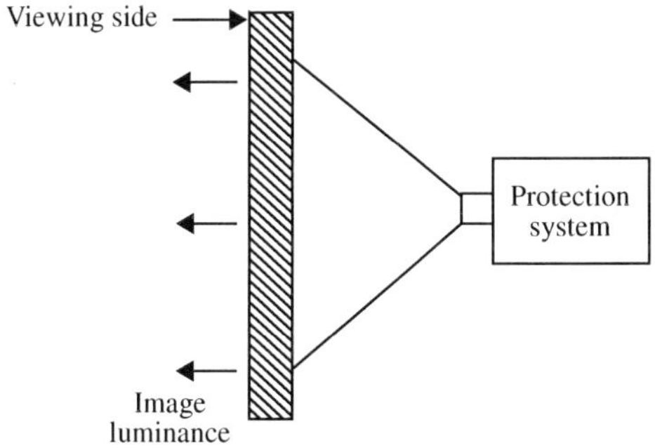

Figure 5.21 (d) Rear projection.

The following are different types of rear-projection televisions, which differ based on the type of projector.

5.5.2.1 CRT Rear-Projection Television

Small cathode ray tubes create the image in the same manner that a traditional CRT television does, which is by firing a beam of electrons onto a phosphor-coated screen and then the image is projected to a large screen. The cathode ray tubes can be arranged in various ways. One arrangement is to use one tube and three phosphor (red, green, blue) coatings. Alternatively, one black-and-white tube can be used with a spinning color wheel. A third option is to use three CRT's, one for red, green, and blue.

5.5.2.2 LCD Rear-Projection Television

A lamp transmits light through a small LCD chip made up of individual pixels to create an image. The LCD projector uses mirrors to take the light and create three separate red, green, and blue beams, which are then passed through three separate LCD panels. The liquid crystals are manipulated using electric current to control the amount of light passing through. The lens system takes the three color beams and projects the image.

5.5.2.3 DLP Rear-Projection Television

A DLP projector creates an image using a digital micro mirror device (DMD chip), which on its surface contains a large matrix of microscopic mirrors, each corresponding to one pixel in an image. Each mirror can be rotated to reflect light such that the pixel appears bright, or the mirror can be rotated to direct light elsewhere and make the pixel appear dark. The mirror is made of

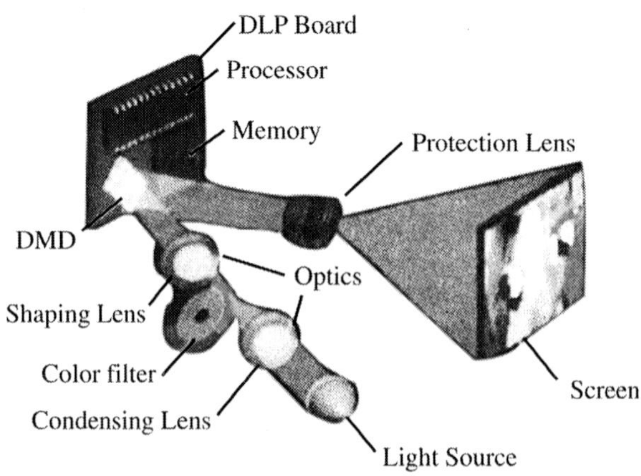

Figure 5.21 (e) Rear projection television system.

aluminum and is rotated on an axle hinge. There are electrodes on both sides of the hinge controlling the rotation of the mirror using electrostatic attraction. The electrodes are connected to an SRAM cell located under each pixel, and charges from the SRAM cell drive the movement of the mirrors.

Advantages

- Significantly cheaper than flat-panel counterparts.
- Projectors that are not phosphor-based (LCD/DLP) are not susceptible to burn-in.
- Rear-projection is not subject to glare.

Disadvantages

- Rear-projection televisions are much bulkier than flat-panel televisions.
- Lamp may need to be replaced after heavy usage.
- Rear-projection has smaller viewing angles than those of flat-panel displays.

5.5.3 Comparison of Different Types of Rear-Projection Televisions

CRT projector

Advantages

- Excellent black level and contrast ratio.
- Achieves excellent color reproduction.
- It has very long lifetimes.
- Greater viewing angles than LCD's.

Disadvantages

- Heavy and large, especially depth-wise.
- If one CRT fails the other two should be replaced as well to maintain color and brightness balance.
- Susceptible to burn-in because CRT is phosphor-based.
- Needs to be 'converged' about every year.
- It has focus problems.

LCD projector

Advantages

- Smaller than CRT projectors.
- LCD chip can be easily repaired or replaced.
- Is not susceptible to burn-in.

Disadvantages

- The screen-door effect, Individual pixels may be visible on the large screen, giving the appearance that the viewer is looking through a screen door.
- Possibility of defective pixels.
- Poor black level.
- Uses lamps for light, lamps may need to be replaced.
- Fixed number of pixels.
- Limited viewing angles.

DLP projector

Advantages

- Slimmest of all types of projection televisions.
- Achieves excellent black level and contrast ratio.
- DMD chip can be easily repaired or replaced.
- Is not susceptible to burn-in.
- Better viewing angles than those of CRT projectors.
- Image brightness only decreases due to the age of the lamp.
- defective pixels are rare.
- Does not experience the screen-door effect.

Disadvantages

- Uses lamps for light, lamps need to be replaced on average once every year.
- Estimated lifetime of LED lamps is over 100,000 hours.
- Fixed number of pixels.

5.6 FLAT PANEL DISPLAY TV RECEIVERS

A flat panel television receiver is one in which the conventional picture tube is replaced by flat panel display. Because the size and shape of a receiver is largely governed by the dimensions of the picture tube (CRT) the adoption of flat screen make the receiver much compact and produce a large screen display. Various thin screen technologies have been developed, they are:

(i) Liquid crystal display (LCD).
(ii) Plasma display (PDP).
(iii) Digital light Processing (DLP).

These technologies can produce large-screen televisions that are much thinner.

There are two methods to achieve the flat panel display:

(i) Simple dot matrix method with tri-colour film-dots.
(ii) Active dot matrix method.

The pixels are arranged in a matrix at the intersection of several hundred rows and columns. The popular standards for screen resolution defines the number of rows and columns matrix for video graphic displays are as follows:

(i) CGA: 640 * 200 (2 colours)
(ii) EGA: 640 * 350 (16 colours)
(iii) VGA: 640 * 480 (16 colours) or 320 * 200 (256 colours)

5.6.1 Liquid Crystal Display

LCD is based on the simple dot matrix approach. It causes ghosting and blurring during the display of fast-moving images, so the quality of the display is not fine. It has poor black level and contrast ratio, to overcome this difficulties, active dot matrix flat panel displays are used.

5.6.2 Active Matrix Display

The active matrix drive display is shown in Figure 5.22. The active display panel measures 370 × 355 × 99 mm. It has an array of 3000 very small

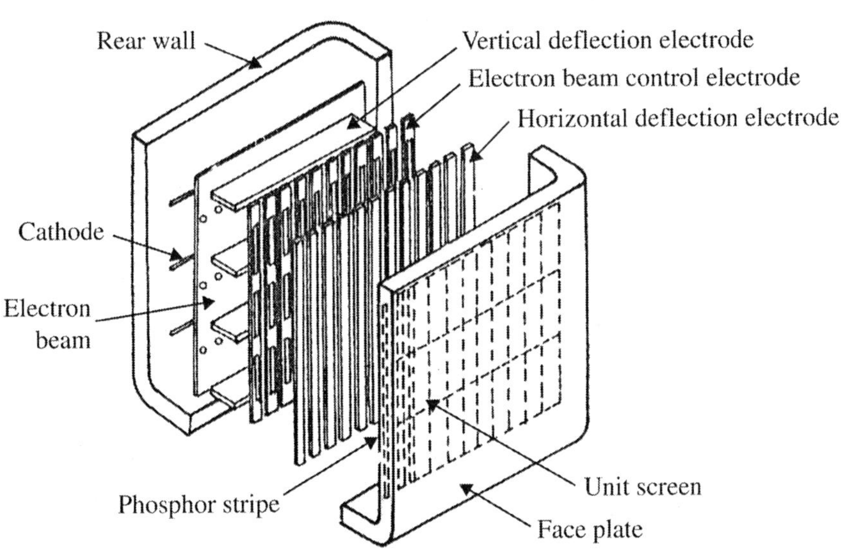

Figure 5.22 Elements of an active matrix drive and deflection flat panel picture display system.

electron tubes are used to form an effective resolution in scanning lines, here 270 scanning lines are employed.

It is made up of lattice electrodes of 0.1mm thick layered between the insulating board. Each panel has 15 fine wire cathodes and 200 electron beam electron electrodes set, deflected in six stages. In order to maintain the uniformity over the surface, the output is controlled by microprocessor.

Advantages

- Slim profile.
- Lighter and less bulky.
- Is less susceptible to burn-in, the low-luminosity areas become permanently visible.
- LCDs reflect very little light, allowing them to maintain contrast levels in well-lit rooms and not be affected by glare.
- Low power consumption.
- Can be wall-mounted.

Disadvantages

- Poor black level and contrast level.
- It is nearly impossible to use an LCD without some image warping occurring.
- It causes ghosting and blurring during the display of fast-moving images, so the quality of the display is not fine.

5.6.3 Plasma Display

Advantages

- Slim profile.
- Can be wall mounted.
- Lighter and less bulky.
- Achieves better and accurate color reproduction than LCDs.
- Produces deep, true blacks allowing for superior contrast ratios (up to 1:1,000,000).
- Absence of motion blur, because of very high refresh rates and faster response times (up to 0.001 milliseconds) make plasmas ideal for fast motion video (films or sports viewing).

Disadvantages

- Susceptible to screen burn-in and image retention.
- Phosphors lose luminosity over time, resulting in gradual decline of absolute image brightness (newer models are less susceptible to this,

having life spans exceeding 60,000 hours, far longer than older CRT technology).

- Generally do not come in sizes smaller than 32 inches.
- Susceptible to reflection glare in bright rooms.
- High power consumption.
- Heavier than LCDs due to the requirement of a glass screen to hold the gases.
- It is difficult to repair than an LCD.

5.7 EXTENDED DEFINITION TELEVISION (EDTV)

EDTV is mainly intended for direct broadcasting satellite system. To improve the performance of TV systems it requires different transmission standards but it maintain the present line numbers and field rates.

In the conventional PAL/NTSC system the luminance and chrominance information are transmitted simultaneously for each line in interleaved form. For sound information there is separate carrier and additional carrier is used for modulating audio signal. Due to this it creates a problem of interference beat called cross luminance and cross colour interference.

In order to avoid the interference problems, a new encoding system called multiplexed analog component system (MAC) was developed. In MAC encoding system the luminance and chrominance signal are time division multiplexed and they are transmitted in a time sequence, as separate components in the time of a horizontal line.

5.7.1 Compression of Signals

In the MAC system, the luminance and chrominance signals are separately time compressed to accommodate in 64μs horizontal line scanning period.

$$\text{Compression factor for luminance signal} = 3:2$$

$$\text{Time period of compressed luminance signal} = 35\mu s$$

$$\text{Compression factor for chrominance signal} = 3:1$$

$$\text{Time period for chrominance signal} = 17.5\mu s$$

It is not possible to transmit sound, line sync pulse signal and other information along with compressed signal. So therefore, the time compressed luminance information is transmitted on each line along with one of the compressed chrominance components (R–Y) or (B–Y) and the complementary signals are being transmitted on next line. As shown in Figure 5.23 and a 10μs period is used for digital sync and sound/data signals.

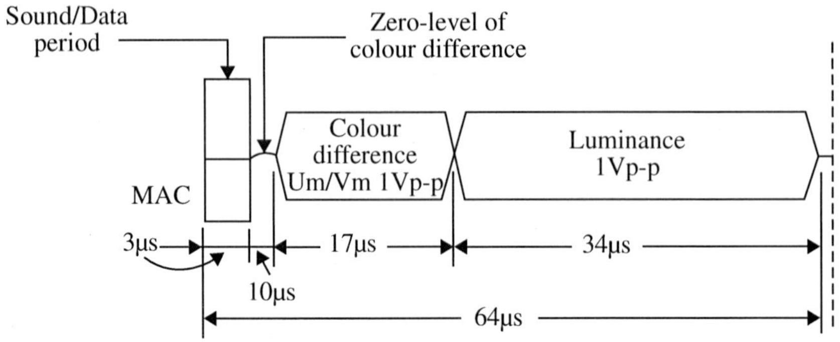

Figure 5.23 MAC encoding format.

The digital signals are converted back into analog format and transmitted in frequency modulation (FM) form. At the receiver side, one line delay is required to store *V* chrominance information of a line, until the *U* chrominance information is available from the next line for decoding colours.

Different versions have developed such as B-MAC, C-MAC and D-MAC, based on the way in which the digital data and sound information's are multiplexed.

5.8 VIDEO TAPE RECORDING SYSTEM

In video recording, we have to record both the video and audio signals simultaneously on the same tape. In this video recording is some what complex due to its high frequency and large bandwidth. Here the basic idea is similar to audio tape recording. But there are many practical difficulties.

The gap in the head must have minimum physical size to place sufficient field strength on the magnetic tape. This gap must be larger than the smallest wave length. If maximum signal frequency is 5 MHz, gap width is 100 micro inch. Then the velocity of the tape

$$V > 100 \times 10^{-6} \times 5 \times 10^{6}$$

But velocity *V* must have least two times greater than this value, to avoid the signal cancellation. Therefore

$$V > 2 \times 100 \times 10^{-6} \times 5 \times 10^{6}$$

$$V > 1000 \text{ inches/sec}$$

That is the constant velocity of tape should be at least 1000 inches/sec. The above value is very high and it is very difficult to maintain the constant speed and large tape length. The relative speed is obtained by moving the tape head in opposite direction to the tape movement.

Actually, the head is mounted on the peripheral of a drum, which rotates around an axis with respect to the direction of tape movement, because of this arrangement the speed of tape is now reduced nearly to 7.5 inches/second. The two methods employed for video tape recordings are:

(i) Quadruple-head method.
(ii) Helical scan methods.

5.8.1 Quadruple (transverse)-Head Method

In this method of recording, four recording head, spaced each at 90° between them are mounted on the rotating drum. Figure 5.24 shows the arrangement of head and tape movement in this method.

As shown in Figure 5.24 the head rotates transversely across the tape, and the tape is pulled slowly from one reel to another. One head comes in contact with the tape as the previous one leaves the tape. Due to this arrangement a large relative motion is achieved between the head gaps and tape surface. In order to achieve good contact between tape and head, the tape is made to move in a curvature as shown in Figure 5.24.

Because of the relative movement of tape and head, the recorded track is some what slanted in the direction of the tape travel this is shown in Figure 5.24. Sync and audio signal are recorded at the top and bottom edges of the edge. A switching arrangement is used to transfer signal to

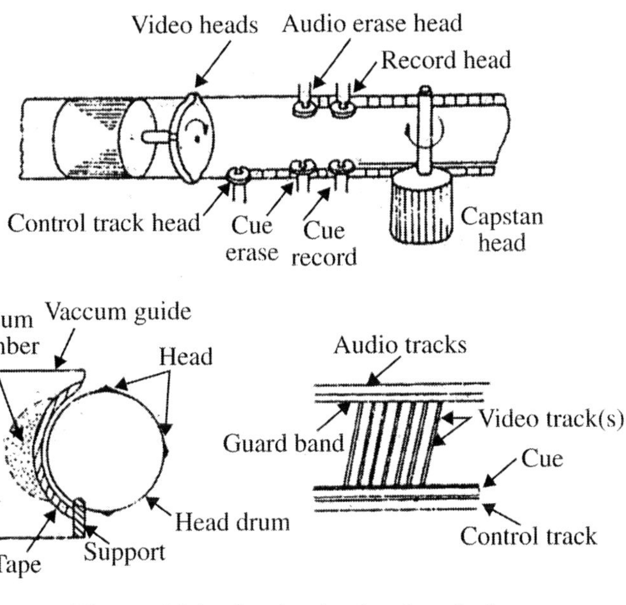

Figure 5.24 Quadruplex-head method.

head which is in contact with tape. Between any two slanted recording tracks, a small guard band is provided. Before the tape goes to head drum, it is passed through a full track erase head.

The audio signals is recorded along the length in a narrow track, near one edge of the tape. To synchronize the record and replay tape motors, cue signal and control pulses are recorded in the other edge in a longitudinal track as shown in Figure 5.24.

5.8.1.1 Block Diagram of Recording Mode

The block diagram arrangement for recording in quadruple head method is shown in Figure 5.25.

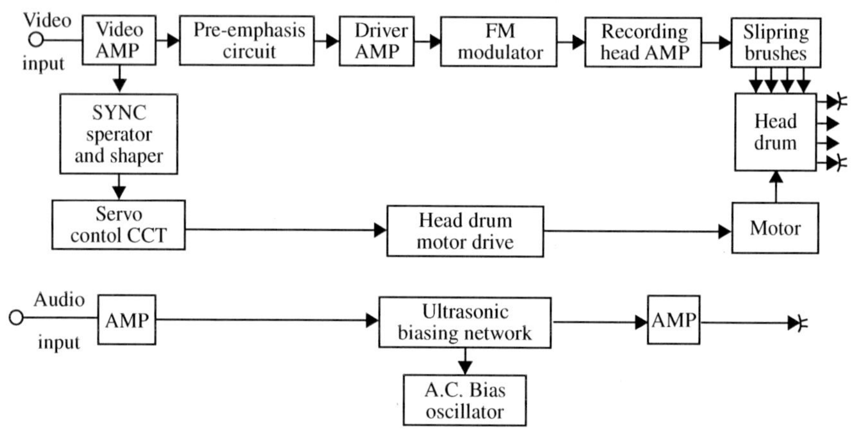

Figure 5.25 Block diagram for recording.

Here the composite video signal fed into wide band video amplifier. The output of this amplifier is fed to the FM modulator through a pre-emphasis network and drive stages. Then recording head amplifier is used to amplify the modulator output through the slip ring brushes, the modulated and amplified signals are fed to recording head. The corresponding audio signals are also amplified and fed to the audio head.

Across the output of the wide band amplifier a sync separator circuit is connected. After separation, the sync pulse are properly shaped and used to synchronize and control the head drum motor speed. Servo control technique is used to control the motor speed.

5.8.1.2 Block Diagram of Play Back Mode

In play back mode, the video outputs are collected through slip-ring brushes from the video head. The block diagram arrangements are shown

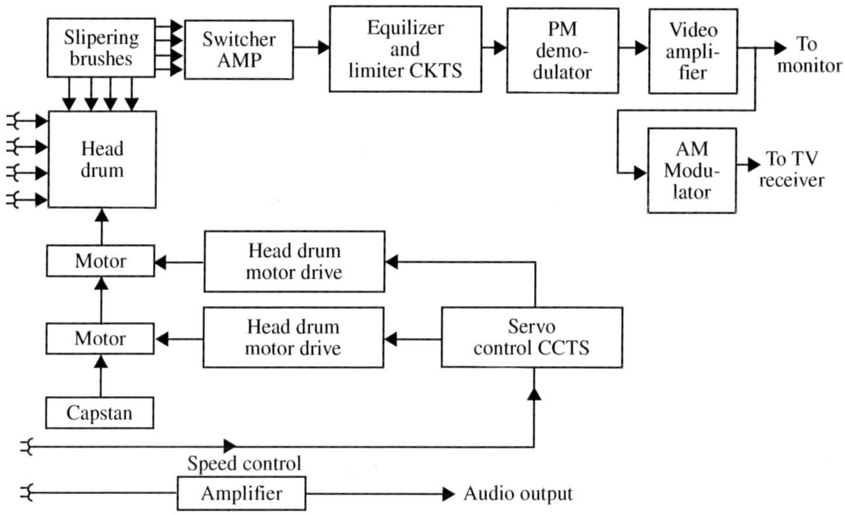

Figure 5.26 Block diagram for play back mode.

in Figure 5.26. The output from the video head is given to an electronic switch, which will select and amplify the head signal.

The selection of signal takes place during horizontal retrace and blanking intervals. So they are not visible. Then selected signal is passed to a FM demodulator through an equalising network and limiter stages. A four stages amplifier is used to amplify the demodulated signal. Then this FM demodulated signal is AM modulated with a carrier frequency of any one band channels, so that the programme is viewed on the TV receiver.

5.8.2 Helical Scan Recording

In this method the tape is wrapped tightly half-way around a drum of about 6.5 cm in diameter. Inside the drum, two recording heads spaced 180° apart are fixed. The arrangements of head and tape for helical scan method are shown in Figure 5.27.

As shown in Figure 5.27 the video head produces through a horizontal slit in the drum and contact is made on the tape. The tape comes in contact at the upper end and leaves the lower end of the drum. That is the tape is wrapped like of half helix round the drum.

The angle of inclination of the tape is about 3.5° the lower section of the drum is stationary and guard band guides the tape around the drum in helix and this method known as 'helical scan recording'.

Two recordings head (mounted on opposite sides are made) contact with the tape with two small rectangular slits. Because of this arrangement,

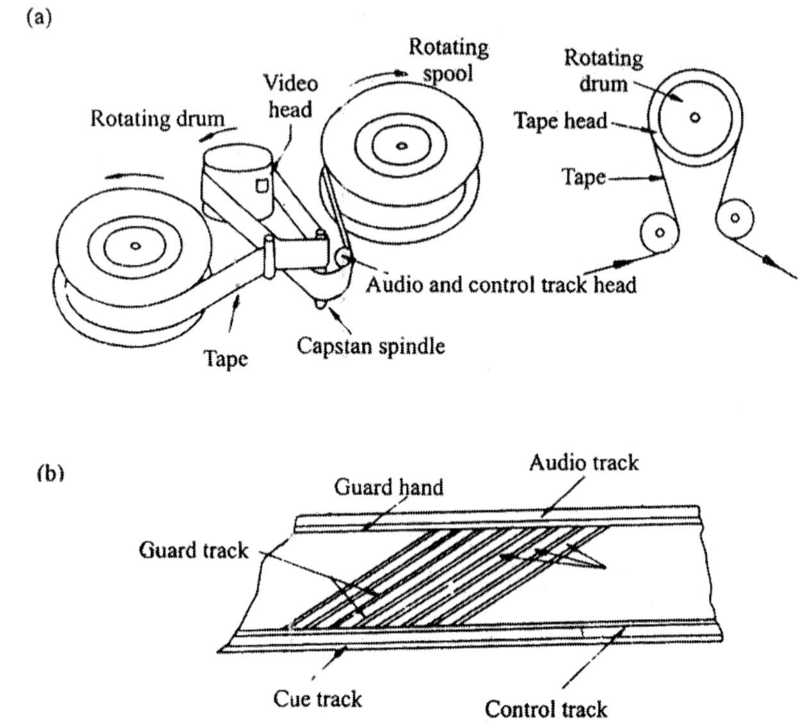

Figure 5.27 Helical scan recording.

the heads trace out diagonal track across the track. The angles of inclination of the tape depend upon the diameter of the drum and speed of the tape movement. Usually the drum is rotated at a speed of 25 revolutions per second. So one complete revolution takes a time 40 ms that is equal to two field period.

As shown in Figure 5.27(b) the tape consist of an audio track guard band at the top and control track cue track, at the bottom.

5.8.3 Steps Involved in VCD Making Process

The video-compact disc working principle is as video tape recorder. There are two types of video disc system to record and playback the pictures in VCD.

(i) Laser disc system.
(ii) Based on capacitance principle.

5.8.3.1 Laser Disc System

Laser (optical) disc is used to store the audio and video signals. The optical disc uses reflected light from the microscopic pits in the disc surface, which corresponds to video and audio signals.

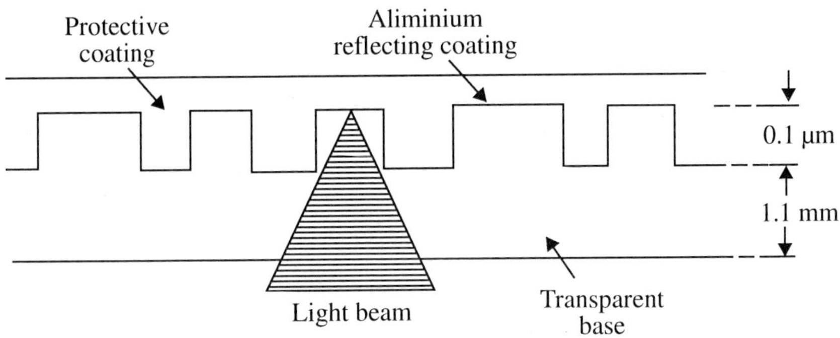

Figure 5.28 VCD making process.

The original recording is done on master disc which consist of a glass substrate. A photo resistive coating is given on one side of the glass substrate.

If the disc rotates, the laser beam exposes the light on a spiral track and it consist of pits and flats. The depths of pits are fixed. But the length of the pits can be varied depending on the information stored on it.

The silver and nickel coating are given on the surface of the disc. This process is called stamping process. This is used to produce the copies of the record. To improve the reflectivity aluminum coating is also given over the nickel coating. Also protective layer of transparent plastic coating is given to avoid the disc from scratches.

The spirals are spaced 1.6μm from one another and they have 0.4μm wide. According to the intensity of the beam the width and spacing of the pits are varied.

The video disc are made in two different formats they are:

(a) Active play.
(b) Long play.

Active Play: In active play, speed of disk rotation is kept constant and angular velocity is also constant. For every 360° rotation, two fields of video recording are produced. Each side will take half an hour. This is shown in Figure 5.29(a). This system is useful for picture reversing, rapid searching, frame freeze and slowing down to permit slow motion display.

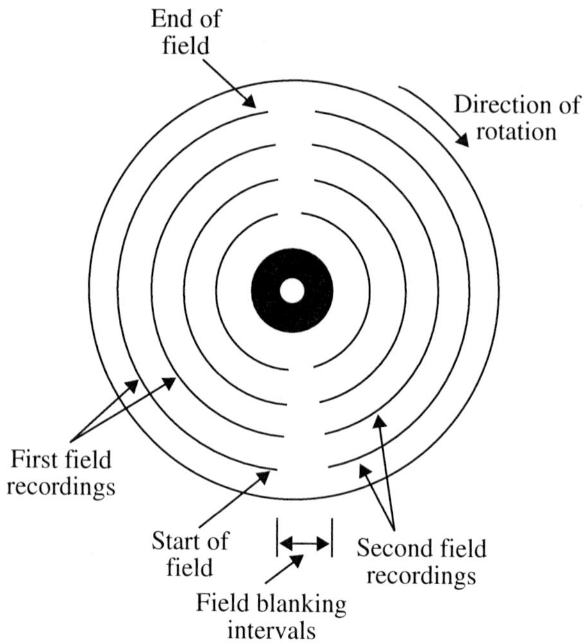

Figure 5.29 (a) Active play.

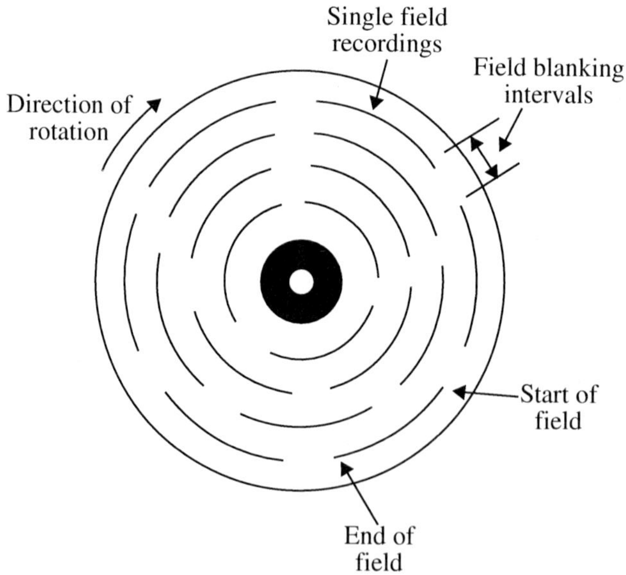

Figure 5.29 (b) Long play.

Long Play: The speed of the motor will be increased continuously as the scanning lasers head comes closer to the centre of the disc. This produces a constant linear velocity to the beam. This enables longer playing time of one hour per side. This is shown in Figure 5.29(b).

5.8.4 Block Diagram of VCD Player in Playback Mode

5.8.4.1 *Digital Signal Reproduction*

The signal on the disc is stored as digital data in the form of 1's and 0's. These 1's and 0's are stored on the disc surface in the form of pits and lands. The flat surface on a spiral track originalities from the centre of the disc and going towards the outer side. The digital 1 is stored as start or end of the pit on the disc surface and 0 is stored as no change in the current position of the pit of flat surface. The digital signal is converted into smooth analog signal by various circuits inside the VCD player.

The complete block diagram of a VCD player is shown in Figure 5.30. The VCD player is known as a Constant linear velocity ("CLV") system. In CLV device such as the VCD player, the rotational speed of the disc player is adjusted with the movement of the reading mechanism on the disc surface. This speed is changed to maintain a constant linear velocity, *i.e.*, the signal on the disc surface always moves at a constant speed of 1.3 m per second under the pickup head.

The CLV is maintained by the VCD player by changing the rotational speed of the disc from 500 rpm (rotations per minute) at the inner surface to around 300–350 rpm, when the reading mechanism moves to the outer edges of the disc.

5.8.4.2 *Decoding VCD*

Let us see how the signal, on the disc surface is processed and reproduce within the VCD player. During the decoding, the digital data on the disc surface is read by the decoding circuit and is converted into analog signal.

5.8.4.3 *Optical Pickup*

As shown in block diagram Figure 5.30, the digital signals stored on the VCD surface as pits and flats. And it is picked up by the optical pickup made of lens assembly, prism, photo detectors and laser diode assembly in the optical pickup unit.

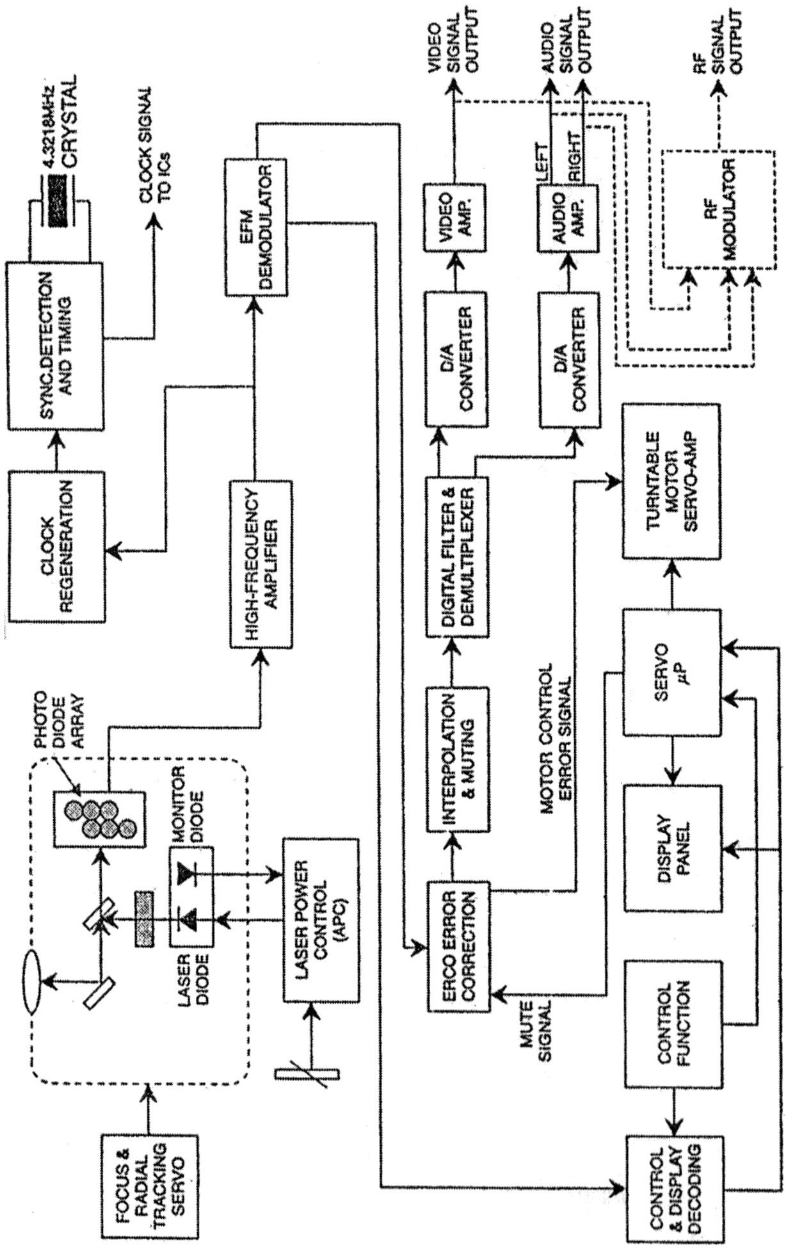

Figure 5.30 Block diagram of VCD player in play back mode.

5.8.4.4 High-Frequency Amplifier

This signal is very weak so it is amplified by a high-frequency RF-amplifier circuit to bring the signal to a proper level. This amplified and filter high-frequency signal contains audio and video signal as well as synchronization signals in 14 bit (8 to 14 modulated formats) EFM.

5.8.4.5 EFM Demodulator

The EFM demodulator separates the modulated data and timing signals from input signal. It also remove the additional coupling bits and convert the 14 bit EFM symbol to actual 8 bit data.

The amplified and filtered EFM signal from the high -frequency amplifier is fed to the clock regeneration circuit, synchronization detecting and timing circuit. These circuits are used to recover the 8 bit clock and the sync pattern from the data. The timing signal separated by this circuit is used to provide timing signal to the VCD system.

5.8.4.6 ERCO Circuit

Demodulated data from the EFM demodulated is given to an error correction ERCO circuit. This demodulated data signal is also given to control and display decoding circuit, which recovers the control and display signals multiplexed into the signal received from the VCD. The ERCO circuit is also responsible for maintaining constant linear velocity of the VCD motors.

5.8.4.7 De-Interleaving

Signal from the ERCO contains the audio signal in the interleaved format before during any further operation on this signal. To restore the original sequence of the information the signal is de-interleaved in the interpolation and muting section.

5.8.4.8 Digital Filter and De-Multiplexer

The de-interleaved and regenerated signal is fed to a digital filter and de-multiplexer, where it is filtered and separated into video and audio signal.

5.8.4.9 D/A Converter

The output from the digital filter and demultiplexer circuit is given to a D/A converter. The video signal from right and left channel signals are processed by separate D/A converter. These converters convert the 16-bit digital signal into analog signal.

5.8.4.10 Audio/Video Amplifier

The analog output from the D/A converter is passed through a sample hold circuit and a low pass filter circuit to provide smooth analog signal output. And then it is fed to a video amplifier and stereo audio amplifier.

5.8.5 DVD

DVD stands for digital versatile disc. It is similar to CD but it has larger data capacity. The laser assembly is used to pass the laser beam on the surface of the disc to read the data stored in it. Due to difference in NTSC and PAL standards the encoding of MPEG format will be different. The player also decodes the audio stream and transmits the signal to Dolby decoder before passing the audio signal to the speakers.

The fundamental components of a DVD player are:

(i) *A drive motor to spin the disc:* It is controlled to rate between 200 and 500 rpm depending on the track area to be read.

(ii) *A laser and lens system to focus in on the bumps and read them:* It allows the DVD laser to focus on the smaller pits.

(iii) *A tracking mechanisms:* It moves the laser assembly so that the laser beam can follow the spiral track.

(iv) *Electronic circuitry:* It processes the digital signals and converts them into analog signal by using DAC and then it is passed to the input of TV receiver.

5.8.5.1 Signal processing in DVD Player

The functional diagram of shown in Figure 5.31 represents how the digital data is decoded into data blocks and converted into analog form by digital to analog converter (DAC).

The laser assembly is used to pass the laser beam on the surface of the poly carbonate layer on the disc, to read the data stored in it. And the reflected signal from the disc is passed to the opto-electronic device, which is used to detect the changes in lights in the from of bits and then these bits can be converted into bytes. After conversion it processes the digital signals and converts them into analog signal by using DAC and then it is passed to the input of TV receiver.

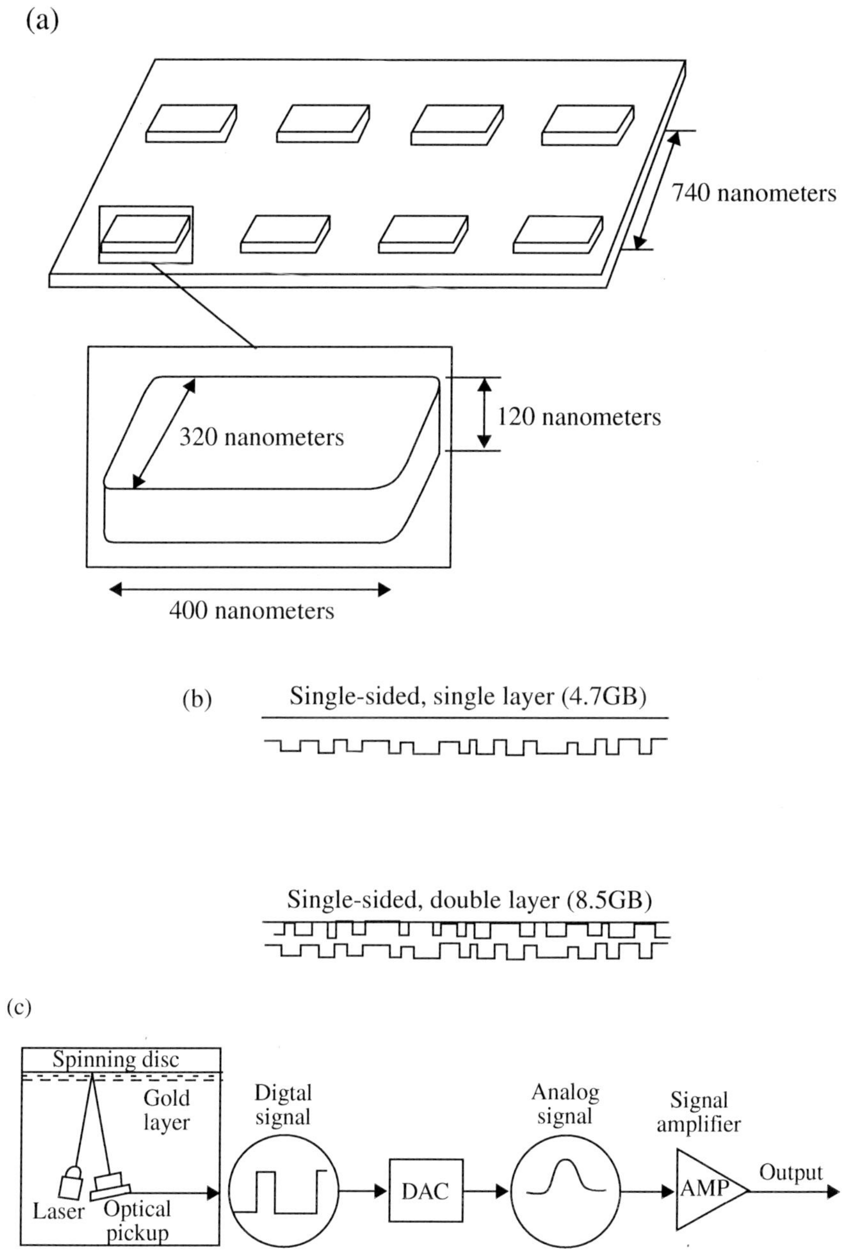

Figure 5.31 Black diagram of a DVD player.

5.8.5.2 Block diagram of DVD player in Play Back Mode

The block diagram of DVD player in playback mode is shown in Figure 5.32.

The main blocks of the DVD player are:

(i) Servo motor.
(ii) Stream demultiplexer.
(iii) Digital decoders for audio and video.
(iv) Descrambler *etc.*

The output signal from the DVD disc is given to the pre-amplifier with AGC circuit. Then the data processor unit processes the audio and video signals. Then the output of the processor is given to stream demultiplexer. This demultiplexer produces different streams of outputs. These outputs are

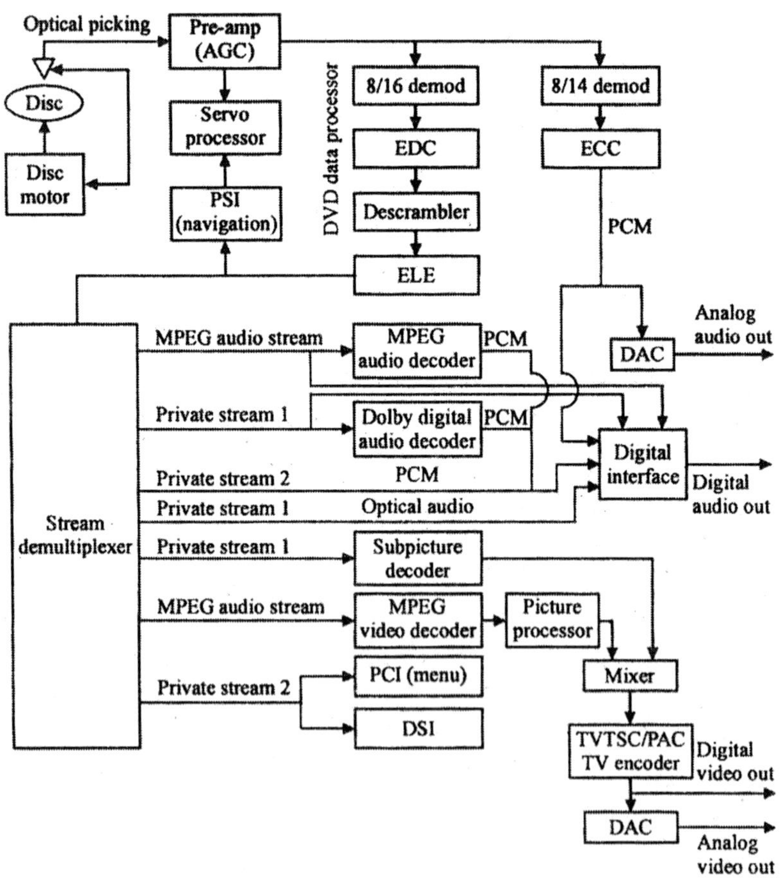

Figure 5.32 DVD – video player of block diagram.

passed to MPEG audio and video decoder. To produce Dolby sound, one stream of Dolby digital decoder is used. This output contains PCM output. Then these pulse code modulation output is fed into the digital interface.

After processing the picture signal is fed into the picture processor for producing the standard television picture. This is fed into the NTSC/Pal decoder.

But the digital signals cannot be directly passed to the TV receiver. Because, the television receiver accepts only the analog signal. So, both the audio and video signal should be converted into analog form before fed into the receiver.

5.8.6 Advantages of DVD Video Over Video CD

5.8.6.1 Capacity

Programmes on DVD can be over four times longer than the video CD. A single-layer DVD video holds over two hours of material per side and a dual-layer disc holds over four hours. A video CD holds 74 minutes and has only one side. Both DVD and video CD can support still frames with audio but DVD has the potential for hundreds of more pictures.

5.8.6.2 Convenience

Since video CDs cannot hold much more than an one hour the disc, must be changed one or more times during a movie. A DVD-video can easily hold a four hour movie on one side.

5.8.6.3 Audio

DVD-video has up to eight digital audio tracks. Each track can hold uncompressed audio with quality better than audio CD.

5.8.6.4 Video

The quality of DVD-video vast exceeds that of video CD, with four times the resolution in normal mode and 5.5 times in wide screen mode. DVD uses MPEG-2 compression used by video CD. Video in MPEG-2 looks better than in MPEG-1. DVD supports both MPEG-1 and MPEG-2.

5.8.7 Advantages of DVD-video Over Video Tape

5.8.7.1 Capacity

For typical use, such as prerecorded movies, both VHS and DVD have sufficient capacity. For longer playing times, a double-sided, dual-layer

DVD can hold over 8 hours of high-quality video or 33 hours at VHS quality.

5.8.7.2 Features

Beyond the basic VCR features of play, pause, step, slow, fast, fast forward, and rewind, DVD adds instant rewind, high speed scan, instant search, multiple camera angles, video menus interactivity and more.

5.8.7.3 Convenience

Discs can be more compact stored and can be transmitted through the mail very easily and cheaply. DVD players can be portable and battery-powered, at a size only slightly larger than the disc.

5.8.7.4 Durability

Videotapes are subject to degradation from wear and stretching, erasure from magnetic fields, and damage from heat. DVDs never wear out, are impervious to magnetic fields and are less susceptible to heat damage discs can be scratched, but as with CDs, only large scratches will cause noticeable playback problems.

5.8.7.5 Audio

Videotape audio is analog. The amount of tape dedicated to the audio track in monophonic VHS and linear stereo VHS is only one-half of one percent. HI–FI, VHS uses more tape and helical scan to get close to CD-quality audio.

5.8.7.6 Video

The video quality of VHS tape is much lower than broadcast or cable signals. Tape dropouts (poor or missing magnetic particles) cause dots and small flashes in the picture. Each generation of a videotape copy has low quality. A digital copy of a DVD is perfect and has high quality.

5.8.7.7 Price

Discs are cheaper than tapes and can be mass-produced faster and more easily.

5.8.8 Advantages of DVD-video Over Laser Disc

5.8.8.1 Features

DVD-video has the same basic features as CLV laser disc (such as scan, pause, search) plus most of the added benefits of CAV laser disc (such as, freeze, slow, fast).

DVD goes beyond laser disc with multistory branching, multiple camera angles, video menus interactivity, and more.

5.8.8.2 Capacity

Programs on DVD can be over four times longer than laser disc at equivalent quality. a single-layer DVD-video holds over two hours of material per side, and a dual-layer disc holds over four hours, a CLV laser disc holds one hour per side, a CAV laser disc holds only one half hour.

DVD-video supports still frames with audio, allowing for hundreds or thousands of picture accompanied by hours of surround sound. In laser disc still frames have no audio.

5.8.8.3 Convenience

Laser disc are large and can be cumbersome to handle. The disc size also makes the players longer and nosier than DVD players. DVD disc can be handled easily and can be send through the mail at lower cost. DVD players can be portable. Because laser discs cannot hold more than one hour per side, the disc must be changed one or more times during a movie.

5.8.8.4 Audio

DVD-video has a up to eight audio tracks. Laser disc has two stereos audio tracks. They are analog and digital.

5.8.8.5 Video

DVD almost always has better video quality than laser disc. Technically, the resolution of DVD-ideo is approximately one third better than laser disc and two third better in wide screen mode. Laser disc suffers from degradation.

5.8.8.6 Noise

Most laser disc players make a whirring noise that can be heard during quiet segments of a movie.

5.8.8.7 Price

DVD players and discs are cheaper than laser disc players and laser disc.

5.9 TELETEXT BROAD CAST RECEIVER

5.9.1 Introduction

Doordarshan of India started a teletext service in the year 1986 and named it as INTEXT meaning India's teletext. In tele text display, the video signal has two logic states, logic-1 represent the screen bright and logic-1 represent the screen dark. The information is reproduced on the screen in the form of pages of alphanumeric text (containing both letters and numbers).

5.9.2 Tele Text Picture

During interlaced scanning 574 lines are required to represent picture information in TV screen, but here 480 lines are used to represent a teletext picture. The horizontal time duration of a row of the teletext character is limited to 40μs trace period and thereby leaving 6μs vertical margin on each side of the picture. This is shown in Figure 5.33(a). A vertical margin of about 10% at the top and bottom and nearly 11% horizontal trace time

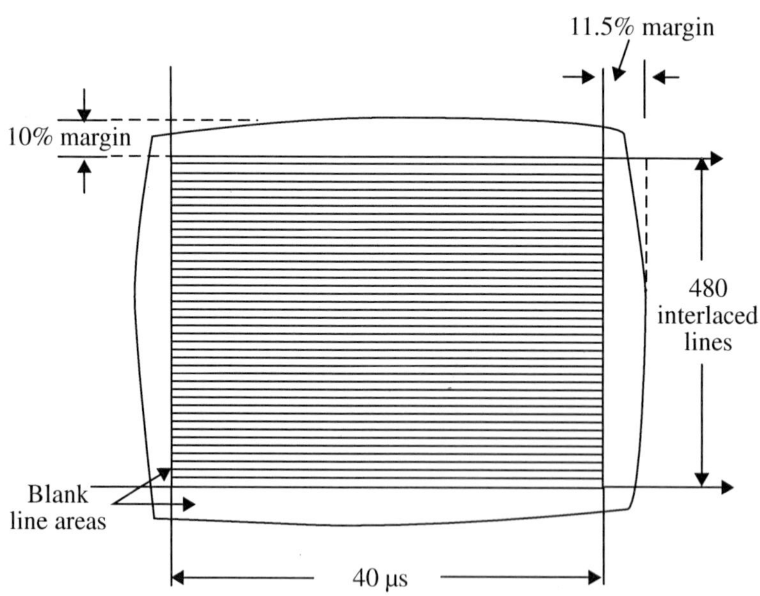

Figure 5.33 (a) Teletext picture area.

is required to enable the clear area on the screen for displaying the tele text data.

5.9.3 Teletext Data Lines

The tele text data's are coded in digital form and they are transmitted during the retrace time period and out of the 12.5 lines are allowed for retrace and other lines are used for test and signaling. But at present line numbers 15,16,17,18 of the even numbered field and line numbers 328, 329, 330 and 331 of odd numbered fields are used for transmitting teletext signals so, it is called as teletext data lines.

The method of digitally representing character shape by logic '1' for maximum amplitude and logic '0' for zero amplitude is called as binary pulse coding. This is shown in Figure 5.33(b).

5.9.4 Teletext Page

The teletext information is displayed by a page at a time, on the receiver screen. A full page contains 24 rows of 40 characters. The word row is used to avoid confusion with scanning lines,

$$\text{No. of characters in a page} = 24 \times 40 = 960 \text{ characters.}$$

To accommodate 24 rows, 240 scanning lines are required

$$\text{Interlaced scanning for each field} = 240 \times 2 = 480 \text{ lines.}$$

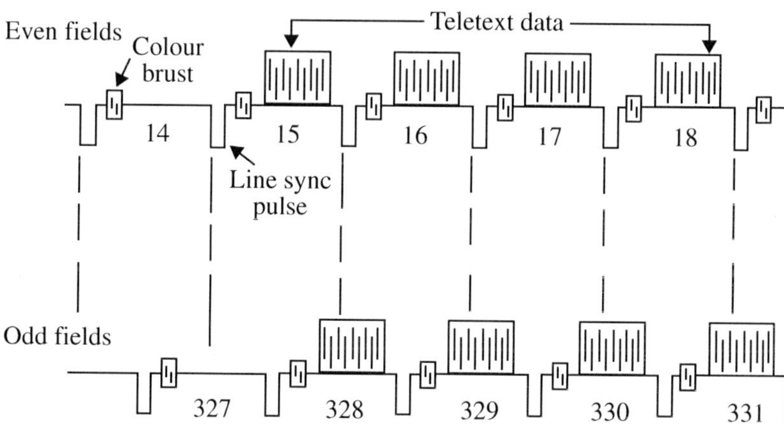

Figure 5.33 (b) Teletext data locations.

For each field 10 scanning lines are required to scan each row in a page. Total no. of lines per row is equal to

10 lines for even field + 10 lines for odd field = 20 lines.

5.9.5 Time for Each Teletext Page

For transmitting teletext information of 24 rows in a page requires 6 fields *i.e.*, $(6 \times 4 = 24)$, but it uses 4 lines in each (odd/even) field. They are (15, 16, 17, 18/328, 329, 330, 331) lines are used. Since the field repetition rate is 50 per second.

Thus the time required to transmit one full page of teletext is approximately $= \frac{1}{0.12} = \frac{1}{8}$ per second.

It means the transmission rate is 8 full pages per second.

5.9.6 Teletext Signal Coding

The digital pulses are used to convey the teletext information which form a teletext line and are arranged in coded groups, and they are decoded in the teletext decoder.

For example, every coded group of pulses is called as word and it represent an 8 bit binary code word(10110001). The binary '0' level is represented by black level (0% of peak value). Binary '1' level represented by white level (66% of peak white). This is shown in Figure 5.33(c).

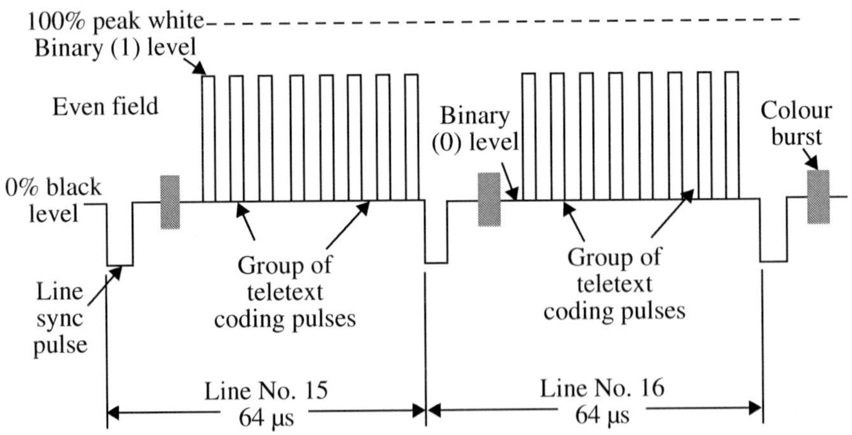

Figure 5.33 (c) Teletext signal coding.

5.9.7 Teletext TV Receiver

The block diagram of teletext receiver is shown in Figure 5.33(d). The main blocks of teletext receiver are:

(i) Vision IF sub system.
(ii) Teletext decoder.
(iii) Data storage RAM and 7-bit words.
(iv) Character generation ROM.
(v) 6 bit serializer.
(vi) Teletext key pad.
(vii) Character rounding.
(viii) Receiver operation and control.

5.9.7.1 Vision IF Sub System

The signal received from the antenna has very low strength and to amplify and process the signal it is fed to RF and vision IF sub system.

5.9.7.2 Teletext Decoder

The pages of information are converted into digital form and transmitted along with the composite video signal sequentially and during allotted line

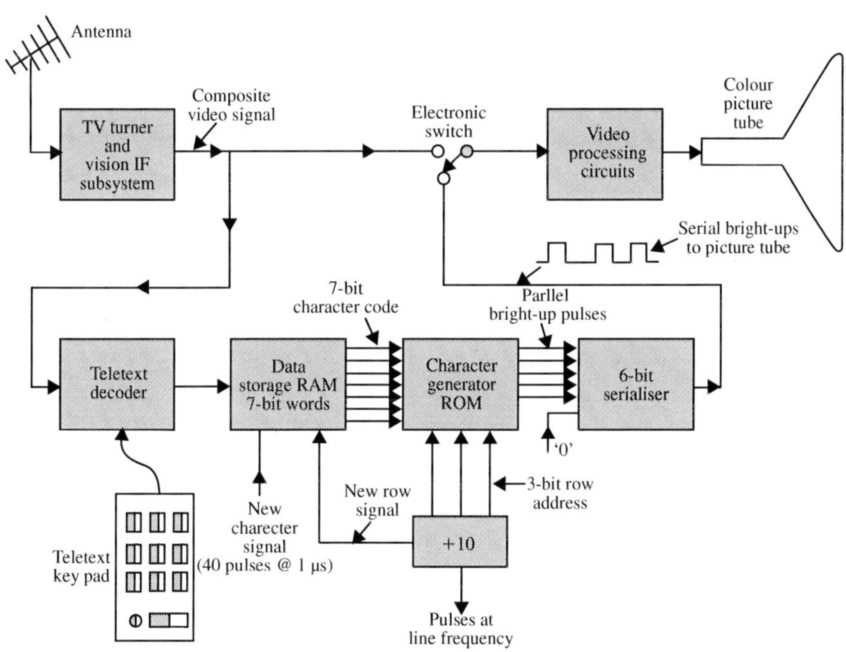

Figure 5.33 (d) Teletext colour receiver.

numbers. The teletext decoder is used to decode the page information and store them in the storage RAM. An electronic switch is used to select the normal picture signal or teletext information to the video processing circuits.

5.9.7.3 Data storage RAM

The teletext decoder decodes the page information in 7 bit binary digital form and they are stored in a data storage RAM.

5.9.7.4 Character Generation

In order to display the teletext signals the decoder is used to recollect the teletext signal stored in a data storage ram and a character generation is used to generate the teletext character according to the 7 bit binary information.

Teletext characters are constructed from a dot matrix of scanning line bright-ups controlled by the data store and character generation. Bright-ups are nothing but bright spots created on screen for very short duration. To display a single character 9 × 5 bright-ups are used. 9 bright-ups are used to build the character height. 5 bright-ups are used to build the character width.

As shown in Figure 5.33(e) the height restriction provides a minimum of one line spacing between the adjacent rows of characters in every even and odd field. Because of interlaced scanning, 9 lines are used to construct a character height appear to be formed from 18 lines of half-thickness bright ups, with the width unchanged at 5 full width bright ups.

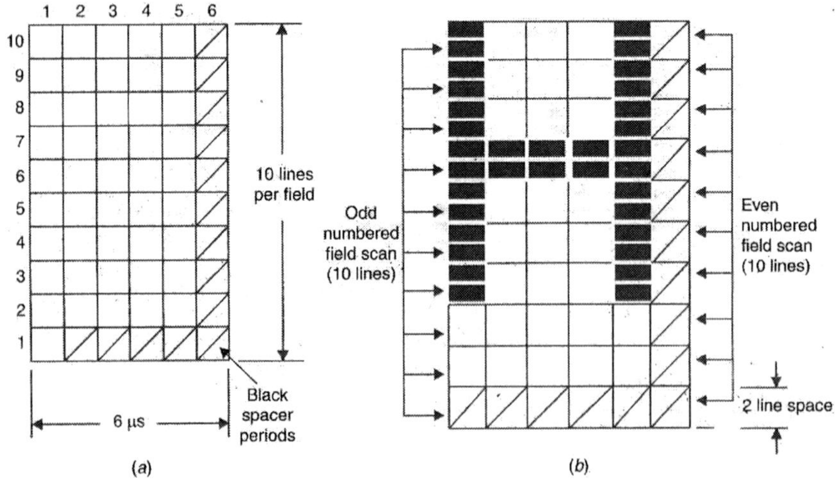

Figure 5.33 (e) Teletext character generation.

To have a sufficient horizontal spacing between the adjacent characters, the 1 μs period is divided into 6 equal periods of $= 1/6^{\text{th}}$ of a microsecond, 5 μs are used for character-width bright-ups and 1 μs as blank space as shown in Figure 5.33(e).

5.9.7.5 Character Rounding

In order to produce a legible display a character rounding technique is used. It reduces the step like abruptness and provides a grater degree of smoothness. The rounding is achieved by adding half-period (1/12 μs) bright-ups either just before or after a normal bright-up, at every point on the character where the character shape start to change diagonally. This is shown in Figure 5.33(f).

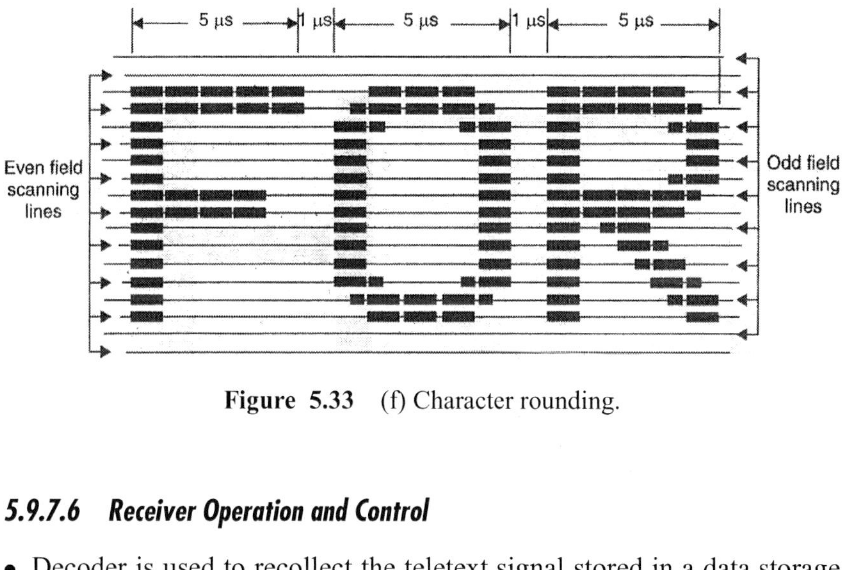

Figure 5.33 (f) Character rounding.

5.9.7.6 Receiver Operation and Control

- Decoder is used to recollect the teletext signal stored in a data storage ram and a character generation is used to generate the teletext character according to the 7 bit binary information.
- The RAM contains all the available characters in the form of 7-bit binary word and it has the word capacity to store all the characters of one full page (24 rows $\times$ 40 characters $= 960$).
- When instructed, RAM supplies the character generator with a sequence of 40 different character codes as the first scanning line of the first row moves from left to right across the screen.
- It then repeats the same sequence for 8 times until the whole 9-line rows of characters has been constructed. The entire sequence is repeated during the next scanning field also, so that at the end of two successive fields a complete interlaced character is produced.

5.9.7.7 6-bit Serializer

The character generator produces a five bright up pulses in parallel at its output. This parallel bright-up pulses is converted into serial data by the serializer and to provide a blank spacing pulse between each character, a sixth binary '0' pulse is applied to the serializer at the end of the 5th parallel data .so it converts parallel data into serial form and passes the information to the picture tube via chroma processor IC one after the other at (1/6 μs) time interval.

5.9.7.8 Teletext Keypad

The teletext keypad is used to select the desired page in the teletext magazine. Represent the Digital equipment for television studios at the end of the unit.

5.10 3D DISPLAY

The 3D television is yet to arrive at the commercial scene. The main reason for this is the delay in development of a picture tube that can directly display three-dimensional pictures. However, considerable research work is going on in this area and encouraging results are forthcoming. The success so far has been in the technique of using the two video signals in different ways to create the illusion of a 3D picture through the existing colour picture tubes. The methods evolved are briefly described:

(i) In one method that was developed earlier, all the phosphor stripes of green colour of the picture tube are excited by the 'G' video signal that originated at one of the two cameras, say the right side camera. Similarly, all the red phosphor stripes are excited by the 'R' video signal that was generated by the left side camera. The blue colour phosphors of the picture tube are not excited at all and the corresponding beam is kept at cut-off. All viewer wearing green filter glass for the right eye and red filter glass for the left eye will see the right view of the scene as produced by the green phosphor stripes and left side view as generated by the red phosphor stripes. The brain perceives a combined effect and thus sees a 3D picture of the scene in various shades of yellow, since green and red lights combine to produce yellow.

(ii) Another technique developed later by which 3D pictures can be seen in natural colours employs two video display units (VDUs). One unit is fed by the right channel RGB video signals and the other by the left channel primary colour signals. A polarizing film is fitted in from of each video display unit (picture tube) and the two light outputs

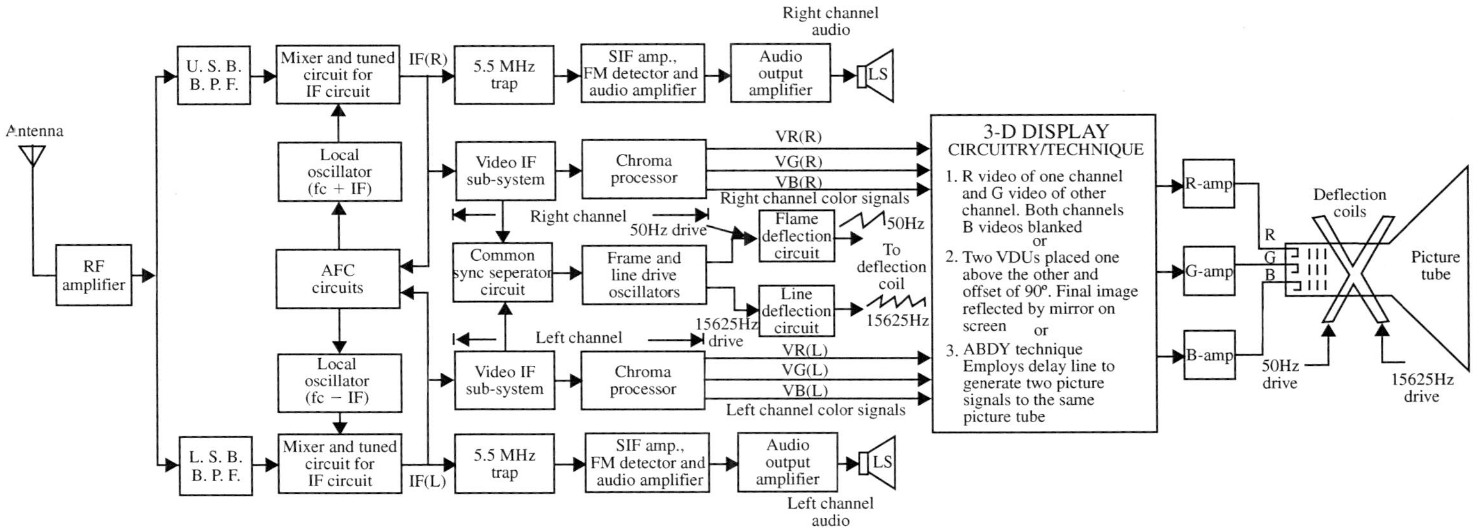

Figure 5.34 Block diagram of 3D display tubes.

thus obtained are reflected together through a mirror arrangement on a screen. The brain combines the two pictures and perceives a 3D picture of the scene being televised in its natural colours. However, this 3D display method is quite cumbersome and complicated needing both electrical and mechanical adjustments.

(iii) A third technique called ABDY 3D system which appears more promising has been developed in West Germany and in this system the video signal containing red information is delayed by nearly 600 μs by additional circuitry in the signal path. The result is a double image formed on the screen of picture tube which viewed through special glasses makes it possible for the viewer to perceive the third dimension of depth. That is, ABDY method of 3D viewing has since been patented and is finding acceptance in Japan and many other countries. Despite this, the final technique enabling 3D pictures is perhaps yet to come.

(iv) In yet another technique, the television has an LCD screen which shows both the left and right images in alternate columns of pascals (picture elements). Covering the screen are corresponding columns of double convex lenses which direct images to the proper eyes.

5.11 DIGITAL EQUIPMENT FOR TELEVISION STUDIOS

The formal available TV cameras and other studio equipments have several limitations. Their digital counterparts that are now available enable impressive operations and this has revolutionized the quantity of television programmes. Some of the devices and instruments that contribute most to the quality of programmes are briefly descried.

The major digital components used in TV studios are:

(i) Digitally controlled cameras.
(ii) Electronic control character.
(iii) Digital art or paint box.
(iv) Electronic controlled studio lights.
(v) Digital audio recorders.
(vi) Computer assisted video and audio editing system.

5.11.1 Digitally Controlled Cameras

Conventional cameras require large multi conductor cables to send and receive picture information and operating commands between the camera-head and the CCU, whereas with digital cameras are much smaller and less bulky cables serve the purpose. Digital camera systems transfer all analog signals into computer like digital pulses or coded signals. These digital signals are not only more precise than conventional analog signals but are also impervious to electrical interference over long distance cables.

5.11.1.1 Advantages

- The signal output from the camera can be stored in a digital memory for any period of time and read and written at different speeds.
- It can also be stretched, delayed, compressed and read in a form different from the one in which it was written. All this has made possible to create special effects by which signals of scenes taken at different times and locations can be mixed as desired.
- This is done with the help of a control computer which is programmed accordingly for a particular presentation.
- It is possible for the producer to call pre-recorded background scenes of jungles, rivers, fields, flowers *etc.* And mix anyone or more of these with the programme to add colour to the whole show.

5.11.2 Electronic Character Generator

The Character Generator (CG) has become as standard a production device as a camera or microphone because it requires absolutely no artistic talent or ability to create perfect lettering. It is used to create lettering and to control the position of characters on the screen.

The character generation has two monitors:

(i) Line monitor or program monitor.
(ii) Set-up monitor or preview.

The CG has two monitors. The first called 'programme' or 'line' displays the message which is currently available to be placed on the 'air'. The second monitor called 'pre-view' or 'set-up' displays another message which the CG operator is preparing. This enables the operator to set up one message while another is available to the programme director during production.

5.11.2.1 Operating Principle

The electronics needed to incorporate all the functions assigned to the character generator is quite complex but essentially digital in nature and controlled by a set of microprocessors.

Each function as indicated on the keyboard is assigned a binary code word, the length of which is determined by the total number of functions the generator is designed to perform. The depression of any key completes the electrical circuit and the assigned binary is generated.

It is either used directly or stored for delay, recall, change of speed, retrieval *etc.*, as necessary for the graphics under production. Once the necessary manipulations and mixing are over, it is superimposed on the main text.

It is then converted by a D/A converter and recorded on a tape. This tape is later used as input to the transmitter at the assigned time of programme.

5.11.3 Digital Art or 'Paint Box'

The most significant and exciting advance in television graphics has been the development of digital art system called 'Paint Box'. The digital art system combines the flexibility of digital video processing with the power of a computer memory to enable an artist to develop desired graphics complete with illustrations and lettering totally electronically.

The paint box system consists of a keyboard, electronic touch board (tablet) with pressure-sensitive stylus and viewing monitor which displays both the artwork that is being done and a chart of possible art effects. The lines, colours and photographs can be varied by the amount of pressure applied on the electronics stylus.

The complete graphics when ready is stored in the computer memory from where it can be recalled whenever necessary. It is also possible to create an animation effect by recalling a stored graphic brush stroke just as the artist draw it originally.

The paint box can also use existing camera ready photographs, maps charts or drawings which are photographed by a camera and captured by a system memory. Once the picture has been captured, the artist can integrate it with electronically produced illustrations and lettering to complete the graphic.

5.11.3.1 Principle Paint Box

The functions that the paint box is designed to perform and their control with microprocessors and computer memory makes its circuitry highly complicated. The tablet (board) is divided into a vertical and horizontal line matrix with a very large number of contact points like picture elements.

Each tiny point is insulated from the other and is designated by a large binary word code that describes its location, nature and choice of colour at that point. As the artist moves the electronic stylus to draw any shape, an electric circuit is completed with various contact points one by one.

Depending on the contact location in the matrix at any instant, the assigned code for that element is generated.

It is both stored in the computer memory and simultaneously processed to produce corresponding image in the monitor's screen. Through a complicated code generation arrangement the pressure on the stylus enables generation of additional code necessary to produce different colours in the drawing. The output of the built-in character generator can also be used to super any lettering *etc.*

5.11.3.2 Frame Storage Facility

Paint box for producing television graphics electronically with speed and versatility, electronic graphics offers another important advantage that free studio cameras and the film chain from having to photograph any graphic material.

Another electronic device which is frequently used along with TV graphics equipment is the 'Still Frame Storage Unit'. It enables the operator to 'capture' and store a complete video frame in its memory to be recalled and used whenever necessary during production.

The 'still-store' unit can store which is made-up of two or more video sources such as a camera picture and character generator super.

5.11.3.3 Television Advertisement Graphics

Paint box offers excellent means of creating any figures, cartoons and animations in different colours to highlight any feature and thus brighten the advertisement. All this together with the frame store facility and a computer to mix various effects perfectly can be used to create any imaginative advertisement graphics that catches the eyes of the viewers.

In fact, television advertisement is now a specialised profession which needs familiarity with the use of modern instruments that go into the making of such graphics. The 'ADS' that we see now were not possible earlier when only analog equipments were available.

5.11.4 Electronic Control of Studio Lights

In a television studio the most accurate and flexible means of controlling light intensity is by Semiconductor-Controller-Rectifier (SCR) controlled dimmers. These dimmers employ solid-stage circuitry and a small "pilot" voltage to control the flow of electric power to various light fixtures focused on performers and instruments.

5.11.5 Digital Audio Recorders

When we use a microphone to pick-up a sound, the acoustical sound vibrations are transferred into electrical energy. Such a conversion is called an analog process because an electrical 'analogy' of the original sound is created. However, the problem with an analog system is that during recording, processing and reproducing some distortion and noise build-up is inevitable which results in a deterioration of the sound quality.

These problems can be eliminated through the use of digital technology. In a digital system the original electric signals produced by the microphone are converted (digitised) into a computer like numerical code which represents the original sound. The coded signals thus obtained are stored on the recording tape. Because only numbers are recorded on the tape and not the complex wave-form itself as is done for analog recording, noise and distortion are virtually eliminated.

Further, it is possible to process and manipulate the digital 'signal' and produce multiple generation copies of the original recording without any loss of quality. During playback the digital audio recorder transforms the numerical code back into analog waveform to reproduce the sound with a quality as rich as the original sound source.

5.11.6 Computer Assisted Video and Audio Editing System

Computer assisted editing systems offer the ultimate in sophisticated video tape editing operation. Such a system combines a data processing computer with television production equipment enabling the operator to accomplish easily a wide range of complex editing in a relatively short period of time. As more and more productions are being shot 'film style' with single camera/single VTR approach, literally hundred of edits are necessary to assemble a finished production and only the computer's ability to keep track of the vast amount of information keeps the editor from being overburdened or confused with the task.

KEY TERMS AND CONCEPTS

Uplink frequency	The signal to be transmitted normally originates at studio from where it is fed to earth station which converts it into the assigned uplink microwave frequency 5.9 to 5 GHz.
Downlink frequency	The receiver is designed to amplify and change the signal to another base frequency as down link frequency 3.7 to 4.2 GHz.
Transponders	A transmitter and receiver combination in satellite is called transponder.
Horizontal polarization	A 24 channel capability is achieved with same bandwidth of 500 MHz by polarization techniques. A signal that is transmitted with vertical polarization does not received on an antenna that is horizontally polarized.
Vertical polarization	Horizontal polarized signal that does not received on an antenna that is vertically polarized.

Frequency reuse	To increase the signal capacity and bandwidth the most commonly used technique is called 'frequency reuse'.
Geo stationary satellite	Satellites appear stationary upon the horizon (approximate height 36000 km).
Domestic broadcast satellite	Domestic broadcasting from geo-stationary satellite is another way that is used to supplement the microwave links and co-axial cable system for TV signal distribution all over the country.
Communication satellite	Geostationary satellites used for communication are called communication satellites. These satellites can offer services like point to point communication.
INTELSAT	The international telecommunication satellites (INTELSAT) are also communication satellites.
DBS	Satellite broadcast directly to the TV receivers, they are called direct broadcast satellites, all these national and international channels which we observe today on the TV and cable network are brought to us by these DBS system.
Translator	It is a frequency converter which mixes the UHF channel frequencies down to a VHF channel.
CATV	Community antenna television systems.
MATV	Master antenna television systems.
CCTV	Closed circuit television.
LNBC	In order to minimize the losses in the co-axial cable that carry dish antenna signals to control room, the collected signals are first translated to a lower frequency range. This is done by LNBC.
Power divider	It divides the signal into independent paths; the signal splitter is commonly called as 'power divider' because it enables equal division of signal power at its output ports.
Camcorder	A cam corder (camera recorder) is essentially a video-com-sound camera and a VCR combination.
Digital television	Digital television means digital processing of TV signals both at the transmitter and in the receiver.

EDTV	Extended Definition Television (EDTV) is mainly intended for new delivery media such as Direct Broadcasting Satellite (DBS) system.
3-D Display	Three dimensional display.
TELETEXT	Doordarshan of India started a teletext service in the year 1986 and named it as INTEXT meaning India's teletext. In tele text display, the video signal has two logic states, logic-1 represent the screen bright and logic-1 represent the screen dark.
Video disc system	The video disc is essentially a gramophone record with pictures on it.
FM–FM stereo multiplex system	The stereo sound broadcasting is done by FM–FM stereo multiplex system, which uses only one FM transmitter to broadcast audio signals on both (left and right) channels.
Pilot carrier	To identify the transmitted signal is mono or stereo or dual sound. The 2nd IF carrier has the identification signal called pilot carrier.
Projection television	A projection television uses a projector to create a small image from a video signal and magnify this image onto a viewable screen.
Flat panel display system	A flat panel television receiver is one in which the conventional picture tube is replaced by flat panel display.
EDTV	EDTV is mainly intended for direct broadcasting satellite system. To improve the performance of TV systems it requires different transmission standards but it maintain the present line numbers and field rates.

REVIEW QUESTIONS

Part A

1. What do you understand by satellite TV?
2. What do you understand by geostationary orbit?
3. Why is uplink and downlink frequencies chosen to be in GHz range and kept faraway part from each other?
4. State the limitations of satellite TV.
5. List some applications of satellite TV.

6. Mention the reason for frequency translation in satellites.
7. What do you mean by translator?
8. Why transponder is necessary in satellite communication?
9. Define foot print area of satellite or coverage area.
10. Define cable TV system.
11. Define global beam.
12. What is mean by scrambling?
13. What are the different methods of scrambling?
14. What is mean by descrambling?
15. Define FM–FM stereo multiplex system.
16. What are the two types of video disc system?
17. List the fundamental components of DVD player.
18. Write short notes on quadruplex (transverse) scan recording.
19. What do you mean by longitudinal video recording?
20. Explain helical scanning in video tape recorder.
21. How a DVD is different from a CD?
22. State the applications of video tape recorders.
23. What is cam recorder?
24. What is projection television and its types?
25. How to avoid the cross interference problem in EDTV?
26. What do you understand by teletext receiver?
27. List the merits of digital TV receives.
28. What do you under stand by extended definition TV?
29. What do you mean by character rounding?
30. State the methods used for projecting images on the screen.
31. What do you understand by EDTV?
32. How to avoid the cross interference problem in EDTV?
33. What are the digital equipments used in television studios?

Part B

1. Draw the block diagram of TVRO earth station receiver and explain how the converted down link IF signal of a TV channel is processed to obtain both video and audio signals.
2. Write a short notes on satellite TV system.
3. With a help of a neat functional diagram, explain the operation of digital TV receivers. And present its merits.
4. Describe how the associated signals of 3D pictures can possibly be transmitted and received and viewed on present-day receivers.
5. Describe the new TV standards for HDTV service.
6. Draw the functional block diagram of a digital television system and explain the operation.

7. Draw a simplified block diagram to represent cable TV system and explain its working principle. Explain the scrambling and descrambling employed in it.
8. Briefly present the satellite based TV broadcast system.
9. Present the working details of laser video disc.
10. Describe the basic distribution CATV network.
11. Explain the operation of projection television system.
12. Draw the block diagram of flat panel display and explain its working principle briefly.
13. Write short notes on EDTV.
14. Explain the working principle of stereo sound in television system.

QUESTIONS AND ANSWERS

1. **What do you understand by satellite TV?**
During television signal transmission, the difficulty of line of sight restriction is overcome by satellites. Here the transmitting station sends the information signal to the satellite and in turn it receives and transmits the signal to the receiving station. Therefore, the better the receiving equipment, the higher the quality of the reception.
2. **What do you understand by Geostationary orbit?**
Geostationary orbit is an orbit at which the satellites appear stationary upon the horizon. The height of the orbit from the earth will be approximately equal to 36000 kms.
3. **List some of the applications of satellite TV.**
Some of the applications of satellite TV are:
   - The shear range of programmes currently available on satellite channel is very much impressive such as 24-hour music videos, news, and feature films.
   - A variety of general entertainment programmes, sports, children's programmes, foreign language broadcasts and cultural programmes are all available for the keyboard dish owner.
   - Some of these come through subscription channels and others by free to watch channels which are sponsored by advertising.
4. **State the merits of satellite TV.**
The merits of satellite TV are:
   - The picture quality from satellite systems is very good and compares well with conventional land based TV transmissions. In addition unlike terrestrial broadcasts it is free from the spectrum of picture ghosting.
   - During television signal transmission, the difficulty of line of sight restriction is overcome by satellites.

5. **Mention the limitations of satellite TV.**
The limitation of the satellite TV has varying picture and audio quality. By Adopting de-emphasis circuits we can avoid the variations.
6. **Mention the reason for frequency translation in satellites.**
   - By using different frequency for uplink and downlink, frequency interference can be avoided.
   - The transponder cannot transmit and receive on same frequency, because the transmitter signal can overload the receiver and block the small uplink frequency signal and thereby prohibiting the communication
7. **Why transponder is necessary in satellite communication?**
The total power radiated by the earth transmitter should be above 10^{-12} watts and 400 dB *i.e.*, $[(-12 + 40) = 10^{-28}]$ watts. Thus it makes impossible to achieve. This problem can be solved by providing a transmitter–receiver combination in satellite called transponder. The satellite transponder receives uplink frequency and provides high amplification and frequency translation. And retransmits the signal to down link frequency to earth stations.
8. **What do you mean by translator?**
Translator is a frequency converter which heterodynes the UHF channel frequencies down to a VHF channel.
9. **Define uplink frequency in satellite communication.**
The frequency at which the information signals are transmitted from earth station to the satellite is called 'uplink frequency'. Most the communication satellite uses 'C' band frequencies.

$$\text{Uplink frequency range} = 5.925\,\text{GHz to } 6.425\,\text{GHz}$$
$$\text{Bandwidth} = 500\,\text{MHz}$$

10. **Define down frequency in satellite communication.**
The frequency at which the information signals are transmitted from satellite to the earth receiving station is called as downlink frequency.

$$\text{Downlink frequency range} = 3.7\text{ to}4.2\,\text{GHz}$$
$$\text{Bandwidth} = 500\,\text{MHz}$$

11. **Define foot print area of satellite or coverage area.**
The satellite communication does not require cable or repeater stations. Within the satellite itself it is possible to cover a specific limited area for transmission. This area is called foot print area of satellite or coverage area.

12. Define global beam.

The satellite antenna is designed for a wide angle beam called 'global-beam' which extends an angle of 18°, and covers maximum possible surface area of the earth. The coverage area is about 1/3rd of surface area of earth.

13. Why uplink and down link frequencies are chosen to be in the GHz range and kept fairly apart from each other?

- The transponder cannot transmit and receive on same frequency, because the transmitter signal can overload the receiver and block the small uplink frequency signal and thereby prohibiting the communication
- The frequencies used for this purpose are in microwave frequency range from (3 to 30 GHz). During UHF, the atmosphere does not act as a barrier and signals can able to travel out into space and return back without any absorption or deflection.

14. Write short notes on CATV.

CATV stands for community antenna television systems. CATV was initially developed for the benefit of communities in rural areas on shadow zones, which distribute a good quality television signal to a very large number of television receivers. This system provides may different channels to subscribers who pay the fee for this service. With this CATV developed into a complex multi-channel system expanded in a big way in all the countries.

15. How does a HDTV differ from conventional TV?

The distinguish feature of HDTV, is its size of the image and additional picture information on the screen and there is no picture sharpness that makes HDTV its overall visual impact.

16. Define bullet amplifier.

The LNA low noise amplifier is provided in the middle of co-axial cable to compensate any losses in it. This LNA is often called as bullet amplifier

17. Define front end converter.

The feed horn and LNBC unit is called front end converter (FEC) and also referred as outdoor equipment because it is located in the open & close to the dish antenna where it is mounted.

18. Define camcorder.

A cam-corder (camera recorder) is essentially a video-com-sound camera and a VCR combination. The recorder which becomes available on the tape are edited and transferred on a standard tape for playing on a VCR for distribution to subscribers.

19. What is mean by scrambling?

Scrambling means, making the picture unintelligible on the screen. Some cable companies offer programs for a minimum charge, but they collect fee for telecasting the special channels. So, that particular signal is scrambled at the transmitter side and it should be descrambled at the receiver restored to its normal form with a signal supplied by the cable operator at the subscriber's request with additional payment.

20. What are the different methods of scrambling?

There are five different types of scrambling:

(i) Trap method.
(ii) SYNC suppression scrambler.
(iii) Base band scrambling.
(iv) Encryption method.
(v) Addressable converter.

21. What is mean by descrambling?

Descrambling means, to restore to its normal form with a signal supplied by the cable operator at the subscribers request with additional payment. It is the reverse process of scrambling.

22. Write short notes on video disc system.

The video disc is essentially a gramophone record with pictures on it. The first video disc was the laser optical and the second video disc system is based on the capacitance principle.

23. What do you mean by longitudinal video recording?

A method in which video signals are recorded on at least several tracks along the length of the tape.

24. What do you mean by Quadruplex (transverse) scan recording?

In transverse scan recording, four recording heads are spaced 90 apart and are mounted on a rotating drum and the tape moves past it, transversely. Each head comes in contact with the tape as the previous one leaves it.

25. What do you mean by helical scan recording?

In helical scan recording, the two recording heads 'look at' the tape surface as it is drawn past them through two tiny rectangular slits mounted on opposite sides of the drum. The heads thus trace out diagonal tracks across the tape, one track per head.

26. What are two types of video disc system?

Laser or optical disc system and capacitance disc system are two types of video disc system.

27. What are the advantages of DVD players over VCR's?

The quality of picture and sound in a DVD is better than on a video tape, and DVD's maintain their high quality over time, because there is no physical contact with the disc as it revolves.

28. List the fundamental components of DVD player.

The fundamental components of DVD player are:

- A drive motor to spin the disc.
- A laser and lens system to focus in on the bumps and read them.
- Tracking mechanism that can move the laser assembly so that the lasers beam can follow the spiral track.
- Electronic circuitry.

29. List four merits of digital TV receivers.

The four merits of digital TV receivers are:

- Reduced ghosts.
- Reduction of 50 Hz flicker.
- High resolution pictures.
- Slow motion action.

30. Give the applications of video tape recorders.

The applications of video tape recorders are:

- Smaller and lower priced video tape recorders using ½ inch tape are available for closed type circuit TV or for use in the home.
- They can record and playback programs on a television receiver in color and monochrome.
- In addition to that small portable cameras are provided for a complete television system with the recorder.
- These portable systems are also employed for taping television programs from a remote are also employed for taping television programs.

31. Define FM–FM stereo multiplex system.

The stereo sound broadcasting is done by FM–FM stereo multiplex system, which uses only one FM transmitter to broadcast audio signals on both (left and right) channels. Two channels signals are combined in a predetermined manner in the FM multiplex transmitter before transmission. At the transmitter side the FM station transmits two signals they are $(L + R)$ and $(L - R)$.

32. Define pilot carrier.

To identify the transmitted signal is mono or stereo or dual sound. The 2nd IF carrier has the identification signal called pilot carrier (54.68 kHz),which is amplitude modulated with 117.5 MHz for stereo or 274.1 Hz for dual transmission.

33. What is projection television and its types?

A projection television uses a projector to create a small image from a video signal and magnify this image onto a viewable screen. The projector uses a bright beam of light and a lens system to project the image to a much larger size. It can be classified into two types they are:

(a) Front-projection television system and

(b) Rear-projection television system.

34. What is flat panel display system?

A flat panel television receiver is one in which the conventional picture tube is replaced by flat panel display. Because the size and shape of a receiver is largely governed by the dimensions of the picture tube (CRT) the adoption of flat screen make the receiver much compact and produce a large screen display.

35. What do you understand by EDTV?

EDTV is mainly intended for direct broadcasting satellite system. To improve the performance of TV systems it requires different transmission standards but it maintain the present line numbers and field rates.

36. How to avoid the cross interference problem in EDTV?

For sound information there is separate carrier and additional carrier is used for modulating audio signal. Due to this it creates a problem of interference beat called cross luminance and cross colour interference. In order to avoid the interference problems, a new encoding system called multiplexed analog component system (MAC) was developed.

37. What do you understand by teletext receiver?

Doordarshan of India started a teletext service in the year 1986 and named it as INTEXT meaning India's teletext. In tele text display, the video signal has two logic states, logic-1 represent the screen bright and logic-1 represent the screen dark. The information is reproduced on the screen in the form of pages of alphanumeric text (containing both letters and numbers).

38. Define teletext data lines.

Out of the 12.5 lines are allowed for retrace and other lines are used for test and signaling. But at present line numbers 15, 16, 17, 18 of the even numbered field and line numbers 328, 329, 330 and 331 of odd numbered fields are used for transmitting teletext signals so, it is called 'teletext data lines'.

39. How much characters contained in teletext page.

The teletext information is displayed by a page at a time, on the receiver screen. A full page contains 24 rows of 40 characters. The word row is used to avoid confusion with scanning lines:

$$\begin{aligned} \text{No. of characters in a page} &= 24 \times 40 \\ &= 960 \text{ characters} \end{aligned}$$

40. What is the need for character rounding?

In order to produce a legible display a character rounding technique is used. It reduces the step like abruptness and provides a grater degree of smoothness.

41. How the character rounding is done?

The rounding is achieved by adding half-period ($1/12\mu$s) bright-ups either just before or after a normal bright-up, at every point on the character where the character shape start to change diagonally.

42. What are the digital equipment used in television studios?

The digital equipment used in television studios are:

- Digitally controlled cameras
- Electronic character generator.
- Digital art or paint box.
- Electronic controlled studio lights.
- Digital audio recorders.
- Computer assisted video and audio editing system.

UNIVERSITY QUESTION PAPERS

B.E./B.Tech. DEGREE EXAMINATION, NOVEMBER/DECEMBER 2007
Seventh Semester
(Regulation 2004)
Electronics and Communication Engineering

EC 1011 TELEVISION AND VIDEO ENGINEERING

Time: Three hours Maximum:100 marks

Answer ALL questions

PART-A ($10 \times 2 = 20$ Marks)

1. What is the significance of choosing the no. of lines as 625 and 623 or 627 and the frame reception rate 25 and not 24 as in motion pictures?
   - In motion illusion the minimum refresh rate is 24 frames/sec.
   - To avoid flicker and brightness blends smoothly by showing each frame twice.
   - Therefore $25 \times 2 = 50$ views/sec.
2. Why is a portion of the lower sideband of the AM picture signal transmitted along with the carrier and full VSB?
   In AM, very low frequencies contain useful information. These frequencies give rise to sidebands that are very close to the carrier frequency and suppression of lower side band will leads to objectionable phase error. To avoid these problems we need to transmit the full upper side band and part of the lower side band.
3. What are the merits of using an RF amplifier before the frequency converter?
   Refer Chapter 2 two marks
4. Why is it necessary to employ an L–C instead of R–C filter to removed IF ripple from the detected output?

To avoid over attenuation of the video signal while filtering out the carrier components. So L–C component is employed.

5. Why is the colour signal band width requirement much less than that of Y signals?
By conducting many experiments with different viewers, it is found that, only the colour region having an area more than 1/25th of screen width will give useful colour information. So for colour signal large bandwidth is not necessary.

6. Why is the colour burst signal transmitted after each scanning line and why PAL colour burst is often called the swinging burst?
Refer Chapter 4 two marks

7. Justify the choice of 3.579545 MHz as the sub carrier frequency in the NTSC system.
Refer Chapter 4 two marks

8. List the functions of the following stages of a PAL-D colour receiver mentioning their input and output. (a) Burst-phase discriminator. (b) Colour killer.
Refer Chapter 4 two marks

9. Why are up-link and down-link frequencies chosen to be in the GHz range and kept fairly apart from each other?
Refer Chapter 5 two marks

10. List the merits of digital TV receivers that are not achievable in a analog receiver.
Refer Chapter 5 two marks

PART-B ($16 \times 5 = 80$ Marks)

11. (a) (i) Show that the highest modulating frequency that needs to be handled in the 625 TV system is 5 MHz. discuss the effect number of lines, interlaced scanning and field frequency on band width (9).
Refer Chapter 1 page number 11 to 12.
(ii) For the composite video signal explain the following.
Refer Chapter 1 page number 23 to 28
1. Why are the horizontal pulses not used as sync Pulses? (3)
Refer Chapter 1page number 14 to 15
2. Why is it necessary to separate the broad vertical sync Pulse and why are they notched at 32 micro interval and not at 64 micro interval. (4)
Refer Chapter 1 page number 15, 145 to 147

Or

(b) (i) Discuss the merits and demerits of positive and negative amplitude modulation and justify the choice of negative modulation in most TV system. (8)
Refer Chapter 1 page number 29 to 32

(ii) Explain how the inherent smear effect in a vidicon is overcome in a plumbicon. Explain with a suitable sketch the mechanism by which the video signal is developed from the P-I-N structure of its target. (8)
Refer Chapter 1 page number 43

12. (a) (i) Describe briefly the factors that influenced the choice of picture IF= 38.9 MHz and sound IF= 33.4 MHz in the 625-B monochrome television system. (8)
Refer Chapter 2 page number from 117 to 119

(ii) what is electronic tuning? Explain with circuit diagram how different channel can be selected by this method. (8)
Refer Chapter 2 page number from 109 to 111 and 115

Or

(b) (i) Give the constructional details of a turnstile antenna and explain by drawing radiation pattern its suitability for television transmission. (8)
Refer Chapter 2 page number 93

(ii) Explain with a circuit diagram how the high voltage pulses, induced in the output transformer windings, are used to generate EHT supply. (8)
Refer Chapter 2 page number 166

13. (a) (i) Explain how by frequency inter leaving the colour information is accommodated with in the same channel band width of 7 MHz. (8)
Refer Chapter 3 page number from 221 to 223

(ii) Explain the need for pincushion distortion correct and the method employed to obtain a distortion free raster? (8)
Refer Chapter 3 page number from 217 to 219

Or

(b) (i) What are weighting factors and why these apply to colour difference signals before quadrature modulation? Draw chrominance signal wave shape for a colour bar pattern and justify the scaling down of (R–Y) and (B–Y) signals. (10)
Refer Chapter 3 page number from 228 to 230

(ii) Describe the gun structure and focusing details of Trinitron picture tube. (6)
Refer Chapter 3 page number 210

14. (a) (i) Explain how the differential phase-error is continuously corrected in the PAL system while affecting QAM of the colour difference signals. (8)
Refer Chapter 4 page number 247

(ii) Show the colour killer circuit and explain how the chrominance signal remains interrupted during monochrome transmission and how indent signal is generated? (8)
Refer Chapter 4 page number 257, 259

Or

(b) (i) Explain the delay line method of separating U and V signals in a PAL receiver. (6)
Refer Chapter 4 page number 247

(ii) Describe the circuit of colour burst blanking and first stage of Chroma band-pass amplifier. Explain the circuit of the saturation control in the chroma band-pass amplifier. (10)
Refer Chapter 4 page number from 252 to 284

15. (a) (i) Draw the block diagram of a TVRO earth station receiver and explain how the converted down-link IF signal of a TV channel is processed to obtain both video and audio signals. (8)
Refer Chapter 5 page number 289

(ii) Describe how the associated signals of 3-D pictures can possibly be transmitted, received and viewed on present day receivers. (8)
Refer Chapter 5 page number 348

Or

(b) (i) Describe the new TV standards for HDTV service. (6)

(ii) Draw the functional block diagram of a digital colour receiver and explain how the audio signal is separated and processed to obtain analog sound output. (10)
Refer Chapter 5 page number 311

B.E./B.Tech. DEGREE EXAMINATION, APRIL/MAY 2008
Seventh Semester
(Regulation 2004)
Electronics and Communication Engineering

EC 1011 TELEVISION AND VIDEO ENGINEERING

Time: Three hours Maximum: 100 marks

Answer ALL questions

PART-A ($10 \times 2 = 20$ Marks)

1. What is meant by interlaced scanning?
   Refer Chapter 1 two marks
2. List the bandwidth required for a VHF TV channel and UHF TV channel in Indian TV broadcasts.
   Refer Chapter 1 two marks
3. Why is AM preferred for picture signal transmission?
   Refer Chapter 1 two marks
4. Mention the reason for DC reinsertion in a TV receiver.
   Refer Chapter 2 two marks
5. Sketch the camera output voltage waveforms for a colour bar pattern consisting of red, green, blue, yellow and white in cascade.
   Refer Chapter 3 page number
6. How is pin cushion distortion adjusted?
   Refer Chapter 3 two marks
7. List four major limitations of NTSC standard.
   Refer Chapter 4 two marks
8. Draw the block diagram of the circuit that separates U and V colour signals.
   Refer Chapter 4 page number
9. Justify the selection of different frequencies for uplink and downlink of a satellite transponder.
   Refer Chapter 5 two marks
10. What do you understand by extended definition TV?
    Refer Chapter 5 two marks

PART-B (16 × 5 = 80 Marks)

11. (a) Justify the choice of 625 lines for TV transmission. Why the total number of lines is kept odd in all TV standards?
Refer Chapter 1 page number from 11 to 12

Or

(b) Sketch the composite video signal waveform and indicate, Blanking level, Pedestal height, back porch and active line period. Why is it necessary to use equalizing pulses and blanking pulses?
Refer Chapter 1 page number from 23 to 28

12. (a) Draw the simplified block diagram of UHF tuner and explain the functions of each section. Why is its output connected to UHF tuner mixer?
Refer Chapter 2 page number from 107 to 109

Or

(b) Explain the principle of operation of a synchronous video detector with the help of a circuit diagram. Present its merits.
Refer Chapter 2 page number 127

13. (a) Draw the block diagram and explain the formation of composite color signal in a PAL encoder. Also draw the phasors to show the effect of their amplitude and angle on the saturation and hue of the color TV signal.
Refer Chapter 3 page number 203, 220, 245

Or

(b) Illustrate the constructional and focusing techniques employed in PIL color picture tube. Explain the need for gray scale tracking.
Refer Chapter 3 page number 208, 219

14. (a) Explain the generation and modulation details of I, Q signals in an NTSC colour TV system. Why is color sub carrier fixed at 3.579545 MHz?
Refer Chapter 4 page number from 240 to 244

Or

(b) Briefly describe the characteristics of NTSC, PAL and SECAM TV standards. Compare them. Explain why India chose PAL-B standard.
Refer Chapter 4 page number from 244, 265, 267, 269, 271

15. (a) Draw the simplified block diagram to represent cable TV system and explain its working principle. Explain the scrambling and descrambling employed in it.
Refer Chapter 5 page number from 297 to 302 and 305 to 306

Or

(b) with the help of a neat functional diagram, Explain the operation of a digital TV receiver. Present its merits.
Refer Chapter 5 page number from 311 to 313.

B.E./B.Tech. DEGREE EXAMINATION, NOVEMBER/DECEMBER 2008
Seventh Semester
(Regulation 2004)
Electronics and Communication Engineering

EC 1011 TELEVISION AND VIDEO ENGINEERING

Time: Three hours Maximum:100 marks

Answer ALL questions

PART-A ($10 \times 2 = 20$ Marks)

1. Define vertical scanning.
   Refer Chapter 1 two marks
2. Define total channel bandwidth using vestigial lower sideband.
   Refer Chapter 1 page number
3. Differentiate co-channel interference and adjacent channel interference.
   Refer Chapter 2 two marks
4. What is EHT and why is it required?
   An anode voltage in order of 15 KV is needed for sufficient brightness in black white picture tubes. This is called as EHT supply. To produce a high voltage with a transformer is impossible and cost effective so EHT is used.
5. Define compatibility and reverse compatibility.
   Refer Chapter 3 two marks
6. Schematically illustrate the formation of the chroma signal for a color bar pattern after the color difference signals have been scaled down in accordance with corresponding weighting factors.
   Refer Chapter 3 page number
7. State the limitations of NTSC system.
   Refer Chapter 4 two marks

8. What is the function of the color killer circuit?
   Refer Chapter 4 two marks

9. What is tele text?
   Refer Chapter 5 two marks

10. Name the various digital equipments required in TV studio.
    Refer Chapter 5 two marks

PART-B ($16 \times 5 = 80$ Marks)

11. (a) (i) With suitable diagrams Explain in detail the interlaced scanning procedure. (8)
    Refer Chapter 1 page number from 11 to 12
    (ii) Draw the constructional detail and explain the operation of vidicon camera tube. (8)
    Refer Chapter 1 page number 38

    Or

    (b) (i) With suitable diagram explain in detail about the composite video signal. (10)
    Refer Chapter 1 page number from 23 to 28
    (ii) Explain the sound signal transmission. (6)
    Refer Chapter 1 page number 34

12. (a) Draw the block diagram of a monochrome television receiver and explain each block in detail. (16)
    Refer Chapter 2 page number from 96 to 101

    Or

    (b) Explain the following in detail.
    (1) DC Re-insertion. (6)
    Refer Chapter 2 page number from 141 to 143
    (2) Receiver antennas. (10)
    Refer Chapter 2 page number from 168 to 173

13. (a) With necessary diagram explain the delta-Gun color picture tube. Describe how purity and convergence are achieved in it. (16)
    Refer Chapter 3 page number from 205 to 208

    Or

    (b) Explain the following
    (1) Pincushion correction techniques. (8)
    Refer Chapter 3 page number from 217 to 219
    (2) Explain the color signal transmission. (8)
    Refer Chapter 3 page number from 220 to 223

14. (a) (i) Draw the simplified block diagram of the NTSC color receiver and explain each block. (8)
Refer Chapter 4 page number from 242 to 244
(ii) Explain the sequence of modulation in the PAL color system and illustrate the color burst swing in PAL system. (8)
Refer Chapter 4 page number from 245 to 247

Or

(b) Describe the following
(1) Separation of U and V signals. (6)
Refer Chapter 4 page number from 261 to 263
(2) Ident and colour Killer circuits (6)
Refer Chapter 4 page number from 259 to 263
(3) Merits and demerits of SECAM system. (4)
Refer Chapter 4 page number from 269 to 271

15. (a) Describe the following in detail with suitable diagram.
(1) Cable TV. (10)
Refer Chapter 5 page number from 297 to 302
(2) VCR. (6)
Refer Chapter 5 page number from 331 to 332

(b) Write notes on:
(1) 3D TV. (8)
Refer Chapter 5 page number 348
(2) EDTV. (8)
Refer Chapter 5 page number 325

B.E./B.Tech. DEGREE EXAMINATION, MAY/JUNE 2009
Seventh Semester
(Regulation 2004)
Electronics and Communication Engineering

EC 1011 TELEVISION AND VIDEO ENGINEERING

Time: Three hours Maximum: 100 marks

Answer ALL questions

PART-A (10 × 2 = 20 Marks)

1. What is vestigial sideband transmission and why is it used for transmission of TV picture signal?
   Refer Chapter 1 two marks
2. Why are blanking pulses not used as sync pulses?
   Refer Chapter 1 page number
3. Where and how is the inter carrier sound signal separated from the video signal?
   The sound signal is separated by a 5.5 MHz trap filter circuit and it is located in the video detector circuit.
4. What is the function of a reflector and a director in a YAGI antenna?
   Refer Chapter 2 two marks
5. What do you understand by degaussing and why is it necessary?
   Refer Chapter 3 two marks
6. Define purity and convergence.
   Refer Chapter 3 two marks
7. Why are different bandwidths assigned to Q and I signal in the NTSC system?
   Refer Chapter 4 two marks
8. Why the color burst signal transmitted after each scanning line in the PAL system?
   Refer Chapter 4 two marks
9. How does a high definition TV differ from a conventional TV?
   Refer Chapter 5 two marks
10. What are the different methods of transmitting 3-D signals over communication lines?
    Refer Chapter 5 two marks

PART-B (16 × 5 = 80 Marks)

11. (a) (i) Discuss the merits and demerits of positive and negative amplitude modulations. (6)
Refer Chapter 1 page number from 29 to 31
(ii) Draw cross-sectional view of an vidicon camera tube and explain how it develop video signal? (10)
Refer Chapter 1 page number 38

Or

(b) (i) Justify the need for pre and post equalizing pulses. (6)
Refer Chapter 1 page number 28
(ii) Show the elements of a picture tube employing electrostatic focusing and magnetic deflection. Explain the function of brightness and contrast control circuit. (10)
Refer Chapter 1 page number from 53 to 60

12. (a) (i) Explain the constructional details of a turnstile antenna and its suitability for television transmission. (8)
Refer Chapter 2 page number 93
(ii) Describe with a block diagram the working principle of a UHF tuner. (8)
Refer Chapter 2 page number from 107 to 109

Or

(b) (i) Describe how EHT is developed from the horizontal output circuit of the sweep amplifier? (8)
Refer Chapter 2 page number 93
(ii) Discuss on the factors that influenced the choice of picture IF and sound IF in the 625B monochrome television system. (8)
Refer Chapter 2 page number from 107 to 109

13. (a) (i) Explain the sequence of operations involved in achieving the luminance and chrominance television system. (8)
Refer Chapter 3 page number from 195 to 203
(ii) Describe the constructional details of a PIL tube and explain its distinguishing features. (8)
Refer Chapter 3 page number 208

Or

(b) (i) Explain how by frequency interleaving the color information is accommodated within in the same channel bandwidth of 7 MHz. (8)
Refer Chapter 3 page number from 221 to 223
(ii) Explain how the grey scale tracking is carried out both for low and high light levels of the video signal. (8)
Refer Chapter 3 page number 219

14. (a) (i) Explain the encoding process in the PAL colour system. (10)
Refer Chapter 4 page number from 245 to 247
(ii) Justify the choice of 3.579745 MHz as the sub carrier in the NTSC system. (6)
Refer Chapter 4 page number from 240 to 243

Or

(b) (i) Explain the delay line method of separating *U* and *V* signals in a PAL receiver. (8)
Refer Chapter 4 page number from 261 to 263
(ii) Describe the encoding and decoding process of SECAM system. (8)
Refer Chapter 4 page number 269

15. (a) Explain the following:
(1) Satellite TV technology. (8)
Refer Chapter 5 page number from 282 to 285
(2) 3D TV. (8)
Refer Chapter 5 page number 348

Or

(b) Explain the following:
(1) Digital TV. (8)
Refer Chapter 5 page number from 311 to 313.
(2) Cable TV. (8)
Refer Chapter 5 page number from 297 to 302.

B.E./B.Tech. DEGREE EXAMINATION, NOVEMBER/DECEMBER 2009

Seventh Semester

(Regulation 2004)

Electronics and Communication Engineering

EC 1011 TELEVISION AND VIDEO ENGINEERING

Time: Three hours Maximum: 100 marks

Answer ALL questions

PART-A ($10 \times 2 = 20$ Marks)

1. Define aspect ratio for a TV image?
   Refer Chapter 1 two marks
2. List the components of composite video signal?
   Refer Chapter 1 two marks
3. Briefly present the concept of AFT?
   Refer Chapter 4 two marks
4. What is the significance of 5.5 MHz sound trap?
   The sound signal is separated by a 5.5 MHz trap filter circuit and it is located in the video detector circuit.
5. Draw the block schematic that generates the Y, (R–Y) and (B–Y) signals from R, G and B outputs of colour camera.
   Refer Chapter 3 page number
6. Justify the use of Bitable multivibrator in the 180° degree PAL switch in a CTV receiver.
   Refer Chapter 4 two marks
7. List out the limitations of the NTSC standard.
   Refer Chapter 4 two marks
8. Give the weighting factors of colour difference signals in color encoding.

$$\text{Weighting factor for } (R - Y) = 0.877$$

$$\text{Weighting factor for } (B - Y) = 0.493$$

9. Present the principle of scrambling of TV signals in a cable TV system.
   Refer Chapter 5 two marks
10. Locate the video and FM sound signals in the 3D stereo sound TV transmission system after modulation.
    Refer Chapter 5 two marks

PART-B (16 × 5 = 80 Marks)

11. (a) Illustrate the interlaced scanning format in detail. Mention the number of active lines traced. (16)
Refer Chapter 1 page number from 11 to 12

Or

(b) Describe the deflection mechanism of a TV picture tube. How do you troubleshoot a fault of 'No brightness' in it?
Refer Chapter 1 page number from 53 to 60

12. (a) Explain with neat block diagram how the video signal is modulated and sketch the output of this modulator.
Refer Chapter 1 page number from 29 to 31

Or

(b) Summarize the design requirements of the vision IF amplifier section. Describe the video and sound signal detection.
Refer Chapter 2 page number from 130–132

13. (a) Explain how the 'C' signal is formed by quadrature modulation of colour difference signals, also present the colour sub carrier interleaving mechanism with the 'Y' signal.
Refer Chapter 3 page number from 220–223

Or

(b) Draw the structure and explain the raster scanning in a Trinitron colour picture tube.
Refer Chapter 3 Page number from 210–211

14. (a) Describe the phase error cancellation process in a PAL TV system, in detail.
Refer Chapter 4 page number from 245–247

Or

(b) Illustrate the basic working of comb filter. Explain fully how the C and Y signals are separated without any loss of bandwidths.
Refer page number 248–251

15. (a) Draw the block diagram and explain the reception and demodulation aspects of a TVRO earth station.
Refer page number from 289–291

Or

(b) How is EDTV different from conventional TV? Explain the MAC technique in detail.
Refer page number 325a

INDEX

S

T

U

V

W

Y